Magnesium-based Hydrogen Storage Materials

Qian Li · Fusheng Pan

Magnesium-based Hydrogen Storage Materials

重庆大学出版社

Springer

Qian Li
Chongqing University
Chongqing, China

Fusheng Pan
Chongqing University
Chongqing, China

ISBN 978-981-96-5530-4 ISBN 978-981-96-5531-1 (eBook)
https://doi.org/10.1007/978-981-96-5531-1

Jointly published with Chongqing University Press Co., Ltd.
The print edition is not for sale in China (Mainland). Customers from China (Mainland) please order the print book from: Chongqing University Press Co., Ltd.

This Springer imprint is published by the registered company Springer Nature Singapore Pte Ltd.
The registered company address is: 152 Beach Road, #21-01/04 Gateway East, Singapore 189721, Singapore

Preface

Materials, telecommunications, and energy form the backbone of human society and industry. The rapid advancement of technology and the global economy has led to an increasing demand for energy. Fossil fuels (e.g., oil, coal, and natural gas) powered the Industrial Revolution in the twentieth century. However, excessive consumption and limited geological reserves of fossil fuels have resulted in a global energy crisis. Additionally, emissions from fossil fuel combustion pose significant environmental challenges. Developing sustainable energy sources to meet the ever-increasing demand has become a critical research focus.

To achieve energy utilization with zero pollutant emissions, renewable energy sources such as solar, tidal, wind, and geothermal energy offer promising alternatives to fossil fuels. However, these sources are inherently limited by time and geographical limitations. As a result, high-quality secondary energy carriers are necessary to complement them and enable "peak-load shifting" in power supply systems. Hydrogen energy, with its advantages of high energy density, reversibility, zero emissions, non-toxicity, abundance, and environmental compatibility, is an ideal secondary energy carrier for storing primary energy. It is expected to play a crucial role in replacing fossil fuels.

The hydrogen energy industry consists of three key sectors: hydrogen production, hydrogen storage/transportation, and hydrogen utilization. High-density hydrogen storage and transportation serve as a crucial link between hydrogen production and utilization, yet this remains a bottleneck in the global hydrogen energy infrastructure. Hydrogen storage techniques can be broadly classified into three categories: low-temperature liquid hydrogen storage, high-pressure gaseous hydrogen storage, and solid-state ambient-pressure hydrogen storage. Solid-state hydrogen storage materials are further categorized into physical adsorption materials and chemical hydrides based on the form of hydrogen within the material structures. Since the discovery of hydrogen storage alloys in 1968, these materials have been considered ideal for hydrogen storage due to their advantages such as safety, high efficiency, high hydrogen storage capacity, and ease of use. To date, four major types of hydrogen storage alloys have been developed: lanthanum (rare-earth) series alloys, titanium-iron series alloys, magnesium (Mg)-based alloys, and zirconium (vanadium) series alloys. Among these, Mg-based hydrogen storage materials stand out due to their abundant raw material supply, high hydrogen storage density, moderate hydrogen absorption/desorption platform, and good reversibility. Consequently, they are considered one of the most promising large-capacity hydrogen storage materials with excellent development prospects. However, challenges such as activation difficulty, high desorption temperature, slow hydrogen absorption/desorption rates, and poor cycling stability hinder their widespread application. Magnesium is a highly reactive metal that readily oxidizes, making hydrogen dissociation on its surface difficult. How can we address these challenges in Mg-based hydrogen storage materials? In the early 1970s, researchers explored alloying strategies using magnesium with nickel (Ni), copper (Cu), titanium (Ti), aluminum (Al), thulium (Tm), and other rare-earth (RE) elements to enhance hydrogenation/dehydrogenation kinetics in systems like Mg-Ni, Mg-RE, and Mg-TM-RE. In the 1980s, the "nanocrystallization" technique was widely adopted to improve the performance of Mg-based hydrogen storage materials. During the alloying and nano-sizing processes, the catalytic effects of various elements and compounds gained significant attention. Today, alloying, nano-sizing, and catalytic enhancements are the primary strategies for improving the hydrogen absorption/desorption performance of Mg-based materials. Future research should focus on optimizing material composition and refining synthesis techniques, aiming for diversified Mg-based hydrogen storage materials and advanced preparation technologies.

For this highly concerned and fast-developing field, several books have been published domestically and internationally on hydrogen energy utilization and hydrogen storage materials, each offering insights from different perspectives. However, few reference books systematically summarize the research progress on Mg-based hydrogen storage materials. To address this gap and promote further development in the field, we were motivated to write this book, *Magnesium-based Hydrogen Storage Materials*. This book aims to comprehensively summarize recent advancements and emerging trends in Mg-based hydrogen storage materials while serving as a professional reference for researchers and engineers in the field. It emphasizes the components, structures, and hydrogen storage properties of Mg-based hydrogen storage material systems. It covers the preparation methods and characterization techniques of Mg-based hydrogen storage materials, as well as the thermodynamic and kinetic modification methods used to enhance their performance and applications.

The book, co-authored by Professor Qian Li and Academician Fusheng Pan, is structured into five chapters:

- Chapter 1 introduces the structures and properties of Mg-based hydrogen storage phases and hydrides, the thermodynamics of Mg-based hydrides, and expands the discussion from simple systems to multi-component and multi-phase systems.
- Chapter 2 introduces the kinetic models and influence factors of the hydrogenation and dehydrogenation processes of Mg-based materials from macro- and micro-perspectives through both theoretical and experimental studies.

- Chapter 3 explores the relationship between processing technologies, microstructures, and hydrogen storage performance, emphasizing advanced preparation technologies for Mg-based alloys, nano-sized alloys and composites, as well as pulverization and hydrogenation methods.
- Chapter 4 discusses the testing techniques for Mg-based hydrogen storage materials, covering testing methods, procedures, critical considerations for performance evaluation, as well as the characterization techniques for structural analysis and hydrogen absorption/desorption mechanisms.
- Chapter 5 focuses on strategies for regulating the thermodynamics and kinetics of Mg-based hydrogen storage materials, identifying key factors influencing hydrogen absorption and desorption reactions, and highlighting improvement methods. Additionally, it presents specific application examples of Mg-based hydrogen storage materials.

This book strives for conciseness while closely integrating reaction thermodynamics and kinetics with the hydrogen absorption and desorption processes of Mg-based hydrogen storage materials. By centering on the research achievements of our group, it also incorporates the latest theoretical and technological advancements in the field, both domestically and internationally. We endeavor to keep the content current, highlighting the contributions of Chinese researchers while providing valuable insights for professionals and students specializing in this area.

Given the rapid development of Mg-based hydrogen storage materials and the large amount of literature on the topic, we have made every effort to ensure the content is comprehensive, objective, and accurate. However, due to our limited knowledge, there may be inadvertent omissions or errors. We sincerely welcome constructive feedback and suggestions from readers to improve this work. We have also consulted numerous references in compiling this book, and we express our deep gratitude to the authors of these sources. We hope this book provides valuable assistance to researchers, engineers, and students interested in Mg-based hydrogen storage materials. We encourage readers to engage in discussions and share their insights on the topics covered.

Finally, we extend our heartfelt appreciation to Dr. Shuanglin Chen of CompuTherm, LLC, Prof. Qun Luo of Shanghai University, and Prof. Yangfan Lu of Chongqing University for their fruitful discussions and support in writing this book.

Chongqing, China

Qian Li
Fusheng Pan

Contents

1 Structure and Thermodynamic Properties of Mg-Based Hydrogen Storage Alloys and Their Hydrides 1
1.1 Structure and Physical Properties of Mg-Based Hydrogen Storage Phases 1
1.1.1 Structure of the Mg-Based Hydrogen Storage Phase 1
1.1.2 Physical Properties of the Mg-Based Hydrogen Storage Phases 10
1.2 Structure and Physical Properties of Mg-Based Hydrides 11
1.2.1 Structure of Mg-Based Hydrides 11
1.2.2 The Physical Properties of Mg-Based Hydrides 13
1.3 Thermodynamics of Hydrogen Absorption and Desorption Reactions in Mg-Based Hydrogen Storage Alloys 16
1.3.1 Thermodynamics Basis of Dehydrogenation/ Hydrogenation in Mg-Based Hydrogen Storage Alloys 16
1.3.2 Thermodynamic Properties of Hydrogen Absorption and Desorption Reactions for Mg-Based Hydrogen Storage Alloys 19
1.3.3 Methods to Improve the Hydrogen Absorption and Desorption Reaction Thermodynamics in Mg-Based Hydrogen Storage Alloys 22
References 24

2 Kinetics of Hydrogen Absorption and Desorption of Mg-Based Hydrogen Storage Alloys 27
2.1 Macroscopic and Microscopic Processes of Hydrogen Absorption and Desorption Reactions of Mg-Based Hydrogen Storage Alloys 28
2.1.1 Macroscopic Processes and Evaluation Methods of Hydrogen Absorption and Desorption Reactions of Mg-Based Hydrogen Storage Alloys 28

2.1.2 Microscopic Processes and Evaluation Methods of Hydrogen Absorption and Desorption Reactions of Mg-Based Hydrogen Storage Alloys 31
2.1.3 Relationship Between Macroscopic and Microscopic Kinetic Processes of Hydrogen Absorption and Desorption Reactions 33
2.2 Micro-kinetics of Hydrogen Absorption and Desorption Reactions in Mg-Based Hydrogen Storage Alloys 34
2.2.1 Micro-kinetic Mechanism of Hydrogen Absorption and Desorption Reaction in Mg-Based Hydrogen Storage Alloys 34
2.2.2 Kinetic Model of the Hydrogen Absorption and Desorption Reaction in Hydrogen Storage Alloys 39
2.3 Factors that Affect the Kinetics of Hydrogen Absorption and Desorption Reactions in Mg-Based Hydrogen Storage Alloys 47
2.3.1 Effect of Temperature 47
2.3.2 Effect of Hydrogen Pressure 55
2.3.3 Effect of Particle Size 56
2.3.4 Effects of Catalysts 57
References 60

3 Preparation of Mg-Based Hydrogen Storage Materials 63
3.1 Preparation of Mg-Based Hydrogen Storage Alloys 64
3.1.1 Melting Method 64
3.1.2 Powder Metallurgy 72
3.2 The Preparation of Mg-Based Hydrogen Storage Nanomaterials 88
3.2.1 Ball Milling 88
3.2.2 Deposition 99
3.2.3 Liquid-Phase Reductions 110
3.3 The Preparation of Mg-Based Composites for Hydrogen Storage 112
3.3.1 Nano Confinement Method 112
3.3.2 Catalytic Synthesis 116
3.4 Methods of Powdering and Hydrogenation of Mg-Based Hydrogen Storage Materials 120
3.4.1 Powdering Methods 120
3.4.2 Hydrogenation Methods 124
References 131

4 Performance Testing Technique and Characterization of Mg-Based Hydrogen Storage Materials 139
4.1 Performance Indicators of Mg-Based Hydrogen Storage Materials 139
4.2 Performance Testing Techniques for Mg-Based Hydrogen Storage Materials 142
4.2.1 Volumetric Method 143
4.2.2 Gravimetric Method 151
4.3 Crystal Structure and Microstructure Characterization of Mg-Based Hydrogen Storage Materials 157
4.3.1 Crystal Structure Characterization Techniques 157
4.3.2 Microstructure Characterization Techniques 163
References 171

5 Thermodynamic and Kinetic Regulations of Mg-Based Hydrogen Storage Materials and the Related Applications 173
5.1 Thermodynamic Regulation of Hydrogenation Reactions in Mg-Based Hydrogen Storage Alloys 174
5.1.1 Factors Influencing the Thermodynamics of Hydrogen Absorption/Desorption Reactions 174
5.1.2 Alloying and Compositing 176
5.1.3 Nanonization 187
5.2 Kinetic Modification of Mg-Based Hydrogen Storage Alloys for Hydrogenation Reactions 198
5.2.1 The Influence Factors on Kinetics of Hydrogen Absorption/Desorption Reactions 198
5.2.2 Catalyst Addition 199
5.2.3 Alloying and Compounding 205
5.2.4 Nano-crystallization 211
5.3 The Application of Mg-Based Hydrogen Storage Materials 215
5.3.1 Hydrogen Storage 216
5.3.2 Thermal Storage 229
5.3.3 Energy Storage 237
5.3.4 Hydrogen Chromatic Film 245
5.3.5 Other Applications 250
References 255

About the Authors

Qian Li is a Professor at the School of Materials Science and Technology of Chongqing University and the Deputy Director of the National Engineering Research Center for Magnesium Alloy Materials. He received his Ph.D. from the University of Science and Technology Beijing in 2004. He has long been engaged in the research of light alloy thermodynamics and dynamics, and has presided over more than 30 national, provincial, ministerial and enterprise cooperation projects, such as the 863 project of the Ministry of Science and Technology of China, the National Support Program, the National Key Research and Development Special Project, the National Natural Science Foundation of China, and the Talent Program of the Shanghai Science and Technology Commission.

He has published more than 300 SCI-indexed papers in journals Advanced Materials, Angewandte Chemie International Edition, Acta Materialia, Journal of Materials Science and Technology, etc., more than 12,500 citations, and an H-index of 62. He has been selected into the Highly Cited Researchers of Elsevier and the list of world's top 2% of scientists, published two Chinese textbooks/monographs and one chapter of English monographs. Sixty invention patents were authorized. He is an Associate Editor of JMST and an Editorial Board Member of JMA, JMI, IJMMM, and many other journals. He has received many scientific and technological awards, such as the "International Magnesium Science and Technology Award" Annual Excellent Academic Paper Award, the China Nonferrous Metals Industry Science and Technology Award (ranked 1st), and the

China Nonferrous Metals Innovation Award (individual award). e-mail: cquliqian@cqu.edu.cn

Fusheng Pan is a Professor of Chongqing University, an Academician of Chinese Academy of Engineering, the President of Chongqing Institute of Chinese Engineering Science and Technology Development Strategy, the Chairman of Chongqing Association for Science and Technology, and the President of Chongqing Institute of New Energy Storage Materials and Equipment. He is Editor-in-Chief of the Journal of Magnesium and Alloys, Chairman of the Committee of ISO/TC 79/SC 5 "Magnesium and Alloys", Chairman of the International Magnesium Society, Vice-President of Chinese Materials Research Society (C-MRS), etc. He has an experience of study and work at the University of Oxford, the University of Stuttgart, Queen Mary, Chiba University, the University of California, Los Angeles, and the University of Queensland. His work is focused in the research and application of lightweight metal structure materials, energy storage materials, composite materials and biological materials. He has published more than 1000 academic papers and 12 works, won 4 National Technological Invention Award and Scientific and Technological Progress Award, and more than 170 authorized invention patents, and has formulated and published more than 10 international standards and national standards. He is the winner of "HLHL Award" and "(USA) DuPont Science and Technology Innovation Award". He was the Chairman of the International Conference on Magnesium for four consecutive terms, and the Co-chairman of the European Materials Conference and the World International Advanced Materials Conference. e-mail: fspan@cqu.edu.cn

Chapter 1
Structure and Thermodynamic Properties of Mg-Based Hydrogen Storage Alloys and Their Hydrides

1.1 Structure and Physical Properties of Mg-Based Hydrogen Storage Phases

Since the discovery of hydrogen storage materials in 1968, these materials have evolved from pure metals to alloys and from binary to multi-component alloys. Among the four types of hydrogen storage systems—rare-earth, TiFe-based, Zr(V)-based and Mg-based alloys—the Mg-based alloys show clear advantages and have attracted broad attention. These alloys possess the merit of low density (Mg: 1.74 g/cm^2), high hydrogen storage capacity (MgH_2: 7.6 wt.%), abundant resources (the content of Mg in the earth's crust is about 2.3%) and low price (2~3 USD/kg). However, the slow hydrogenation/dehydrogenation kinetics, high operation temperature, and rapid degradation of hydrogen storage capacity have limited their practical applications. Various strategies, including alloying, nanosizing, adding catalysts, and employing the external energy field, have been developed to address these issues. Among these strategies, alloying is one of the most widely used methods to improve hydrogen storage functionalities by optimizing the structure and phase constitution of Mg. In this section, the crystal structure and physical/chemical properties of multiple Mg-based hydrogen storage alloys are introduced.

1.1.1 Structure of the Mg-Based Hydrogen Storage Phase

Mg is a silvery-white metal with high chemical activity, easily oxidized in air and reacting with acids to produce hydrogen. As one of the alkaline earth metals, Mg belongs to the third period and Groip IIA. The crystal of Mg possesses the hexagonal structure with the space group of $P6_3/mmc$ (194), as shown in Fig. 1.1. The cell parameters of Mg at room temperature are $a = b = 3.1954$ Å and $c = 5.1872$ Å. There are few slip systems in the Mg lattice. Mg can form alloys with various heteroatoms,

Q. Li and F. Pan, *Magnesium-based Hydrogen Storage Materials*,
https://doi.org/10.1007/978-981-96-5531-1_1

including main group and rare earth elements (Al, Zn, Sc, Gd, Y, etc.). Meanwhile, the solubility of transition metal elements (Ti, V, Fe, Co, Ni, Zr) in Mg is relatively low. The melting point of Mg is 923 K, and its density is only about 2/3 that of Al. Mg is one of the most abundant elements in nature and an essential element in the human body. The physical properties of Mg are shown in Table 1.1.

Mg and its alloys have the following characteristics:

(1) **Low density**. The density of Mg alloy is 1.75~1.85 g/cm^3, which is about 2/3 that of Al-based alloys, 3/4 that of steels, and 17/10 that of plastics.
(2) **High specific strength and specific stiffness**. The specific strength of Mg is higher than that of aluminum alloy and steel, but slightly lower than that of fiber-reinforced materials. The specific stiffness and specific elastic modulus of Mg are comparable to high-strength aluminum alloy and alloy steel, and much higher than that of fiber-reinforced materials.
(3) **Low modulus of elasticity and good damping performance**. When Mg alloys are subjectd to same external impact, the stress may distribute more evenly, avoiding stress concentration. Mg alloys can absorb over 50% more energy than

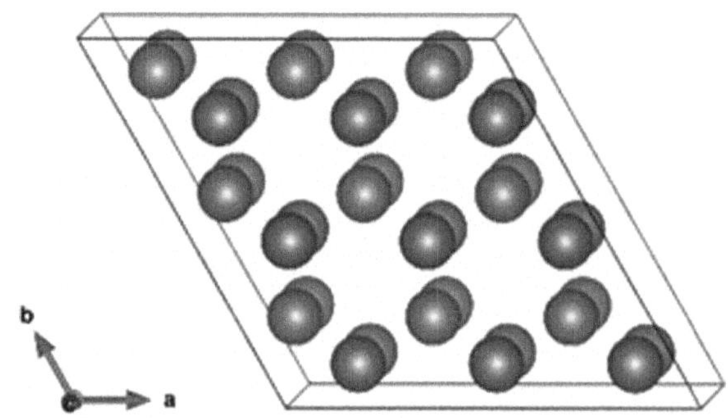

Fig. 1.1 The crystal structure of Mg

Table 1.1 The fundamental properties of Mg

Characters	Properties	Characters	Properties
Atomic number	12	Melting point (K)	923 ± 1
Number of valence electrons	2	Boiling point (K)	1380 ± 3
Relative atomic mass (g/mol)	14.305	Latent heat of vaporization (J/g)	5150~5400
Volume (cm^3/mol)	14.0	Heat of sublimation (J/g)	6133~6238
Atomic diameter (nm)	0.320	Heat of combustion (J/g)	24,900~25,200
Poisson's ratio	0.33	Specific heat capacity of Mg vapor [J/(g·K)]	0.8709
Density (g/cm^3)	1.74	Room temperature electrical conductivity (Ω/m)	23 × 10^6
Resistivity/(nΩ·m)	47	Room temperature thermal conductivity [W/(m·K)]	153.66

aluminum when subjected to an impact load within the elastic range. Therefore, they are suitable for the production of earthquake-resistant components.

(4) **Good castability and excellent dimensional stability**. Mg remains inert toward iron, reducing the melt corrosion during die-casting process. This property extends the service life of the mold. The high casting precision of Mg alloys can meet the requriements of high-speed machining.
(5) **Good cutting processability**. Due to the small cutting force and fast heat conductivity, the high-speed and large-feed cutting is feasible.
(6) **High thermal stability and good thermal conductivity**.
(7) **Good anti-electromagnetic interference and shielding performance**.
(8) **Recyclability**. Mg can be easily recycled at low cost, which is beneficial for reducing the exploitation of mineral resources and protecting the environment.
(9) **Prone to violent oxidation and burning**. Molten Mg is prone to violent oxidation and burning.

When the additive elements are oversaturated, intermetallic compounds with different structures can form, some of which are capable of storing hydrogen. After hydrogenation, these compounds transform to the associated hydrides. The well-studied binary Mg-based hydrogen storage systems include Mg-Ni, Mg-Co, Mg-Ti, Mg-Fe, and Mg-RE. The intermetallic compounds for hydrogen storage materials are summarized in Table 1.2.

As shown in Fig. 1.2, the Mg-Ni binary system has two intermetallic compounds: Mg_2Ni and $MgNi_2$ [1]. Among them, Mg_2Ni is one of the hydrogen storage materials. The melting point of Mg_2Ni is approximately 1032 K. When the Ni content is lower than 31.57 at.%, the eutectic microstructure of Mg_2Ni and α-Mg can form. The eutectic temperature of α-Mg and Mg_2Ni is approximately 781 K. Thus, Mg-Ni

Table 1.2 The crystal structure and hydrogen storage capacity of binary Mg-based hydrogen storage materials

Phases	Pearson symbol/structure	Space group	Lattice parameters (Å)	Hydrogen storage capacity (wt.%)
Mg_2Ni	$hP18/Mg_2Ni$	$P6_222$	$a = 5.19$ $c = 13.21$	3.62
Mg_3La	$cF16/BiF_3$	$Fm\overline{3}m$	$a = 7.494$	2.89
$Mg_{17}La_2$	$hP38/Th_2Ni_{17}$	$P6_3/mmc$	$a = 10.359$ $c = 10.240$	4.87
$Mg_{12}Ce$	$tI26/ThMn_{12}$	I4/mmm	$a = 10.33$ $c = 5.96$	5.68
Mg_3Ce	$cF16/BiF_3$	$Fm\overline{3}m$	$a = 7.428$	3.62
Mg_3Nd	$cF16/BiF_3$	$Fm\overline{3}m$	$a = 7.37 \sim 7.41$	1.95
$Mg_{12}Nd$	$tI26/ThMn_{12}$	I4/mmm	$a = 10.310$ $c = 5.930$	5.63
$Mg_{24}Y_5$	cI58/αMn	$I\overline{4}3m$	$a = 11.25$	5.34
Mg_3Pr	$cF16/BiF_3$	$Fm\overline{3}m$	$a = 7.425$	2.58

alloys can be prepared via conventional melting method. The crystal structure of Mg_2Ni is shown in Fig. 1.3, and a small amount of hydrogen can be dissolved in the interstitial sites of Mg_2Ni to form $Mg_2NiH_{0.3}$. When the hydrogen pressure is higher than the equilibrium pressure of metal/hydrides, Mg_2NiH_4 hydride can form, with a structure different from $Mg_2NiH_{0.3}$. The Mg-Ni binary alloy is a commonly used hydrogen storage alloy, and the theoretical hydrogen storage capacity of Mg_2Ni is approximately 3.6 wt.%. Another compound in the Mg-Ni binary system is $MgNi_2$. However, $MgNi_2$ cannot absorb hydrogen when the temperature is lower than 623 K and the pressure is lower than 40 MPa H_2 [2]. It is worth emphasizing that the Mg-Ni binary system possesses intrinsic catalytic function groups that can enhance the thermodynamic and kinetic properties of Mg-based alloys, without affecting the theoretical hydrogen storage capacity of Mg (7.6 wt.%) and Mg_2Ni (3.6 wt.%).

Unlike the Mg-Ni system, stable binary intermetallic compounds cannot be formed in Mg-Fe, Mg-Co, and Mg-Ti systems. Moreover, the maximum solubilities of these transition metals in α-Mg are lower than 2 at.%. Mg and Fe powders need to be sintered at high temperature (~773 K) and under high hydrogen pressure (2~12 MPa) for several days, or undergo mechanical alloying in a hydrogen atmosphere to form Mg_2FeH_6. The Mg_2FeH_6 has a relatively high gravimetric (5.4 wt.%) and volumetric hydrogen storage capacity (150 kg H_2/m^3) [3–5]. The hydrogen storage capacity of Mg_2COH_5 and $Mg_6CO_2H_{11}$ obtained after ball milling of Mg and Co powders is approximately 4.5 wt.% and 4 wt.%, respectively. For Mg-Ti, a mechanical alloying strategy can be employed to prepare oversaturated solid solutions of $Mg_xTi_{100\text{-}x}$ alloys. The relationship between Mg content and the crystal

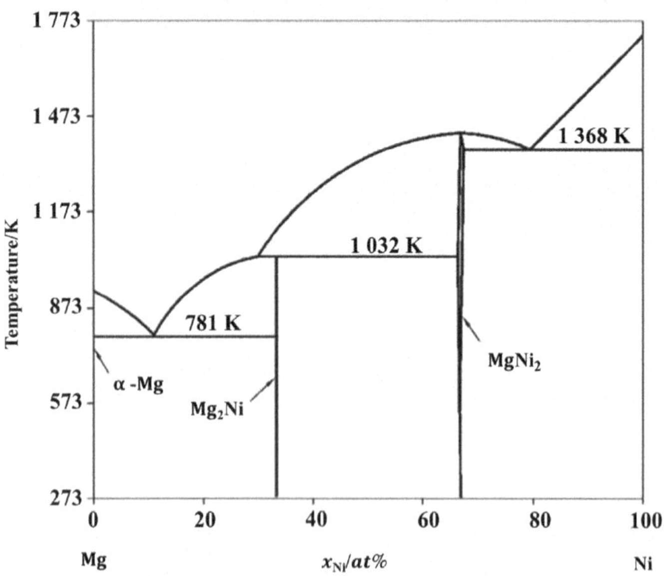

Fig. 1.2 The phase diagram of Mg-Ni system [1]

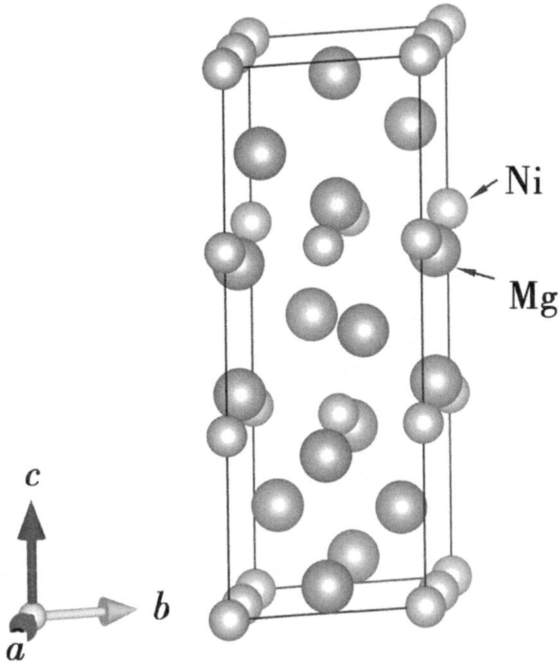

Fig. 1.3 The crystal structure of Mg_2Ni

structure of Mg_xTi_{100-x} can be expressed as $25 \leq x \leq 65$ (BCC), $35 \leq x \leq 80$ (FCC) and $65 \leq x \leq 80$ (HCP) [6–8]. It is important to note that these oversaturated solid solutions will decompose when the temperature exceeds 473 K.

The Mg-RE system contains many intermetallic compounds. Rare-earth elements, such as La, Ce, Nd, Y, and Pr, can form a series of intermetallic compounds with similar structures when combined with Mg, such as $REMg_{12}$, RE_2Mg_{17}, RE_5Mg_{41}, $REMg_3$, $REMg_2$, and REMg. Except for the Mg-Pm binary system, which still remains to be explored, the rest of the Mg-RE-based intermetallic compounds are summarized in Fig. 1.4. The REMg phases in almost all the Mg-RE alloy systems possess a simple CsCI structure. The crystal structure of $REMg_2$ can be classified into two types, $MgZn_2$ and $MgCu_2$. The $REMg_3$ phase usually crystallizes in a BiF_3 structure, which is one of the most commonly observed precipitation-hardening phases and can be stable in the Mg-La/Ce/Pr/Nd/Sm/Gd/Tb/Dy systems as an intermetallic compound. The higher the Mg content, the fewer Mg-RE phases can form due to increased structural complexity. For example, RE_5Mg_{41} phase exists only in Mg-Ce/Pr/Nd/Sm systems; RE_2Mg_{17} phase is limited to Mg-La/Ce/Eu systems; $REMg_{12}$ phase occurs solely in the Mg-La/Ce/Pr systems, and $NdMg_{12}$ is a metastable phase. After hydrogenation, Mg-RE alloys decompose into REH_x and MgH_2, as described by the reaction in Eq. (1.1). Considering that the equilibrium pressures of REH_x are relatively low, and that the decomposition of REH_x is difficult under ambient temperature, the MgH_2 phase contributes the most to the recyclable hydrogen storage capacity in Mg-RE alloys. The crystal structures of well-studied hydrogen storage

Mg-RE alloys are summarized in Fig. 1.5.

$$Mg_aRE_b + \left(\frac{2a+bx}{2}\right)H_2 \rightarrow aMgH_2 + bREH_x \tag{1.1}$$

Most ternary Mg-based hydrogen storage alloys consist of α-Mg and binary hydrogen storage phases, and their hydrogen storage properties are usually expressed as the combination of of α-Mg and binary intermetallic compounds. Among numerous Mg-based hydrogen storage alloys, the Mg-TM-RE (TM: transition metals; RE: rare-earth elements) ternary systems have drawn extensive attention due to the various ternary intermetallic compounds they contain, such as long-period stacking ordered (LPSO) and quasi-crystalline phases. It has been reported that

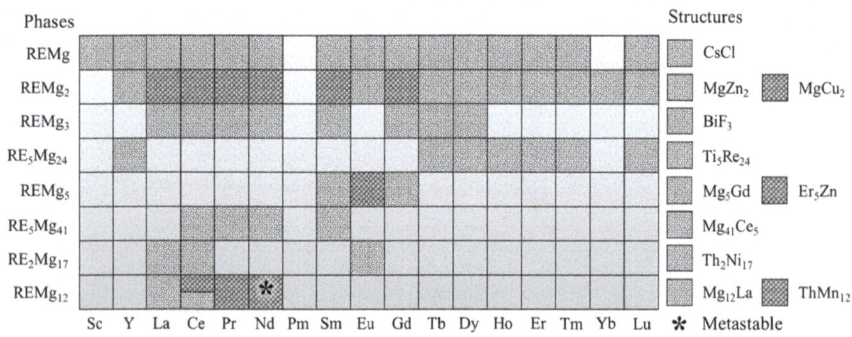

Fig. 1.4 The crystal structures of Mg-RE intermetallics [9]

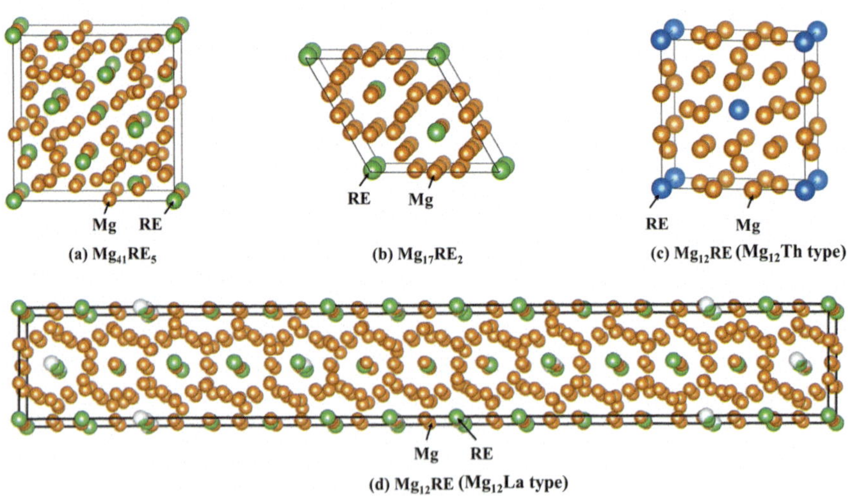

Fig. 1.5 The crystal structures of Mg-RE intermetallics

adding transition metals and rare-earth elements enhances the thermodynamic and kinetic properties of dehydrogenation /hydrogenation in these alloys. The structural information and hydrogen storage capacity of certain ternary intermetallic compounds are listed in Table 1.3.

Table 1.3 The crystal structures and hydrogen storage capacities of ternary Mg-based hydrogen storage phases

Phases	Pearson symbol/ structure	Space group	Lattice parameters (Å)	Hydrogen storage capacity (wt.%)
18R-LPSO ($Mg_{85}Zn_6Y_9$)	mS144/ $Mg_{29}Gd_4Al_3$	C2/m	$a = 11.1$ $b = 19.3$ $c = 16.0$	4.6~5.3
18R-LPSO ($Mg_{12}NiY$)	–	–	$a = 11.12$ $b = 19.26$ $c = 46.89$	4.6~5.1
14H-LPSO ($Mg_{92}Cu_{3.5}Y_{4.5}$)	–	$P6_3/mcm$	$a = b = 11.1$ $c = 36.58$	5.5
10H-LPSO	hP120 / $Mg_{23}Zn_3Y_4$	$P6_3/mcm$	$a = b = 11.1$ $c = 26.0$	–
12R-LPSO	–	P1	$a = 11.28$ $b = 11.13$ $c = 30.68$	–
$Nd_4Mg_{80}Ni_8$	tI92/$Mg_{20}NdNi_2$	I41/amd	$a = b = 11.274$ $c = 15.917$	4.9
$Nd_{16}Mg_{96}N_{12}$	oS124/ $Mg_{24}Nd_4Ni_3$	Cmc21	$a = 15.342$ $b = 21.675$ $c = 9.486$	3.9
$LaMg_2Ni$	oS16/$MgCuAl_2$	Cmcm	$a = 4.223$ $b = 10.303$ $c = 8.360$	1.9
$CeMg_2Ni$	oS16/$MgCuAl_2$	F–43 m	$a = 4.863$	1.8
$PrMg_2Ni$	oS16/$MgCuAl_2$	F–43 m	–	1.8
$NdMg_2Ni$	oS16/$MgCuAl_2$	F–43 m	–	2.1
$NdMgNi_4$	cF24/$MgCu_4Sn$	F–43 m	–	1.1
$SmMgNi_4$	cF24/$MgCu_4Sn$	F–43 m	$a = 7.057$	0.9
$LaMg_2Ni_9$	hR36/$LaMg_2Ni_9$	R–3 m	$a = 4.924$ $c = 2.387$	1.9
$CaMg_2Ni_9$	hR36/$LaMg_2Ni_9$	R–3 m	$a = 4.925$ $c = 23.787$	1.5
$La_2Mg_2Ni_9$	hR36/$LaMg_2Ni_9$	R–3 m	$a = 5.031$ $c = 24.302$	1.5
$Sm_2Mg_2Ni_9$	hR36/$LaMg_2Ni_9$	R–3 m	$a = 4.969$ $c = 24.253$	1.4

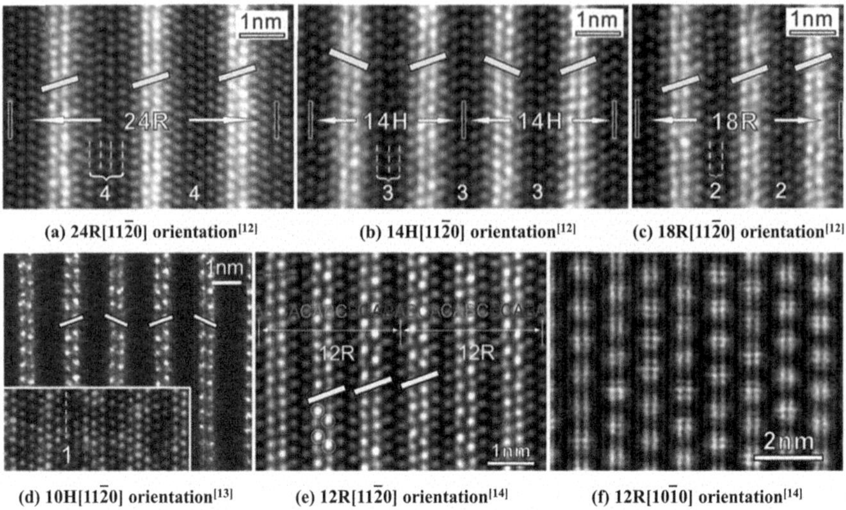

(a) 24R[11$\bar{2}$0] orientation[12] (b) 14H[11$\bar{2}$0] orientation[12] (c) 18R[11$\bar{2}$0] orientation[12]

(d) 10H[11$\bar{2}$0] orientation[13] (e) 12R[11$\bar{2}$0] orientation[14] (f) 12R[10$\bar{1}$0] orientation[14]

Fig. 1.6 The scanning transmission electron microscopy (STEM) images of LPSO phases in the Mg-TM-RE systems [12–14]

In Mg-TM-RE (TM = Al, Zn, Cu, Ni, Co) systems, LPSO phases such as 24R, 14H, 18R, 10H, and 12R have been discovered, where the number indicates the stacking period, and H and R represent the different symmetry structures. Figure 1.6 shows the high-angle annular dark-field imaging of several representative LPSO structures, where the bright spots with a periodic arrangement indicate the stacking layers of transition metals or rare earth elements. During the dehydrogenation/hydrogenation reactions, the LPSO phases in-situ transform to Mg_2Ni and REH_x. These Mg-rich LPSO phases are considered potential constituents of high-performance hydrogen storage materials. For example, the theoretical capacity of 18R-$Mg_{12}NiY$ is 4.6~5.1 wt.% of hydrogen at 573 K [10, 11], 14H-$Mg_{92}Cu_{3.5}Y_{4.5}$ phase, with higher Mg content, has a hydrogen storage capacity of 5.5 wt.%.

So far, the LPSO phases were not found in the Mg-Ni-Nd system. However, several kinds of intermetallic compounds can form in the Mg-rich region of Mg-Ni-Nd system, e.g., $Nd_4Mg_{80}Ni_8$, $Nd_{16}Mg_{96}Ni_{12}$, $NdMg_5Ni$, $NdMg_2Ni$, etc. The crystal structures of $Nd_4Mg_{80}Ni_8$ and $Nd_{16}Mg_{96}Ni_{12}$ were determined by synchrotron radiation and high-resolution TEM characterizations (Table 1.3). The hydrogen storage capacities of these two phases areapproximately 4.8 wt.% and 3.9 wt.%, respectively [15]. Additionally, the $Nd_4Mg_{80}Ni_8$ phase shows excellent cycling properties, retaining 80% of its hydrogen storage capacity after 819 cycles at 573 K. This improved cycling property originates from its stable phase structure. During hydrogenation, the $Nd_4Mg_{80}Ni_8$ phase decomposes into NdH_2, α-Mg, and Mg_2Ni nanoparticles. Notably, the average particle size of NdH_2 is quite small, only about 32 nm. The stable NdH_2 particles prevent the growth of α-Mg particles during the dehydrogenation/hydrogenation cycles. Furthermore, NdH_2 nanoparticles also catalyze

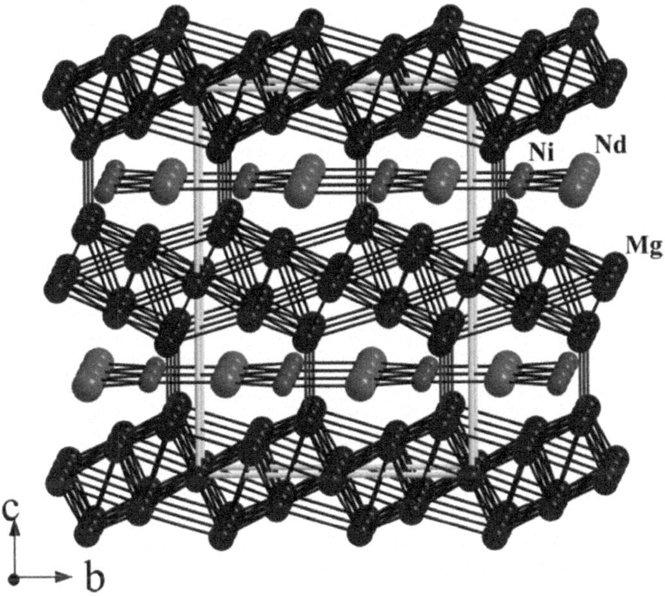

Fig. 1.7 The $NdMg_5Ni$ structure observed along the *a*-axis direction [17]

the formation of α-Mg and MgH_2, enhancing the overall kinetic performance of the alloys. $NdMg_5Ni$ has a laminar structure conposed of several Mg layers and a NiNd layer along the *c*-axis, similar to that of LPSO (Fig. 1.7). However, due to the low Mg content, the hydrogen storage capacity of $NdMg_5Ni$ is quite low, at approximately 3.2 wt.% [16].

$REMg_2Ni$ compounds, e.g., $LaMg_2Ni$, $CeMg_2Ni$, $NdMg_2Ni$, and $PrMg_2N$, can also searve as hydrogen storage materials. However, the low Mg content (~50 at.%) in these compounds significanly reduces their the hydrogen storage capacities really low, which typically range between 1.7 and 2.0 wt.%. It has been reported that $REMg_2Ni$ phases spontaneously transform to REH_2 and Mg_2NiH_4 after hydrogenation. $REMgNi_4$ and $(RE_xMg_{1-x})_3Ni_9$ are extraordinary phases in the RE-Mg-Ni system and capable of reversible hydriding and dehydriding at room temperature. After hydrogenation, the two phases transform to $REMgNi_4H_4$ and $(RE_xMg_{1-x})_3Ni_9H_{12}$. However, their theoretical capacities are low, ~1 wt.% for $REMgNi_4$ and ~1.5 wt.% for $(RE_xMg_{1-x})_3Ni_9$. The ternary or multicomponent hydrogen storage phases typically form through the solid solution of other components in mono- and binary compounds. However, multicomponent compounds have been rarely discovered in recent years. Most of hydrogen storage phases are the α-Mg, binary, and ternary phases. Combination of these multicomponent compounds exhibit varting hydrogen storage properties.

In addition to Mg-rich systems, Ni-rich compounds also contain intermetallic phases for hydrogen storage materials, such as $REMgNi_4$, and superlattice compounds of AB_3, A_2B_7, A_5B_{19}, and AB_4. The superlattices are formed by stacking

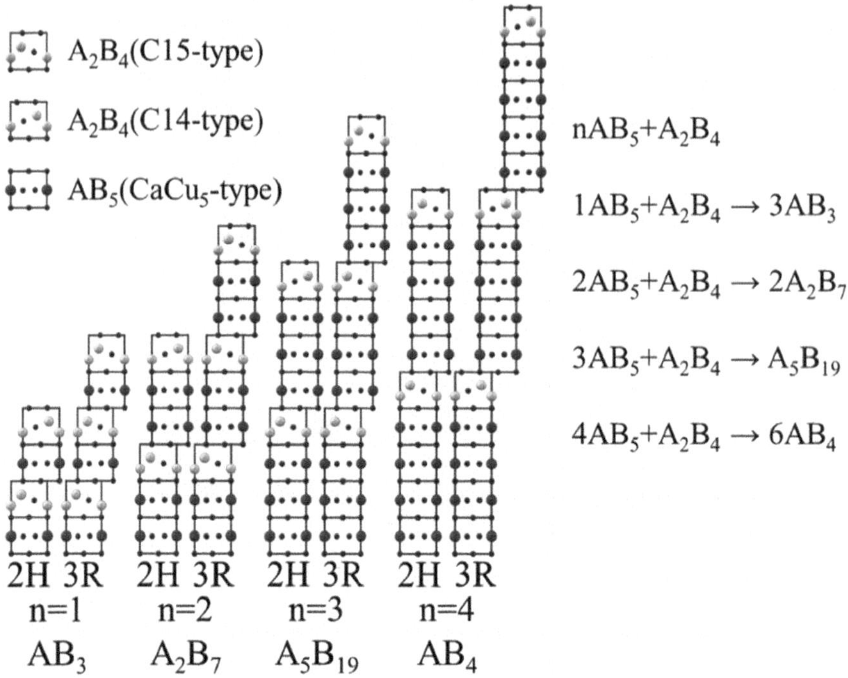

Fig. 1.8 The periodic stacking structure of Mg-TM-RE systems

[AB_5] and [A_2B_4] substructures along the *c*-axis, with ratios between different units of 1:1, 2:1, 3:1 and 4:1, respectively. The crystal structure of the above-mentioned four superlattices are shown in Fig. 1.8. The [A_2B_4] substructure unit comprises two distinct structures, known as C14 and C15-type structures. These superlattices give rise to various intermetallic compounds, including AB_3, A_2B_7, A_5B_{19}, and AB_4 types, which crystallize in 2H and 3R configurations, respectively. The AB_3-type compounds, such as $REMg_2Ni_9$, can form $REMg_2Ni_9H_{12\sim13}$ hydrides at room temperature through dehydrogenation/hydrogenation reactions, making them ideal candidates for supplying hydrogen to the anode of fuel cells.

1.1.2 Physical Properties of the Mg-Based Hydrogen Storage Phases

Mg-based alloys are used as solid-state hydrogen storage materials due to their advantages of low hydrogen storage pressure and high bulk hydrogen storage density. However, they also have some shortcomings, such as low powder flowbility, volume expansion, and low thermal conductivity of metal hydride powder, which may negatively impact both the material and hydrogen storage devices. Although the thermal

Table 1.4 The physical properties of Mg-based hydrogen storage materials

Phases	Density (g/cm^3)	Melting point (K)	Hydrides
Mg	1.74	923	MgH_2
Mg_2Ni	3.8	1043	Mg_2NiH_4
MgCo	4.533	–	Mg_2COH_5 $MgCo_2$
$LaMg_2$	3.677	1048	$LaMg_2H_7$
$CeMg_2$	3.907	1023	$CeMg_2H_7$
$LaMg_2Ni_9$	7.11	–	$LaMg_2Ni_9H_{12}$

conductivity of pure Mg is approximately 156 W/(m·K), the effective thermal conductivity of pulverized alloy is reduced to 0.01~0.1 W/(m·K). The hydrogenation process is exothermic, and the poor thermal conductivity can lead to rapid heating up. It is important to note that some Mg-based intermetallics have relatively low melting points and sublimate easily. A significant temperature rise may result in sluggish kinetics and degrade cycling properties. Therefore, the density and melting point of the Mg-based metal storage phases are critical parameters for hydrogen storage alloys. Although a variety of Mg-based hydrogen storage alloys exist, most decompose easily after hydrogenation. Table 1.4 summarizes the densites and melting points of several Mg-based hydrogen storage phases. Stable intermetallic phases are not found in Mg-Ti, Mg-Fe, and Mg-Co systems. The MgCo alloy with a BCC structure can form Mg_2CoH_5 and $MgCo_2$ through ball-milling under hydrogen atmosphere. The melting point of hydrogen storage intermetallics is higher than that of pure Mg, but compounds formed by various mixed hydrogen storage phases tend to have lower melting points. This property can be used to develop more advanced Mg-based hydrogen storage alloys.

1.2 Structure and Physical Properties of Mg-Based Hydrides

1.2.1 Structure of Mg-Based Hydrides

Although various hydrogen storage alloys have been reported in Mg-based hydrides, most of them transform to nanocomposites composed of MgH_2, Mg_2NiH_4, TiH_2, and REH_2 after hydrogenation. Ti and RE elements have a strong affinity with hydrogen, leading to the formation of TiH_2 and REH_2 hydrides. However, these phases remain in the composite after the dehydrogenation process as their dehydrogenation pressure is much lower than that of MgH_2. Table 1.5 provides the structural information of some Mg-based hydrides. The structure of hydrogen storage phases before and after dehydrogenation/hydrogenation processes is different as hydrogen atoms do not occupy the interstitial sites but instead coordinate with Mg via covalent bonds to form hydrides.

Only one hydride, MgH_2, can form in the reaction between Mg and hydrogen. Different from α-Mg (Fig. 1.9) each unit cell of MgH_2 is composed of one Mg atom coordinated with six hydrogen atoms via covalent bonds to form an Mg-centered octahedron. The Mg-H bonds in MgH_2 have different lengths: one is 1.943(8) Å and the other is 1.955(2) Å. As shown in the Mg-H phase diagram (Fig. 1.10a), hydrogen has limited solid solubility in Mg. Figure 1.10b shows the phase diagram of the Mg-rich side under different hydrogen pressures. It can be seen that the hydrogen pressure has a limited but temperature-dependent effect on the solid solubility of hydrogen in α-Mg. At 373 K, the solid solubility of hydrogen in α-Mg is negligible; however, when the temperature increases to 495 K, the solid solubility of hydrogen is increased to 0.0036 at.%. The phase stability of MgH_2 is closely related to the hydrogen pressure and temperature. As shown in Fig. 1.10a, the stability of MgH_2 increases as hydrogen pressure rises. At 0.01 MPa, MgH_2 will decompose to α-Mg and hydrogen at 495 K, but 560 K is needed for the decomposition reaction when hydrogen pressure is 0.1 MPa.

The Mg_2NiH_4 unit cell has two types of configuration, i.e., a low-temperature rhombohedral structure (rt) and a high-temperature face-centered structure (ht), as shown in Fig. 1.11. The transition temperatures of the two structures are approximately 508~513 K, regardless of hydrogen pressure. The phase diagram of the Mg_2Ni-H system is shown in Fig. 1.12. Hydrogen atoms have moderate solid solubility in Mg_2Ni, with a maximum solid solubility of approximately 10 at.% at 689 K and 3 MPa pressure. The hydrogen desorption temperature of high-temperature Mg_2NiH_4 increases with increased hydrogen pressure, reaching 571 K at 0.3 MPa H_2 and 689 K at 3 MPa H_2, respectively.

Table 1.5 Crystal structures of Mg-based hydrides

Mg-based hydrides	Pearson symbols/structural prototypes	Space group	Lattice parameters (Å)
MgH_2	$tP6/TiO_2$	$P4_2/mnm$	$a = 4.5096$, $c = 3.0152$
Mg_2NiH_4 (ht)	$cF36/K_2PtCl_6$	Fm-3 m	$a = 6.4890$
Mg_2NiH_4 (rt)	$mS56/Mg_2NiH_4$	C12/c1	$a = 6.4067$, $b = 6.4941$, $c =$ 14.3667
Mg_2FeH_6	$cF36/K_2PtCl_6$	Fm-3 m	$a = 6.4300$
Mg_2COH_5	$tP16/Mg_2CoH_5$	$P4/nmmO_2$	$a = 4.4800$, $c = 6.\ 6190$
$LaMg_2H_7$	$tP40/LaMg_2H_7$	$P4_12_12$	$a = 6.4054$, $b = 9.\ 5994$
$CeMg_2H_7$	$tP40/LaMg_2H_7$	$P4_12_12$	$a = 6.3663$, $b = 9.5226$

Note ht indicates high temperature type, rt indicates low temperature type

Fig. 1.9 The crystal structure of MgH_2

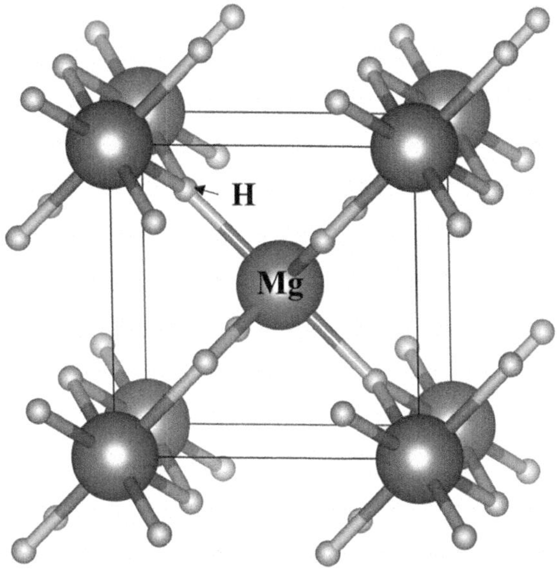

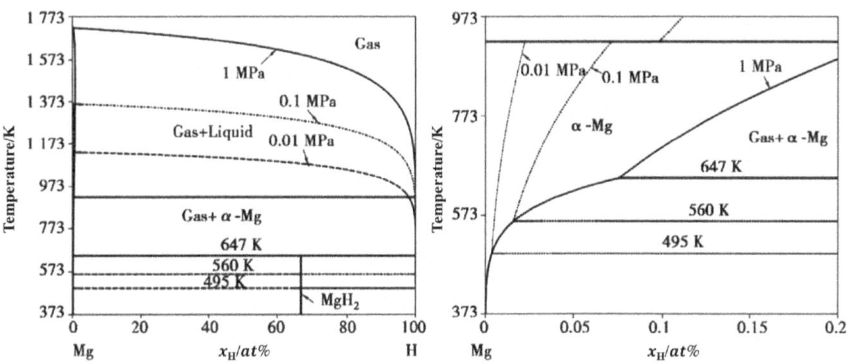

Fig. 1.10 **a** The Mg-H phase diagram under different hydrogen pressure, **b** enlarge view of the Mg-rich region at 0.1 MPa H_2

1.2.2 The Physical Properties of Mg-Based Hydrides

The volume of Mg-based hydrogen storage phases usually expands after hydrogenation, leading to decreased density, as shown in Table 1.6. Pilling and Bedworth proposed the concept of metal-oxide-to-metal volume ratio [18], i.e., the Pilling-Bedworth Ratio (PBR), to explain the volume expansion or shrinkage of the materials after oxidation. This concept has also been introduced to explain the volume change from metals to their hydrides [19, 20]. Among several typical Mg-based hydrogen storage materials, Mg_2NiH_4 has the largest PBR value, indicating the

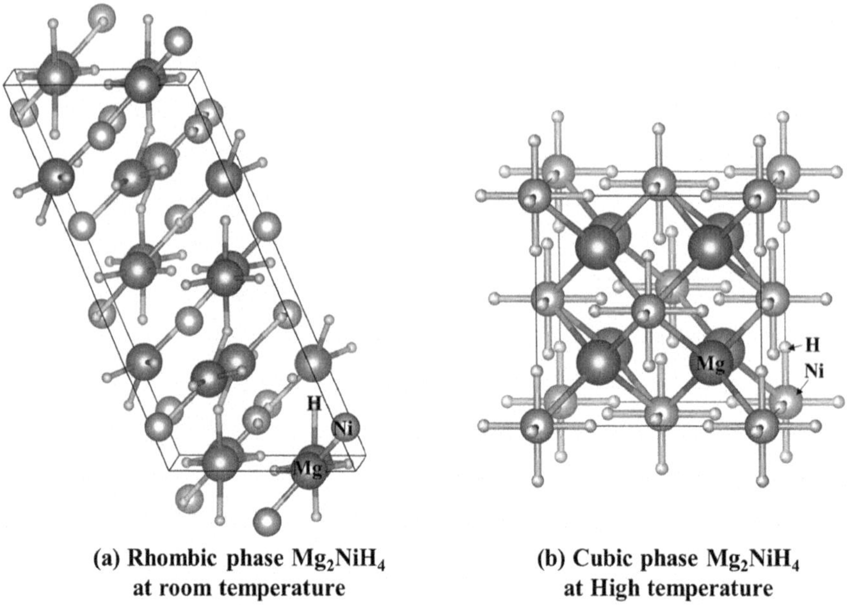

Fig. 1.11 The crystal structures of Mg_2NiH_4

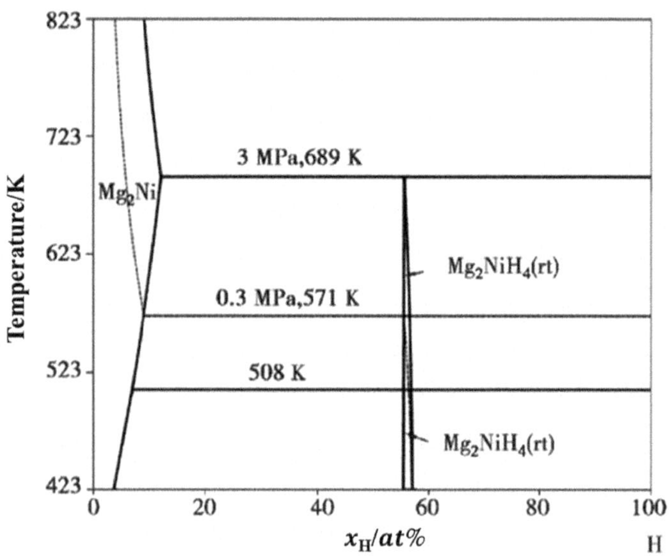

Fig. 1.12 The phase diagram of Mg_2Ni-H with different hydrogen pressures

most significant volume expansion of Mg_2Ni after hydrogenation. Recent studies show that low-temperature Mg_2NiH_4 comes in two configurations: monoclinic LT1 and orthorhombic LT2. Both structures tend to shrink at 494~733 K, with volume shrinkage of approximately 18.7% and 11.3%, respectively. This unusual phenomenon is caused by the phase transition and hydrogen release of LT1 and LT2 during the heating process [21]. In contrast, $LaMg_2$ and $CeMg_2$ show mild volume expansions during hydrogenation, approximately 1.05% and 1.06%, respectively.

MgH_2 has relatively high thermodynamic stability and a high decomposition temperature. Its dehydrogenation temperature is approximately 560 K under 0.1 MPa H_2. However, the decomposition temperature of Mg_2NiH_4 is slightly lower than that of MgH_2 (528 K under 0.1 MPa H_2). Due to the large reaction heat, Mg-based hydrogen storage alloys are considered as one of the most promising high-temperature heat storage materials. Another important factor for high-temperature heat or hydrogen storage material is the thermal conductivity. For example, the effective thermal conductivity of MgH_2 is 0.6~1.6 W/(m·K) [22], and the limited thermal conductivity hinders the hydrogen absorption and desorption to some extent. The thermal conductivity of hydrides varies with hydrogen pressures and powder sizes. For example, the thermal conductivity of ball-milled powders under 0.4 MPa hydrogen atmosphere is approximately 0.03 W/(m·K) and finer powders exhibit even smaller thermal conductivity. Meanwhile, the effective thermal conductivity can be increased to 38 W/(m·K) by adding graphite into MgH_2 [26].

Table 1.6 The physical properties of Mg-based hydrides

Hydride phases	Density (g/cm^3)	PBR	Thermal conductivity	Decomposition temperature
MgH_2	1.44	1.12	0.6–1.6 W/(m·K) [22] 0.03W/(m·K) [23] (Ball milling powder under 0.4 MPa H_2)	560 K (0.1 MPa)
Mg_2NiH_4	2.685	1.36	0.83 W/(m·K) (at 4.5 MPa H_2 and 373 K) [24] 0.65 W/(m·K) (at 3.5 MPa H_2 and 308 K) [25]	528 K (0.1 MPa)
Mg_2COH_5	2.882	–	–	–
$LaMg_2H_7$	3.349	1.05	–	–
$CeMg_2H_7$	3.522	1.06	–	–

1.3 Thermodynamics of Hydrogen Absorption and Desorption Reactions in Mg-Based Hydrogen Storage Alloys

1.3.1 Thermodynamics Basis of Dehydrogenation/ Hydrogenation in Mg-Based Hydrogen Storage Alloys

The most important hydrogen storage phase in Mg-based hydrogen storage alloys is α-Mg. After hydrogenation, Mg-TM (TM: transition metals) or Mg-RE (RE: rare-earth) alloys transform to different phases such as MgH_2, Mg_2NiH_4, Mg_2FeH_6 and RE hydrides varying with their initial composition. As mentioned above, Mg-based hydrogen storage alloys contain numerous ternary hydrogen storage phases, such as 18R-LPSO ($Mg_{12}NiY$), $Nd_4Mg_{80}Ni_8$, and $LaMg_2Ni$. Taking them as examples, their hydrogen absorption and desorption reactions can be expressed as follows:

$$Mg_{12}NiY + 13H_2 \leftrightarrow 10MgH_2 + Mg_2NiH_4 + YH_2 \tag{1.2}$$

$$Nd_4Mg_{80}Ni_8 + 84H_2 \leftrightarrow 64MgH_2 + 8Mg_2NiH_4 + 4NdH_2 \tag{1.3}$$

$$LaMg_2Ni + 3H_2 \leftrightarrow Mg_2NiH_4 + LaH_2 \tag{1.4}$$

Due to the low decomposition pressure, RE hydrides usually do not involve in reversible hydrogen absorption and desorption cycles. Therefore, the hydrogenation reaction of the Mg-based hydrogen storage materials can be simplified as hydrogen absorption and desorption reaction of the Mg and the Mg-based hydrogen storage phase, following the reaction of:

$$M + H_2 \leftrightarrow MH_2 \tag{1.5}$$

and the Gibbs free energy of the reaction can be expressed as:

$$\Delta G = \Delta G^\circ + RT\ln\frac{a_{MH_2}}{a_M P_{H_2}} \tag{1.6}$$

The activities of solid hydrides and metals are usually considered to be one. When the reaction reaches equilibrium, ΔG equals zero. Thus, Eq. (1.6) can be re-written as:

$$\ln P_{H_2}^{eq} = \frac{\Delta H^\circ}{RT} - \frac{\Delta S^\circ}{R} \tag{1.7}$$

where $P_{H_2}^{eq}$ is the equilibrium pressure of hydrogen (MPa), ΔH° is the standard reaction enthalpy (kJ/mol H_2) and ΔS° is the standard reaction entropy (J/(mol · K) H_2).

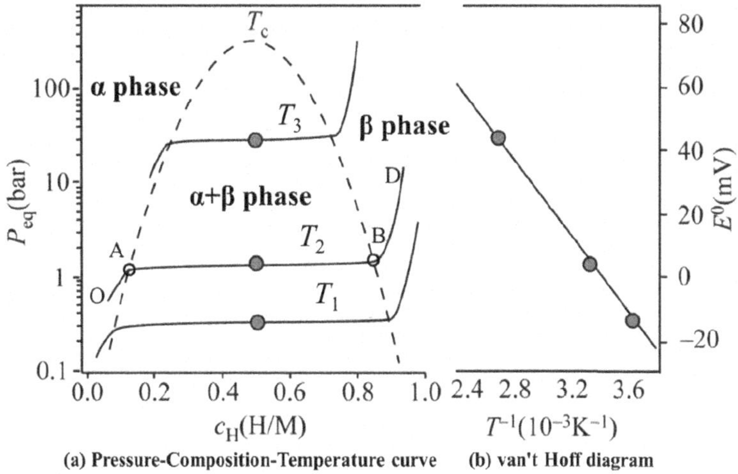

Fig. 1.13 The relationship between the P–C–T diagram and van't Hoff plot

Equation (1.7) is the well-known van't Hoff equation, showing that $P_{H_2}^{eq}$ is a function of temperature, ΔH° and ΔS°.

From the phase transition point of view, the hydrogen absorption/desorption reaction is composed of three stages, i.e., (1) the hydrogen storage phase of α-Mg, (2) the co-existence of α-Mg and hydrides (MgH_2), and (3) MgH_2. Usually, the α phase represents the solid solution of metals and hydrogen-containing metals, while the β phase represents their corresponding hydrides. Thermodynamically, when the metals are exposed to a specific temperature and hydrogen pressure for a sufficiently long time to reach equilibrium, the corresponding phase diagram of hydrogen pressure (P), alloy composition (C), and temperature (T), i.e., the P–C–T curve, can be obtained. Figure 1.13a shows a schematic of the P–C–T curve. Based on the relationship between the plateau pressure and temperature, the van't Hoff diagram can be obtained (Fig. 1.13b). In a certain temperature range, ln P_{H_2} depends linearly on $1/T$.

The P–C–T diagram contains abundant thermodynamic information, such as hydrogen storage capacity and hydrogen absorption/desorption equilibrium pressure at a specific temperature. If the P–C–T curve read as a composition-hydrogen pressure phase diagram, each point on the isotherm corresponds to an equilibrium state. When the alloy reaches equilibrium in a specific hydrogen environment, hydrogen in each phase in the alloy has the same chemical potential (Eq. 1.8).

$$\mu_H^{\alpha} = \mu_H^{\beta} = \mu_H^{Gas} \tag{1.8}$$

The chemical potentials of hydrogen atoms and the hydrogen molecules have the following relationship:

$$\mu_{\mathrm{H}}^{\mathrm{Gas}} = \frac{1}{2}\mu_{\mathrm{H_2}}^{\mathrm{Gas}} = \frac{1}{2}\left(\mu_{\mathrm{H_2}}^{\mathrm{o,Gas}} + RT\ln\frac{P_{\mathrm{H_2}}^{\mathrm{eq}}}{P_{\mathrm{H_2}}^{\mathrm{o}}}\right) \tag{1.9}$$

Where $\mu_{H_2}^{Gas}$ and $\mu_{H_2}^{\circ}$ are the chemical potential of hydrogen in the gaseous state and under standard conditions, respectively (J/mol).

Since the gaseous hydrogen pressure can be detected by a pressure sensor, the chemical potential of hydrogen in each phase of the alloy can be easily obtained by monitoring the gaseous hydrogen pressure. The pressure-composition curve is useful for understanding the thermodynamic hydrogen absorption/desorption properties of the metals or alloys. Figure 1.13a shows a typical isothermal hydrogen pressure curve of the metal/alloy-hydrogen system, where the abscissa and ordinate represent the atomic ratio of hydrogen in the metal/alloy and hydrogen pressure, respectively. The common P–C-T curve can be divided into three stages:

(1) Stage 1: Hydrogen-containing solid solution (α phase) is located at the *OA* section on the left side of the P–C–T platform. Hydrogen atoms are dissolved in the metal or alloy.
(2) Stage 2: The *AB* section corresponds to the two-phase equilibrium of metal/alloy and hydride (α + β phase). The α and β phases co-exist in this region.
(3) Stage 3: The hydride formation is completed on the right side of the P–C–T platform (*BD* section).

During Stage 1, as the hydrogen pressure increases, the hydrogen content in the α phase also increases. When α phase becomes saturated with hydrogen, it begins to react with the excessive hydrogen to form metal hydride (β phase), initiating Stage 2. As this process continues, a two-phase equilibrium of α + β is established in the alloy. At this point, the degree of freedom in the system is one, and the chemical potential of hydrogen is the same in the α phase, β phase and gas phase, as shown in Fig. 1.14. Therefore, the equilibrium hydrogen pressure is independent of the the content ratio of α and β phases, resulting in a plateau characteristic as shown in Fig. 1.13a.

Once the hydrogen pressure continues to increase, the α phase is completely transformed into the β phase, leading to Stage 3. In this stage, the phases are completely hydrogenated, and the chemical potential of hydrogen in the β phase rises with increasing hydrogen pressure, as hydrogen exists only in the β phase hydride and gaseous form. The curves T_1, T_2, and T_3 in Fig. 1.13 illustrate different reaction equilibrium pressures at various temperatures, with the equilibrium hydrogen pressure increasing with temperature. For Mg-based hydrogen storage alloys, the composition intervals of the α and β phases is difficult to observe due to the extremely low solid solubility of hydrogen in α-Mg and MgH_2. Thus, only the broad α-Mg + MgH_2 platform can be identified in the P–C–T curve.

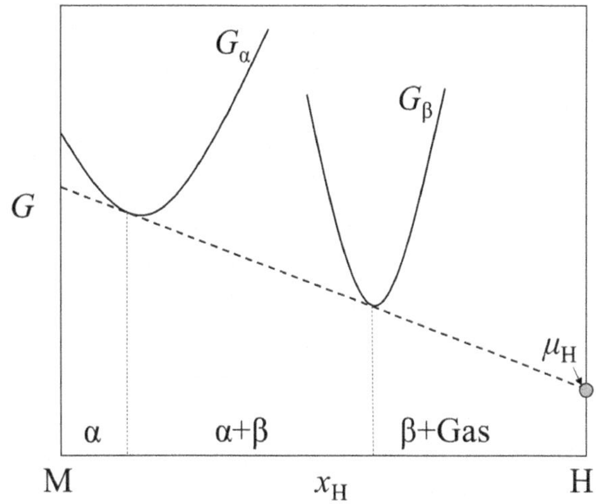

Fig. 1.14 Gibbs free energy curves and equal chemical potential diagram of metal or alloy-hydrogen system

1.3.2 Thermodynamic Properties of Hydrogen Absorption and Desorption Reactions for Mg-Based Hydrogen Storage Alloys

The thermodynamic properties of the hydrogen absorption and desorption reactions of Mg-based hydrogen storage alloys are different from those of metal hydrides, such as AB_5 type, AB_2 type, AB type, and solid solution. Specifically, the hydrogen solubility in Mg is much lower than in other hydrogen storage materials. For example, some metals such as Pd and $LaNi_5$ can store hydrogen in their interstitial positions in the form of solid solutions, while MgH_2 forms a completely different crystal structure from Mg due to the covalent bonding of hydrogen with Mg.

The reaction enthalpy (ΔH°) and entropy (ΔS°) of the hydrogenation reaction can be calculated using the van't Hoff equation Eq. (1.7). By plotting the equilibrium hudrogen pressure $\left(P_{H_2}^{eq}\right)$ as the function of $1/T$, ΔH° and ΔS° are obtained from the slope and intercept, respectively. Figure 1.15 shows the van't Hoff diagram for some Mg-based hydrogen storage alloys, where all reaction enthalpies are negative, indicating the hydrogenation reaction is exothermic. A greater the slope corresponds to a larger ΔH°. A minor addition of In in α-Mg slightly reduces the ΔH° of the hydriding reaction. The reaction enthalpy of Mg_2Ni is significantly lower than that of pure Mg, reducing the hydrogen desorption temperature at 0.1 MPa. ΔS° mainly depends on gaseous hydrogen because hydrogen transforms from gas phase to solid phase. ΔS° is equal to $-S_{H_2}^{\circ}$, where $S_{H_2}^{\circ}$ is the absolute entropy of hydrogen, approximately 130.6 J/(mol·K) H_2 at 298 K [27]. The metals have similar ΔS° values.

Table 1.7 presents the standard enthalpy and entropy of hydrogen desorption reaction for various well-studied hydrides. The standard enthalpies of hydrogen release for AB_5-type $LaNi_5H_6$, AB_2-type $TiMn_2H_3$, and AB-type $TiFeH_2$ hydrides are much

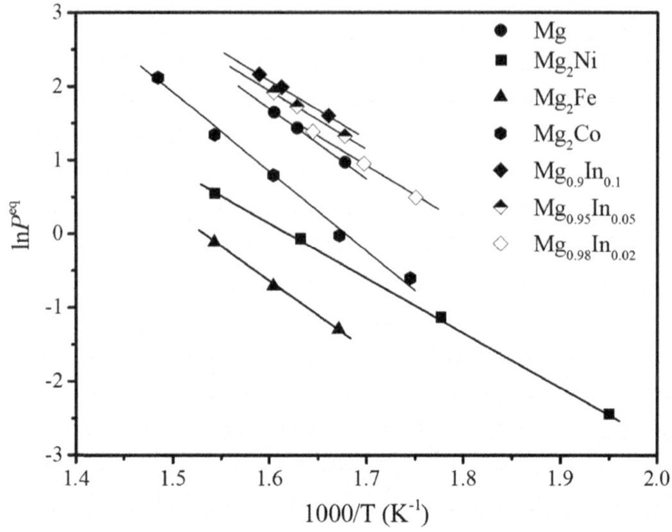

Fig. 1.15 The van't Hoff diagram of Mg-based hydrogen storage alloys

smaller than that of MgH_2, which indicates a lower hydrogen desorption temperature. The high reaction enthalpy (heat absorption and release) of Mg-based hydrogen storage alloys also makes them promising materials for energy storage [22].

The hydrogen absorption and desorption platforms of Mg-based hydrogen storage alloys is smoother and broader than other systems which also indicates the characteristics of large hydrogen storage capacity of Mg-based hydrogen storage alloys. Figure 1.16 shows the hydrogen absorption and desorption P–C–T curves of Mg and Mg_2Ni at different temperatures. MgH_2 has a higher thermal stability with decomposition ΔH° and ΔS° estimated at −74.4 kJ/mol H_2 and 130.1 J/(mol·K) H_2, respectively. This results in a decomposition temperature higher than 573 K at 0.1 MPa hydrogen pressure. The equilibrium pressure of Mg_2NiH_4 is higher than that of MgH_2, allowing hydrogen desorption at 523 K under 0.1 MPa.

Table 1.7 Standard enthalpy and standard entropy of hydrogen desorption of various hydrides

Hydrides	H_2 desorption temperature (K)	ΔH (kJ/mol H_2)	ΔS (J/(mol · K) H_2)
MgH_2	573	−74.4	130.1
Mg_2NiH_6	527	−64.4	122.2
Mg_2FeH_6	583	−80.0	137.0
Mg_2COH_5	590	−79.0	134.0
$LaNi_5H_6$	283	−30.9	109.0
$TiMn_2H_3$	216	−24.6	114.0
$TiFeH_2$	265	−28.1	106.0

Note The hydrogen desorption temperature condition given in the table is at 0.1 MPa H_2

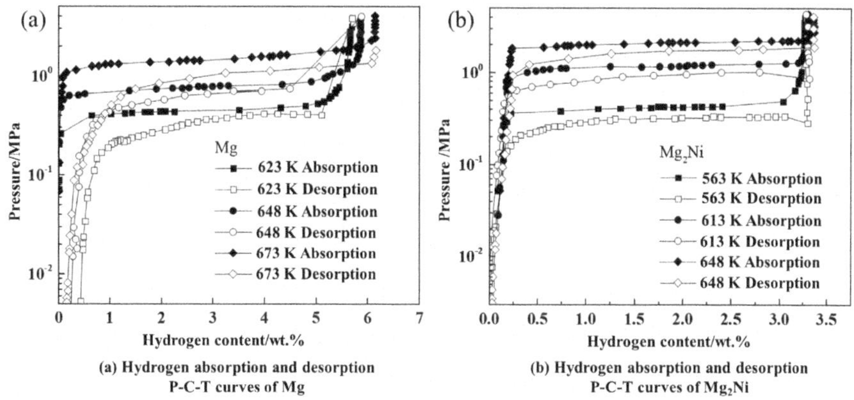

Fig. 1.16 Hydrogen absorption and desorption P–C–T curves of Mg and Mg_2Ni

The hydrogen absorption and desorption reactions are reversible processes, their theoretical equilibrium pressures are comparable. However, hysteresis is often observed between the hydrogen absorption and desorption processes. The equilibrium pressure of hydrogen absorption is generally higher than that of desorption. This phenomenon is called pressure hysteresis (Fig. 1.16). The primary reason for this phenomenon is the expansion of the metal lattice after hydrogenation, which creates compressive stress between the lattices. When hydrogen is released, the hydride remains unaffected by the stress, reducing the decomposition pressure and resulting in hysteresis in the P–C–T curves. The hysteresis coefficient (H_f) can be calculated as follows:

$$H_f = \ln\left(\frac{P_a}{P_d}\right) \tag{1.10}$$

where P_a and P_d are the equilibrium pressures of hydrogen absorption and desorption, respectively (MPa).

A larger hysteresis coefficient indicates a greater plateau pressure difference between the hydrogen absorption and desorption processs. During hydrogen absorption or desorption, energy is exchanged between the alloy and hydride. A larger pressure difference promotes more efficient reactions, but the hydrogen storage capacity and reaction heat cannot be effectively utilized in the systems with large hysteresis. Therefore, a smaller hysteresis coefficient is preferred for practical applications. Table 1.8 compares the hysteresis coefficients of Mg-based hydrogen storage alloys with several other typical hydrogen storage alloys.

Table 1.8 Hysteresis coefficients of several typical hydrides

Hydrides	P_a/MPa	P_d/MPa	Hysteresis coefficient
MgH_2	0.80 (648 K)	0.66 (648 K)	0.19
Mg_2NiH_4	1.15 (648 K)	0.93 (648 K)	0.21
$LaNi_5H_6$	0.20 (353 K)	0.16 (353 K)	0.22
Mg_2FeH_6	3.19 (573 K)	0.69 (573 K)	1.53
$TiFeH_2$	0.99(298 K)	0.37 (298 K)	0.98
$TiMn_2H_3$	0.70 (298 K)	0.44 (298 K)	0.46

1.3.3 Methods to Improve the Hydrogen Absorption and Desorption Reaction Thermodynamics in Mg-Based Hydrogen Storage Alloys

The high hydrogen desorption temperature is a bottleneck in practical applications of Mg-based hydrides. Significant efforts have been made to improve the hydrogen absorption and desorption thermodynamics, such as adding alloying elements (Ni, Cu, Al, Ti, Fe, Nb, La, Ce, Y, Si, etc.) and employing advanced preparation methods (ball milling, amorphization, microwave, and magnetic field heat treatment). The alloying strategy can form new Mg-based hydrides, reducing MgH_2 stability to lower the hydrogen desorption temperature. If the MgH_2 particles are sufficiently small, the influence of surface energy becomes significant, and the nano effect can improve the alloy's thermodynamic properties. External energy fields can introduce additional energy into the alloys, enhancing their hydrogen absorption and desorption properties. These approaches, which include reducing the reaction enthalpy by alloying, increasing the surface energy by nanosizing, and introducing additional energy via external field, effectively reduce the reaction energy in different ways. This section focuses on the theoretical mechanism to improve the thermodynamic properties of alloys. Detailed methods to improve the hydrogen absorption and desorption thermodynamics of Mg-based alloys are beyond this section's scope.

According to Eq. (1.6), when surface energy and additional energy terms are present, the Gibbs free energy difference can be expressed as:

$$\Delta G = \Delta G^{\circ} + RT \ln \frac{a_{\mathrm{MH_2}}}{a_{\mathrm{M}} P_{\mathrm{H_2}}} + \Delta G^{\mathrm{surf}}_{\mathrm{M \leftrightarrow MH_2}} + \Delta G^{\mathrm{ext}}_{\mathrm{M \leftrightarrow MH_2}} \tag{1.11}$$

where $\Delta G^{surf}_{M \leftrightarrow MH_2}$ is the surface energy change (J/mol) between M and MH_2 phases, and $\Delta G^{ext}_{M \leftrightarrow MH_2}$ is the energy change associated with external fields (J/mol).

When the hydrogen desorption reaction reaches equilibrium, Eq. (1.11) can be written as

$$RT \ln P^{\mathrm{eq}}_{\mathrm{H_2}} = \Delta H^{\circ} - T\Delta S^{\circ} + \Delta G^{\mathrm{surf}}_{\mathrm{M \leftrightarrow MH_2}} + \Delta G^{\mathrm{ext}}_{\mathrm{M \leftrightarrow MH_2}} \tag{1.12}$$

This can be obtained by collating Eq. (1.12):

$$\ln P_{H_2}^{eq} = \frac{\Delta H^\circ + \Delta G_{M\leftrightarrow MH_2}^{surf} + \Delta G_{M\leftrightarrow MH_2}^{ext}}{RT} - \frac{\Delta S^\circ}{R} \tag{1.13}$$

Bérubé et al. [28] summarized the effect of particle size on the heat of formation of hydride, where surface energy can be expressed as:

$$\Delta G_{M\leftrightarrow MH_2}^{surf} = \frac{3V_M E_{M\rightarrow MH_2}(\gamma, r)}{r} \tag{1.14}$$

where $E_{M\rightarrow MH_2}$ (γ,r) represents surface energy (J/m^2), determined by the surface energy per unit area γ (J/m^2) and particle radius r (m). The calculation of surface energy can be expressed as:

$$E_{M\rightarrow MH_2}(\gamma, r) = \left[\gamma_{MH_2}(r)\left(\frac{V_{MH_2}}{V_M}\right)^{2/3} - \gamma_M(r)\right] + E_{ads} \tag{1.15}$$

where V_{MH_2} and V_M are the molar volumes of MH_2 and M, respectively. Typically, the volume of metal or alloy expands 10~20% after hydrogenation. E_{ads} is the reduced surface energy due to expansion after the metal bonding with hydrogen.

The influence of an external magnetic field on Gibbs free energy can be expressed as:

$$\Delta G_{M\rightarrow MH_2}^{mag} = \left(G_{MH_2}^{mag} - G_M^{mag}\right) - \left(E_{MH_2}^{mag} - E_M^{mag}\right) \tag{1.16}$$

where $\Delta G_{MH_2}^{mag}$ and ΔG_M^{mag} represent Gibbs free energy (J/mol) of MH_2 and M adjusted by the magnetic field. According to the Hillert-Jahl model [28], $\Delta G_{M\leftrightarrow MH_2}^{mag}$ can be expressed as a function of magnetic moment (m) and curie temperature (T_c). In Eq. (1.16), $E_{MH_2}^{mag}$ and E_M^{mag} represent the static magnetic energy (J/mol) introduced by the external magnetic field, given by:

$$E_\phi^{mag} = \int_0^B m(T, B)dB \tag{1.17}$$

where $m(T,B)$ is the magnetic susceptibility, which is a function of temperature T (K) and magnetic field B (T).

By substituting the surface energy and magnetic Gibbs free energy into the van't Hoff equation [Eq. (1.12)], the effects of standard reaction enthalpy, surface energy, and external field energy on the reaction heat and equilibrium pressure can be obtained. The standard reaction enthalpy is related to the alloy composition, and ΔH° can be adjusted by alloying with different elements. Reducing the particle size of MgH_2 or introducing an external field may also reduce the reaction enthalpy change. Hartree–Fock ab initio calculations and density functional theory have estimated the effect of grain size on the thermal stability of Mg and MgH_2, showing that the stability of both Mg and MgH_2 decreases with cluster size. For cluster smaller than

19 Mg atoms, MgH_2 becomes less stable than Mg. Researchers have predicted that when the MgH_2 cluster is smaller than 0.9 nm, the hydrogen desorption temperature may decrease to 473 K. These theoretical insights provide guidance for designing Mg-based hydrogen storage alloys with high capacity and low hydrogen desorption temperatures.

References

1. JACOBS M H G, SPENCER P J. A critical thermodynamic evaluation of the system Mg-Ni [J]. Calphad, 1998, 22(4): 513–525.
2. REILLY J J, WISWALL R H. Reaction of hydrogen with alloys of magnesium and nickel and the formation of Mg_2NiH_4 [J]. Inorganic Chemistry, 1968, 7(11): 2254–2256.
3. POLANSKI M, PŁOCIŃSKI T, KUNCE I, et al. Dynamic synthesis of ternary Mg_2FeH_6 [J]. International Journal of Hydrogen Energy, 2010, 35(3): 1257–1266.
4. POLANSKI M, NIELSEN T K, CERENIUS Y, et al. Synthesis and decomposition mechanisms of Mg_2FeH_6 studied by in-situ synchrotron X ray diffraction and high pressure DSC [J]. International Journal of Hydrogen Energy, 2010, 35(8): 3578–3582.
5. CASTRO F J, GENNARI F C. Effect of the nature of the starting materials on the formation of Mg_2FeH_6 [J]. Journal of Alloys and Compounds, 2004, 375(1–2): 292–296.
6. ASANO K, AKIBA E. Direct synthesis of Mg-Ti-H FCC hydrides from MgH_2 and Ti by means of ball milling [J]. Journal of Alloys and Compounds, 2009, 481(1–2): L8–L11.
7. ASANO K, ENOKI H, AKIBA E. Synthesis of Mg-Ti FCC hydrides from Mg-Ti BCC alloys [J]. Journal of Alloys and Compounds, 2009, 478(1–2): 117–120.
8. ASANO K, ENOKI H, AKIBA E. Synthesis process of Mg-Ti BCC alloys by means of ball milling [J]. Journal of Alloys and Compounds, 2009, 486(1–2): 115–123.
9. LI Q, CHOU K C. Relations between key phases and their interfaces with properties in rare earth-magnesium aloys [J]. The Chinese Journal of Nonferrous Metals, 2019, 29(9): 1934–1952.
10. ZHANG Q A, LIU D D, WANG Q Q, et al. Superior hydrogen storage kinetics of $Mg_{12}YNi$ alloy with a long-period stacking ordered phase [J]. Scripta Materialia, 2011, 65(3): 233–236.
11. LI Y, GU Q, LI Q, et al. In-situ synchrotron X-ray diffraction investigation on hydrogen-induced decomposition of long period stacking ordered structure in Mg-Ni-Y system [J]. Scripta Materialia, 2017, 127: 102–107.
12. NIE J F, ZHU Y, MORTON A J. On the structure, transformation and deformation of long-period stacking ordered phases in Mg-Y-Zn alloys [J]. Metallurgical and Materials Transactions A, 2014, 45(8): 3338–3348.
13. YAMASAKI M, MATSUSHITA M, HAGIHARA K, et al. Highly ordered 10H-type long-period stacking order phase in a Mg-Zn-Y ternary alloy [J]. Scripta Materialia, 2014, 78: 13–16.
14. LIU C, ZHU Y, LUO Q, et al. A 12R long-period stacking-ordered structure in a Mg-Ni-Y alloy [J]. Journal of Materials Science & Technology, 2018, 34(12): 2235–2239.
15. LUO Q, GU Q, LIU B, et al. Achieving superior cycling stability by in situ forming NdH_2-Mg-Mg_2Ni nanocomposites [J]. Journal of materials chemistry A, 2018, 6(46): 23308–23317.
16. OURANE B, GAUDIN E, LU Y F, et al. The new ternary intermetallic $NdNiMg_5$: hydrogen sorption properties and more [J]. Materials Research Bulletin, 2015, 61: 275–279.
17. OURANE B, GAUDIN E, ZOUARI R, et al. $NdNiMg_5$, a new magnesium-rich phase with an unusual structural type [J]. Inorganic Chemistry, 2013, 52(23): 13289–13291.
18. PILLING N B, MEMBER M S, BEDWORTH R E. The oxidation of metals at high temperatures [J]. Journal of the Institute of Metals, 1923, 29: 529–591.

19. CHOU K C, LUO Q, LI Q, et al. Influence of the density of oxide on oxidation kinetics [J]. Intermetallics, 2014, 47: 17–22.
20. LI T, LI Q, LONG H, et al. Interpretation of negative temperature dependence of hydriding reaction in $LaNi_5$-Mg alloys by modified Chou model [J]. Catalysis Today, 2018, 318: 97–102.
21. LUO Q, CAI Q, GU Q, et al. Negative thermal expansion and phase transition of low-temperature Mg_2NiH_4 [J]. Journal of Magnesium and Alloys, 2023, 11(9): 3338–3349.
22. SHEPPARD D A, BUCKLEY C E. The potential of metal hydrides paired with compressed hydrogen as thermal energy storage for concentrating solar power plants [J]. International Journal of Hydrogen Energy, 2019, 44(18): 9143–9163.
23. CHAISE A, DE RANGO P, MARTY P, et al. Enhancement of hydrogen sorption in magnesium hydride using expanded natural graphite [J]. International Journal of Hydrogen Energy, 2009, 34(20): 8589–8596.
24. SUISSA E, JACOB I, HADARI Z. Experimental measurements and general conclusions on the effective thermal conductivity of powdered metal hydrides [J]. Journal of the Less Common Metals, 1984, 104(2): 287–295.
25. ISHIDO Y, KAWAMURA M, ONO S. Thermal conductivity of magnesium-nickel hydride powder beds in a hydrogen atmosphere [J]. International Journal of Hydrogen Energy, 1982, 7(2): 173–182.
26. POHLMANN C, RöNTZSCH L, HU J, et al. Tailored heat transfer characteristics of pelletized $LiNH_2$-MgH_2 and $NaAlH_4$ hydrogen storage materials [J]. Journal of Power Sources, 2012, 205: 173–179.
27. OSUMI Y. Properties and applications of metal hydrides [M]. Beijing: Chemical Industry Press, 1990.
28. BERUBE V, RADTKE G, DRESSELHAUS M, et al. Size effects on the hydrogen storage properties of nanostructured metal hydrides: a review [J]. International Journal of Energy Research, 2007, 31(6–7): 637–663.

Chapter 2
Kinetics of Hydrogen Absorption and Desorption of Mg-Based Hydrogen Storage Alloys

From a macroscopic point of view, the hydrogen absorption and desorption process in the Mg-based hydrogen storage system involves the storage and release of hydrogen in the metal hydrides. On the micro level, however, it is a reaction process between Mg alloys and hydrogen. The apparent reaction rates correspond to multiple steps in the microscopic reaction process. Therefore, to improve the kinetics of Mg-based hydrogen storage alloys, it is necessary to understand the factors influencing the rate-controlling steps and the kinetic mechanism on hydrogen absorption and desorption. The reaction kinetic mechanism is mainly associated with specific pathways and corresponding reaction rates. To this end, it is important to identify the rate-controlling steps and analyze the relationship between reaction rates and various variables. Macroscopic reaction rates are closely related to fluid flow status and mass/heat transfer characteristics. At the microscopic level, studies on reactions reveal the influence of alloy composition, micro-morphology, reaction temperature, initial hydrogen absorption pressure, and other factors on the hydrogen absorption and desorption reaction of Mg-based alloys. The research methods for macroscopic and microscopic studies mainly include experimental and theoretical methods. The experimental research provides the hydrogen absorption and desorption data under specific conditions. It needs to be measured again if the experimental conditions are changed. The theoretical research mainly relies on the physical models, which are constructed based on experimental data and then combined with existing theories to establish the mathematical relationships between the reaction rate and variables. Finally, a kinetic model is obtained to mimic the reaction process. Some parameters in kinetic models are obtained from experiments, and then the reaction rates in other unmeasured conditions can be predicted via the model.

In this chapter, the relationship between the macroscopic and microscopic processes of hydrogen absorption and desorption of Mg-based hydrogen storage alloys will be discussed. Next, the classic kinetic models of hydrogen absorption and desorption, as well as the influencing parameters, will be introduced. Finally, the

Q. Li and F. Pan, *Magnesium-based Hydrogen Storage Materials*,
https://doi.org/10.1007/978-981-96-5531-1_2

microscopic kinetic mechanisms of hydrogen absorption and desorption reactions of Mg-based hydrogen storage alloys will be summarized.

2.1 Macroscopic and Microscopic Processes of Hydrogen Absorption and Desorption Reactions of Mg-Based Hydrogen Storage Alloys

2.1.1 Macroscopic Processes and Evaluation Methods of Hydrogen Absorption and Desorption Reactions of Mg-Based Hydrogen Storage Alloys

Macro-kinetics study focuses on the influence of fluid flow rate and mass/heat transfer characteristics on the hydrogen absorption and desorption process in the Mg-based hydrogen storage alloys, which are also related to the shape and size of the reactor. Metal hydrides are usually filled into the reactors, which possess the merit of high safety, high volumetric hydrogen storage capacity, mild hydrogen absorption and desorption temperature and low working pressure, etc. The hydrogen absorption and desorption processes in reactors can be categorized as the hydrogen absorption and desorption process in the Mg-based hydrogen storage alloys and the heat and mass transfer in hydrogen storage tanks. These two processes, i.e., heat transfers and mass transfers during hydrogen absorption and desorption, are the major research focuses in macro-kinetics. For example, the hydrogen storage capacity, operation temperature and pressure, reaction rate, and rate-to-capacity ratio are intensively investigated in the macro-kinetic analysis.

Hydrogen absorption in Mg alloy is an exothermic process ($\Delta H = -74.4$ kJ/mol·H_2). On the contrary, external heat is needed to drive the hydrogen desorption reaction. For microscopic reactors, the small-scale reactions result in negligible temperature and pressure gradient at different locations. However, in medium-to large-scale reactors, the reaction heat generated by hydrogen absorption cannot be conducted out promptly due to the poor heat conductivity of powder reaction beds [less than 2 W/(m·K)], leading to the drastic temperature changes and significant temperature gradient. As the temperature increases, the equilibrium pressure of hydrogen absorption rises, decreasing the hydrogen absorption rate obviously. It has been revealed that the heat-transfer process is one of the controlling steps in the hydrogen absorption reaction, especially in the initial stage, and the temperature of the reaction bed varies significantly [1].

The design of reactors needs to consider both operating conditions and heat/mass transfer characteristics. Operating conditions refer to the initial temperature, pressure, hydrogen content, cooling gas/liquid flow rate, etc. These factors directly affect the hydrogen absorption and desorption reaction rates in the reactor. The enhancement of the heat and mass transfer characteristics of the reactor is mainly achieved by improving the heat and mass transfer performance, such as ameliorating the temperature fluctuations and pressure fluctuations, and thus increase the hydrogen absorption and desorption rates of the reactor. Taking hydrogen absorption process of hydrogen storage alloys as an example, the alloy is filled in the reactor and then reacts with the hydrogen gas flowing into the reactor, the reaction generates heat and consumes hydrogen. Without a continuous hydrogen supply to the powder bed, the hydrogen inside the powder bed will be consumed until the pressure drops to the equilibrium pressure. Finally, the hydrogen absorption reaction stops. Therefore, a continuous hydrogen flow is necessary for the hydrogen absorption reaction in the reactor. Furthermore, even with the continuous hydrogen supply, a pressure gradient is inevitably generated when the hydrogen absorption reaction happens, as shown in Fig. 2.1. When the reaction of the alloy is initiated, the hydrogen absorption reaction releases a drastic amount of heat. Thus, the equilibrium pressure is also increased, suppressing hydrogen absorption kinetics. If heat loss is not considered, the temperature inside the reaction bed will continuously increase until it reaches the temperature corresponding to the equilibrium pressure, and the hydrogen absorption reaction is terminated. When the heat dissipation at the wall results in that wall temperature is lower than that inside the powder bed, the temperature gradient between the wall and the powder bed promotes the hydrogen absorption reaction. Heat dissipation is also necessary for the continuous hydrogen absorption reaction in reactors. Therefore, the alloy's hydrogen absorption and desorption reaction should affect heat transfer and hydrogen gas flow in the powder bed.

The two key parameters affecting the heat transfer and hydrogen gas flow rates in the powder bed are the effective thermal conductivity and the permeability, which vary with pressure, particle size, and porosity. The volumetric expansion and contraction phenomena of the alloys during the hydrogen absorption and desorption process

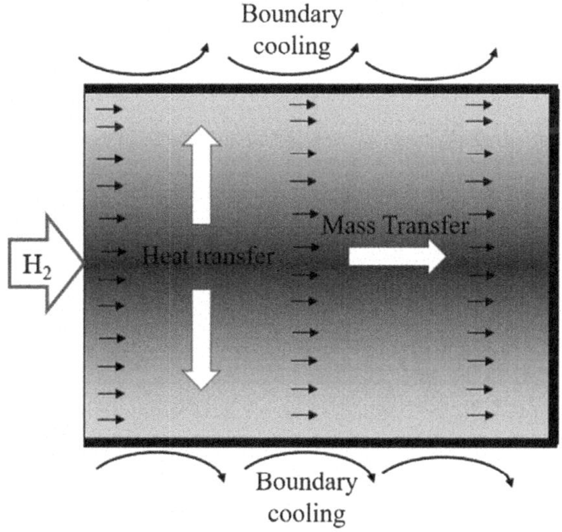

Fig. 2.1 Heat and mass transfer processes in the reactor

cause changes in particle size and porosity, which in turn lead to changes in the effective thermal conductivity and permeability of the powder beds concurrently. These factors affect the hydrogen absorption and desorption reaction rates of the reactor.

The methods for measuring hydrogen absorption and desorption kinetics of hydrogen storage alloys can be classified into three categories: isothermal type, variable temperature type, and constant hydrogen flow rate type. Isothermal hydrogen absorption and desorption refer to the reactions conducted at a constant temperature. Their reaction rates depend on the temperature, initial hydrogen pressure, and cooling gas/liquid flow rate, etc. The variable-temperature reactions indicate that the reactions are conducted at varying temperatures with a specific heating or cooling rate. The reaction rates are controlled by the temperature change rate, initial hydrogen pressure, and cooling gas/liquid flow rate. The reaction rates under the constant hydrogen flow rate are determined by the hydrogen flow rate, temperature, hydrogen pressure, and gas/liquid flow rate.

The research methods of macroscopic hydrogen absorption and desorption reaction kinetics in metal hydride reactors can be categorized by experimental and theoretical studies. Experimental studies are generally based on qualitative or semi-quantitative empirical knowledge, which cannot accurately reflect the comprehensive performance of the reactor. Importantly, it relies more on the empirical design of the reactor. The specific performance of the reactor can only be determined after conducting the hydrogen absorption and desorption reaction experiments, which are costly and time-consuming. Moreover, the temperature and pressure gradients inside the powder bed cannot be obtained. Thus, experimental study is usually used to obtain the real hydrogen absorption and desorption reaction data of the reactor to verify the theoretical studies. Physical simulation and numerical simulation are the two main methods in theoretical studies. The physical simulation method is used to replicate a real physical process in a certain proportion to study the influence of different parameters on the reactor. What needs to be emphasized is that the scaled-down model should be based on realistic conditions, including geometric and operational conditions, etc. However, considering that the hydrogen absorption and desorption reaction performance of the reactor relies on heat/mass transfer, but the heat transfer distance and area of the simulated reactor are different from the original one by scale reduction. Thus, the physical simulation of the reactors can only be applied to qualitative analysis, and this method is rarely applied in the field of reactor design. Numerical simulation is an effective method to study and predict the kinetic performance of hydrogen absorption and desorption reactions in the reactor. The numerical simulation method constructs a mathematical model, including partial differential equations or functions, based on a simplified model for the physicochemical processes of the hydrogen absorption and desorption reactions in the reactor. Under certain initial and boundary conditions, the mathematical model can be solved using finite element method. It provides the variation of temperature, pressure, hydrogen content, and their distribution in the reactor. As a result, quantitative predictions of the hydrogen absorption and desorption reactions in the reactor can be obtained. The numerical simulation has the following advantages:

① It provides a fast and relatively accurate prediction of the hydrogen absorption and desorption reaction performance in the reactor.
② The cost of numerical simulation is lower than that of experiments.
③ Some experimentally unavailable data, such as temperature, pressure, and hydrogen content distribution inside the reactor, can be obtained via the simulation method.

Therefore, numerical simulation has been widely used in the hydrogenation and dehydrogenation kinetic studies in reactors.

2.1.2 Microscopic Processes and Evaluation Methods of Hydrogen Absorption and Desorption Reactions of Mg-Based Hydrogen Storage Alloys

Microscopic kinetics is the basis of chemical reactions. It mainly studies the relationship between the reaction rate and the factors associated with the reaction rate, such as concentration and temperature. Microscopic kinetics assumes that the internal temperature of the object is uniform and excludes the influences of mass, heat, and momentum transfers. It focuses on the reaction rate and intrinsic mechanism of the material.

The essential chemical equation of hydrogen absorption and desorption reaction of Mg alloys can be expressed as follows:

$$Mg + H_2 \leftrightarrow MgH_2 \tag{2.1}$$

As the hydrogen absorption reaction proceeds, the content of MgH_2 gradually increases until all the Mg is fully transformed. The hydrogen storage capacity is typically expressed by the mass fraction of hydrogen in the alloy (wt.%), the hydrogen/metal (H/M) ratio, or the ratio of the absorbed hydrogen relative to the theoretical value (ξ). For practical measurements, the amount of absorbed and desorbed hydrogen is calculated by changing the hydrogen pressure under a constant volume. Under constant pressure, the amount of absorbed and desorbed hydrogen can be obtained by mass change as a function of time or temperature.

The hydrogen absorption processes of alloys are generally composed of six steps, as shown in Fig. 2.2. The hydrogen desorption reaction can be regarded as the reverse of the hydrogen absorption reaction.

The main processes of the hydrogen absorption reaction can be described as follows [2]:

① H_2 physisorption on the particle surface: H_2 molecules are physically adsorbed on the particle surface by the van der Waals force.
② H_2 dissociation and chemisorption on the particle surface: H_2 molecules is dissociated into H atoms and chemisorbed on the particle surface.

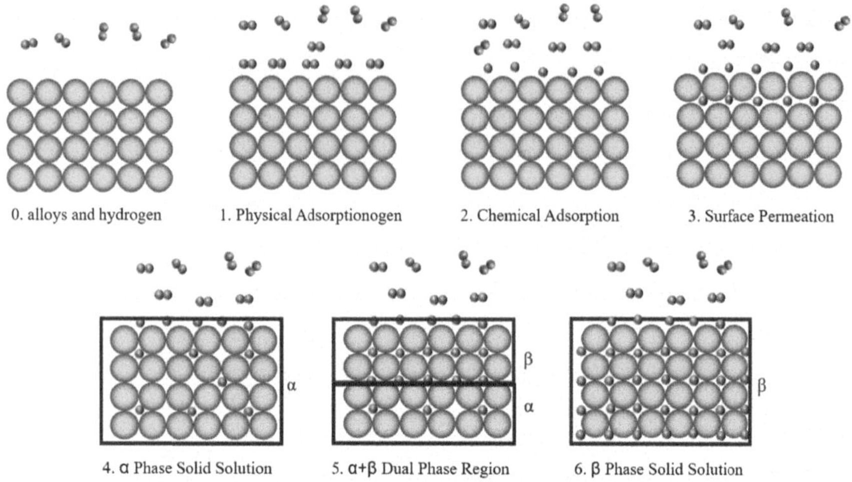

Fig. 2.2 Schematic illustration of the hydrogen absorption process in alloys

③ Surface penetration of hydrogen atoms: H atoms penetrate through the surface into the alloy lattice.
④ α solid solution: Hydrogen atoms gradually diffuse into the lattice to form a solid solution.
⑤ α + β two-phase region: When the concentration of H atoms reaches a certain level, the alloy undergoes a phase transition to form the hydride phase.
⑥ β phase solid solution: After all the α-phase transforms to the β phase, the hydrogen atoms are distributed into the β phase under higher pressure.

Hydrogen absorption kinetics of the alloys can be influenced by environmental conditions, including gas pressure, temperature, and the intrinsic properties of the alloy, etc. When the hydrogen pressure increases, the hydrogen concentration on the surface of the sample increases. However, since the equilibrium pressure is temperature-dependent, the increased gaseous hydrogen pressure raises the gap between gaseous pressure and equilibrium pressure, thus enhancing the diffusion rate and the absorption rate of hydrogen. The increased temperature can enhance the rate of hydrogenation reaction. Meanwhile, since hydrogen absorption is an exothermic reaction, the increased temperature results in a higher equilibrium pressure, which reduces the driving force of the reaction. The intrinsic properties of the alloy include the alloy composition and the morphology and particle size of alloy powder, etc. The addition of some active elements, such as rare earth and transition metals (Ni, Fe, Co, Ti) in Mg significantly enhances the reaction rate. Decreasing the particle size of Mg alloy leads to the increased contact area between the particles and hydrogen, significantly improving the reaction rate.

Similar to the macroscopic study of hydrogen absorption and desorption reactions, microscopic study also can be categorized into experimental study and theoretical model analysis as well. Experimental studies need to measure the hydrogen absorption and desorption rates as a function of time under different reaction conditions, which is time-consuming and costly. In fact, most kinetic processes are studied by combining theoretical and experimental methods. The theoretical study is based on the experimental data. By combining the existing theoretical kinetic models with the experimental data, the relationship between the reaction rate and each influencing parameter can be mathematically obtained. When the kinetic analysis fits well with experimental data, it demonstrates that the hydrogen absorption and desorption reactions of different metal hydride compounds are governed by kinetic laws. However, if some kinetic parameters are missing or the model equations are too complicated to solve, some empirical assumptions are needed, some parameters of the kinetic model also need to be determined by experiment.

2.1.3 Relationship Between Macroscopic and Microscopic Kinetic Processes of Hydrogen Absorption and Desorption Reactions

Some aspects of the macroscopic kinetic processes of hydrogen absorption and desorption reactions differ from microscopic kinetic processes, but they share similarities. From the research perspective, the macroscopic kinetics is associated with large-scale phenomena and mainly studies the influence of heat and mass transfer conditions on the reaction rate in the whole reactor. Meanwhile, microscopic kinetics focuses on the relationship between the intrinsic reaction rate of the material, hydrogen concentration, reaction temperature, and other factors. From the aspect of the research objective, macroscopic kinetics needs to consider the hydrogen absorption and desorption reactions of the alloy, heat, and mass transfer process of the reactor. However, microscopic kinetics mainly considers the reaction rate and reaction mechanism of the material. The research fields of macroscopic studies include the shape of the reactor, the structure of the gas pipeline, the type and arrangement of the heat transfer medium, and the method of reactant molding, etc. The microscopic studies focus on the factors affecting the intrinsic kinetics of the material, including the hydrogen partial pressure, temperature, heating rate, alloy composition, particle size, morphology, etc. Both macroscopic and microscopic studies have experimental and theoretical branches. The macroscopic studies need to measure the temperature distribution in the reactor, hydrogen flow rate, and hydrogen pressure at the inlet/outlet of the reactor. Microscopic studies mainly focus on the change of hydrogen pressure and the amount of absorbed and desorbed hydrogen during the reaction. The simulation of macroscopic kinetics is complicated, but it is more suitable to choose numerical simulation for investigating hydrogen absorption and desorption reaction kinetics, heat transfer, and mass transfer in the reactor. Microscopic studies are usually conducted by employing kinetic models, and the rate-controlling step of the reaction is evaluated by fitting different models. Table 2.1 summarizes the

Table 2.1 Similarities and differences between macro- and micro-reaction kinetic studies

Items	Macroscopic kinetics	Microscopic kinetics
Scale	Large, macroscopic	Small, microscopic
Research subjects	Heat/mass transfer in reactors	Reaction rate and mechanism of materials
Research field	Shape, structure, material, reactant accumulation method of reactors	Hydrogen partial pressure, temperature, temperature rise rate, particle size and morphology
Research methods	Experimental measurement, numerical simulation	Experimental measurements, kinetic model calculations

similarities and differences between macroscopic and microscopic kinetic studies. Cutting-edge kinetic research is mainly focused on microscopic reaction kinetic analysis, and this chapter introduces several commonly used classical kinetic models and their applications on Mg-based hydrogen storage alloys.

2.2 Micro-kinetics of Hydrogen Absorption and Desorption Reactions in Mg-Based Hydrogen Storage Alloys

2.2.1 Micro-kinetic Mechanism of Hydrogen Absorption and Desorption Reaction in Mg-Based Hydrogen Storage Alloys

The hydrogen absorption and desorption of alloys is a gas–solid reaction. To establish a reasonable effective model, it is required to understand the reaction processes. Typically, the hydrogen absorption and desorption reactions of alloys involve in the following steps:

① Physisorption: Hydrogen is first physically adsorbed on the particle surface.
② Chemisorption: Adsorbed hydrogen molecules dissociate into hydrogen atoms that occupy chemical vacancies on the surface of alloy particles.
③ Surface penetration: Hydrogen atoms penetrate into the crystal lattice of the alloy.
④ Hydrogen diffusion: Hydrogen atoms diffuse across the β phase to the interface between the α and β phases when a certain thickness of β phase is formed.
⑤ Nucleation and particle growth: The β phase nucleates and grows in the supersaturated α phase. In this step, no clear interface between the α phase and the β phase is formed because of the large solid solubility of hydrogen in the α phase.
⑥ Interfacial chemical reaction: The β-phase forms at the interface between the α and the β phases.

The schematic illustration of these reaction steps is shown in Fig. 2.3. The hydrogen desorption reaction is the reverse process of the hydrogen absorption reaction.

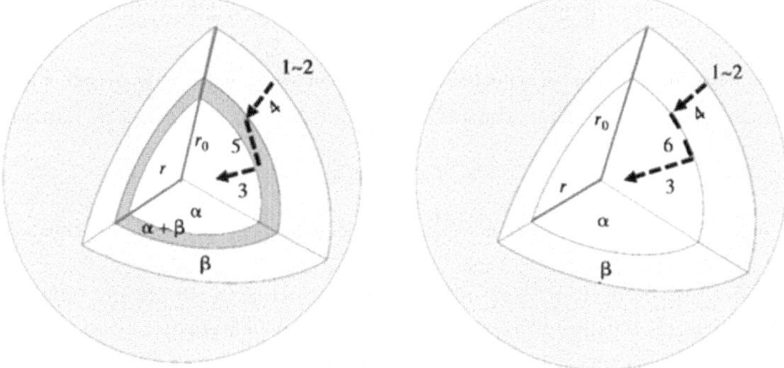

(a) Nucleation and growth process (b) Reaction occurring at the α/β interface

Fig. 2.3 Schematic illustration of the hydrogen absorption process. The reaction steps include ① physical adsorption, ② chemical adsorption, ③ surface penetration, ④ hydrogen diffusion, ⑤ β phase nucleation, and ⑥ interfacial chemical reaction

The micro-scale hydrogen storage kinetics is the basis for the analysis of macroscopic processes. In-situ experiments and numerical modeling of microstructure can elucidate the kinetic mechanism of hydride formation in Mg-based alloys.

Step ① is the hydrogen physisorption process. The major driven force for hydrogen diffusion to the surface of magnesium-based metals is the pressure difference. On the microscopic scale, the interaction between hydrogen and the surface of hydrogen storage alloys occurs in the boundary layer of the fluid. The transport of hydrogen to the surface and its adsorption on the surface usually reach a local equilibrium. The adsorption process of hydrogen on the Mg-based alloy surface is driven by intermolecular forces or hydrogen bonding. A monomolecular layer first forms on the surface when hydrogen is physically adsorbed, followed by multilayer adsorption of hydrogen molecular through intermolecular forces. Considering the kinetics of adsorption and its inverse process, the rate of the multilayer absorption process can generally be described by the following equation:

$$v_{pa} = k_{ab}^{pa}\left(1 - \theta_{pa}\right)P - k_{de}^{pa}\theta_{pa} \tag{2.2}$$

where k_{ab}^{pa} and k_{de}^{pa} are the kinetic coefficients of physical adsorption and its inverse process, which can be obtained by the Arrhenius formulas with variables of temperature and activation energy θ_{pa} represents the surface coverage and P represents the hydrogen pressure.

According to the Arrhenius formula, Eq. (2.2) can be expressed as follows:

$$v_{pa} = k_{ab,0}^{pa}\exp\left(-\frac{E_{ab}^{pa}}{RT}\right)\left(1 - \theta_{pa}\right)P - k_{de,0}^{pa}\exp\left(-\frac{E_{de}^{pa}}{RT}\right)\theta_{pa} \tag{2.3}$$

where E_{ab}^{pa} and E_{de}^{pa} are the activation energy of hydrogen adsorption and its inverse process.

If no activation energy is needed for the physical adsorption and the rate of adsorption is much faster than the reverse process, Eq. (2.3) can be simplified as follows:

$$v_{pa} = k_{ab,0}^{pa}(1 - \theta_{pa})P \quad (2.4)$$

In step ②, hydrogen chemisorption and dissociation occur on the surface of the Mg-based hydrogen storage alloys. The dissociation of hydrogen can be expressed as $H_2 \leftrightarrow 2[H]$. H_2 is converted to H atoms that penetrate and diffuse in alloys and hydrides. Chemisorption is a complex kinetic process, which is associated with chemical interactions between physically adsorbed hydrogen and solid surface atoms. The chemisorption process includes electron transfer, atomic rearrangement, and chemical bond formation/breaking. The process also includes the formation of hydride compounds on the first layer of molecules. Due to surface roughness, only certain areas such as lattice defects, crystal edges, and other positions with low nucleation energy have chemisorption activity and become active sites for chemisorption. A hydrogen molecule dissociates into two hydrogen atoms, each occupying an active site. After completely covering these active sites, the surface becomes saturated with chemisorption of hydrogen atoms. In general, the kinetic rate of chemisorption can be calculated by Elovich equation, which is expressed as follows:

$$\begin{aligned} v_{ca} = {} & k_{ab,0}^{ca} \exp\left(-\frac{E_{ab}^{ca} + a\theta_{ca}}{RT}\right)(1 - \theta_{ca})^2 P \\ & + k_{de,0}^{ca} \exp\left(-\frac{E_{de}^{ca} + b\theta_{ca}}{RT}\right)\theta_{ca}^2 \end{aligned} \quad (2.5)$$

where a and b are the coefficients related to the surface state of the particles. θ_{ca} is the solid surface-active site occupancy, whose value is generally between 0 and 1.

When the value of θ_{ca} is relatively small (approach to zero), the inverse process of chemisorption can be neglected. Therefore, Eq. (2.5) can be simplified as follows:

$$v_{ca} = k_{ab,0}^{ca} \exp\left(-\frac{E_{ab}^{ca} + a\theta_{ca}}{RT}\right)P \quad (2.6)$$

For moderate values of θ_{ca}, the effects of $(1 - \theta_{ca})^2$ and θ_{ca}^2 in Eq. (2.5) are negligible and can be merged into the kinetic coefficients $k_{ab,0}^{ca}$ and $k_{de,0}^{ca}$, which are expressed as follows:

$$v_{ca} = k_{ab,0}^{ca} \exp\left(-\frac{E_{ab}^{ca} + a\theta_{ca}}{RT}\right)P + k_{de,0}^{ca} \exp\left(-\frac{E_{de}^{ca} + b\theta_{ca}}{RT}\right) \quad (2.7)$$

Step ③ is the hydrogen transport in solid, i.e., the hydrogen atoms dissolve and diffuse in the Mg-based material. At the initial stage of the hydrogen storage process, the surface of Mg-based materials is covered by a passivation layer of metal oxides. The hydrogen transportation first needs to pass through the passivation layer and reduce Mg oxides to Mg metal simultaneously. Generally, the hydrogen diffusion in the passivation layer is very slow. Therefore, reducing the oxide is crucial to the incubation period of hydride growth. Hydrogen transport to the metal is faster at the defective or detached passivation layer. Therefore, a localized region of higher hydrogen concentration is generated near the surface oxide layer, which becomes the preferred site for the nucleation and growth of hydride. In addition, localized volume expansion associated with hydride formation may result in cracks in the passivation layer, providing a channel for faster hydrogen transport.

The surface penetration process is the process where hydrogen atoms that are chemically adsorbed on the surface penetrate through the lattice of the surface layer and then enter the interior of the alloy. When the passivation layer is reduced to Mg metal, the hydrogen atoms directly contact the β phase and penetrate into the lattice from the surface-active sites into the bulk of the alloy. This process requires high activation energy, only hydrogen atoms with higher energy can permeate into the alloy. The concentration of hydrogen atoms involved in surface penetration can be described as $C_{sur} = k_{ca}\theta_{ca}$ and the kinetic rate of surface penetration is calculated as:

$$v_{sp} = k_{ab}^{sp} C_{sur} - k_{de}^{sp} C_{\beta} \tag{2.8}$$

where k_{ab}^{sp} and k_{de}^{sp} are the kinetic coefficients; C_{β} is the hydrogen concentration in the β phase.

In step ④, the hydrogen transport process follows the classical diffusion law once the hydrogen atoms are accommodated inside the β phase. The diffusion process consists of the hydrogen diffusion from the gas/β phase surface to the α/β phase interface or the diffusion of hydrogen atoms from the α/β phase interface to the inside of the β grains distributed in the α phase. The diffusion equation can be expressed as follows:

$$\frac{\partial C}{\partial t} = \nabla (D_H \nabla \mu) \tag{2.9}$$

where C is hydrogen concentration, D_H is the diffusion coefficient, and μ is the chemical potential of hydrogen.

The diffusion of hydrogen in the solid phase can be regarded as multiple-phase diffusion, including the hydrogen diffusion in the α and β phases, the redistribution of hydrogen on both sides of the α/β phase interface, and the diffusion of hydrogen along grain boundaries. Therefore, the diffusion coefficient D_H in Eq. (2.9) is affected by the temperature, phase, and grain boundaries at each location. Thus, the diffusion potential expression may differ in different phases. In the initial stage, the β phase usually nucleates and grows near the passivation layer due to the high gaseous hydrogen concentration. However, the diffusion coefficient of hydrogen in the β phase is relatively small, which hinders hydrogen transport and restricts the growth of hydrides towards the interior.

In step ⑤, the nucleation and growth of the β phase is the critical step for the hydrogen storage reaction. This step is generally governed by the generation and movement of the α/β phase interface. As the β phase grows, the α/β interface moving inwards and participates in the chemical reaction. It can also be regarded as the appearance of hydride in the hydrogen supersaturated region. Typically, the generation and movement of α/β interface are influenced by three factors: chemical driving force in the interfacial region, micro-elastic stress caused by the volume expansion, and Laplace pressure. According to the classical Stefan model, the kinetic equation of the interface movement can be expressed as follows:

$$v_{\perp} = \beta(T)\left(-\Delta G^{ca}_{\alpha\to\beta} - f^{ela} - \sigma\kappa\right) \tag{2.10}$$

where $v_{\perp}$ is the velocity of motion to the interface, $\beta(T)$ is the temperature-dependent kinetic coefficient, $\Delta G^{c\alpha}_{\alpha\to\beta}$ is the driving force for the phase transition from the α phase to the β phase, f^{ela} is the micro-elastic stress at the interface, σ is the interfacial energy, and κ is the interfacial curvature.

The chemical driving force $\Delta G^{c\alpha}_{\alpha\to\beta}$ is determined by the temperature, the hydrogen concentration on the interface, and the composition of the Mg-based material. Considering that hydrogen atoms diffuse rapidly along the grain boundaries in metals, the hydrogen concentration at and near grain boundaries tends to be higher than that in the bulk, leading to a higher chemical driving force around grain boundaries. In addition, the kinetic factor of phase transition at grain boundaries, $\beta(T)$, is generally larger than that in bulk. The faster diffusion rate and larger kinetic factor are favorable for the growth of hydrides around the grain boundaries. The β phase hydrides tend to nucleate near the grain boundaries of the α phase and grow into the bulk. The transition from the α phase to the β phase accompanies a specific volume expansion, resulting in localized stress around the β phase. The volume of the β phase generally expands by ~20% compared to that of the α phase. The stress generated by this expansion strain usually suppresses the growth of the expanding phase, i.e., inhibits the development of the expanding phase during the hydrogen storage process. In the case of Mg-based materials, the growth of the β phase is inhibited during the hydrogen storage process, while the growth of the α phase is promoted during the hydrogen release process. In addition, the expansion caused by the formation of the β phase may generate microcracks, forming additional gas-phase channels for hydrogen penetration. Under thermal equilibrium conditions, the plateau region of the P-C-T curve is related to the effect of micro-elastic stress. The Laplace pressure depends on the energy and the curvature of α/β phase interface, which is the origin of the Gibbs–Thomson effect. In particular, in the early stage of the β phase nucleation and growth, the interface curvature is very large and significantly affected by the Laplace pressure.

2.2.2 Kinetic Model of the Hydrogen Absorption and Desorption Reaction in Hydrogen Storage Alloys

Normally, steps ①–③ are very fast, and the hydrogen diffusion, nucleation, or chemical reaction is considered to be the rate-determining step (RDS). To date, various models have been proposed to describe the reaction rate of these steps, such as the classical Jander model [3], Ginstling-Brounshtein (G-B) model [4], Valensi-Carter (V-C) model [5], and Chou model [6] for describing diffusion-controlled kinetic models; Jonhson-Mehl-Avrami-Kolomogorov (JMAK) model [7–9] for nucleation and growth-controlled kinetic models; as well as nucleation self-catalysis model [10–12]. In the following sections, these kinetic models of hydrogen absorption and desorption reactions are described based on the corresponding RDS.

1. Diffusion controlled kinetic model

These models assume a uniform temperature inside the particles and a constant external hydrogen pressure. The hydrogen diffusion in the hydride layer is the slowest and it is assumed to be the RDS.

(1) Jander model

For the hydrogen diffusion process, the hydrogen absorption and desorption rates are inversely proportional to the thickness of the β phase layer. Considering the dimensions (d) of spherical, cylindrical, and plate-like particles as 3, 2, and 1, respectively, the reaction fraction ξ can be expressed as a function of radius r as follows:

$$\xi = 1 - \left(\frac{r}{r_0}\right)^d \tag{2.11}$$

For simplicity, we assume that the area of the diffusion interface is constant, i.e., simplifying three- and two-dimensional diffusion into one-dimensional diffusion [3, 13]. We further assume that the particle volume before and after hydrogen absorption and desorption remains unchanged. According to Fick's first law of diffusion, hydrogen absorption and desorption rates can be expressed as follows:

$$\frac{\rho dr}{dt} = -\frac{D_H \Delta C}{r_0 - r} \tag{2.12}$$

where ρ is the density of the alloy, D_{H} is the diffusion coefficient, ΔC is the difference in hydrogen concentration between the gas/hydride interface (g/β) and the hydride/metal interface (β/α), r_0 is the original radius of the particle. By separating and integrating the variables in Eq. (2.12), we can derive the following equation:

$$\frac{1}{2}r^2 - r_0 r - \frac{1}{2}r_0^2 = \frac{D_H \Delta C}{\rho} t \tag{2.13}$$

Combining Eqs. (2.11) and (2.13), the integral equation of the Jander model is obtained:

$$\left[1-(1-\xi)^{\frac{1}{d}}\right]^2=\frac{2D_H\Delta C}{r_0^2}t=k_d t \tag{2.14}$$

where $\frac{2D_H\Delta C}{r_0^2}$ is replaced with k_d, the reaction rate constant. Its value indicates the speed of the reaction, and a larger k_d indicates a faster reaction rate. The Jander model is simple and widely used in the reaction in which the diffusion process is the RDS. By fitting the data between the reaction rate and time, the reaction rate constant, which is used to compare the magnitude of reaction rates, can be obtained. However, the Jander model only considers the relationship between reaction rate constant and particle size, while many other factors affecting diffusion are ignored.

(2) G-B model

The G-B model is established based on the Jander model, amending the improper assumption that the constant area of the diffusion interface [4, 14]. According to Fick's first law, the diffusion equation for two-dimensional columnar particles can be expressed as follows:

$$\frac{\rho\partial r}{\partial t}=-\frac{D_H\Delta C}{r\ln(r_0/r)} \tag{2.15}$$

For three-dimensional spherical particles, the diffusion-controlled equation can be expressed as follows:

$$\frac{\rho\partial r}{\partial t}=-\frac{D_H\Delta C r_0}{(r_0-r)r} \tag{2.16}$$

Substituting the integrated Eq. (2.15) and Eq. (2.16) to Eq. (2.11), for the reaction fraction versus particle size for two- and three-dimensional particles, the G-B model can be expressed by the following equations:

$$(1-\xi)\ln(1-\xi)+\xi=\frac{4D_H\Delta C}{r_0^2\rho}t=k_d t \tag{2.17}$$

$$1-\frac{2}{3}\xi-(1-\xi)^{\frac{2}{3}}=\frac{2D_H\Delta C}{r_0^2\rho}t=k_d t \tag{2.18}$$

The G-B model is more complicated than the Jander model, but it is more accurate as it accounts the changes in the diffusion interface area.

(3) V-C model

The V-C model is built by further amending the G-B model [5, 15]. The model takes into account the volume change before and after hydrogen absorption and desorption in three-dimensional spherical particles. The equation can be modified as follows:

$$\frac{\rho \partial r}{\partial t} = -\frac{D_H \Delta C}{r - r^2/[z r_0^3 + r^3(1-z)]^{\frac{1}{3}}} \tag{2.19}$$

where z is the volume expansion ratio of the alloy before and after hydrogen absorption and release. The integral of the V-C model is as follows:

$$\frac{z - [1 + (z-1)\xi]^{\frac{2}{3}} - (z-1)(1-\xi)^{\frac{2}{3}}}{z-1} = \frac{2D_H \Delta C}{r_0^2 \rho} t = k_d t \tag{2.20}$$

The V-C model does not make any additional approximations or assumptions, making it one of the most accurate models for describing the hydrogen absorption and desorption behaviors of three-dimensional spheres. However, the model is also more complicated. Although it has not yet been applied to the hydrogen absorption and desorption in metal hydrides, it has been used for oxidation kinetics, which is similar process.

(4) Chou model

When the RDS is the hydrogen diffusion from the gas/hydride interface (g/β) to the hydride/metal interface (β/α), the diffusion rate can be expressed according to Fick's first law [6]:

$$V_d = \frac{D_H}{r_0 - r}\left[C_H\left(\frac{g}{\beta}\right) - C_H\left(\frac{\beta}{\alpha}\right)\right] \tag{2.21}$$

where C_H(g/β) and C_H (β/α) are the hydrogen concentrations at the gas/hydride and hydride/metal interfaces, respectively. Derivation of Eq. (2.21) gives the reaction rate equation for spherical particles:

$$\frac{d\xi}{dt} = -\frac{3r^2}{r_0^3}\frac{dr}{dt} \tag{2.22}$$

where the relationship between reaction rate and metal particle radius can be expressed as:

$$\frac{dr}{dt} = \frac{V}{v_m} \tag{2.23}$$

where v_m is the correlation coefficient depending on the molar mass of the substance. Based on Eq. (2.11) and Eqs. (2.21)–(2.23), the differential equation for reaction rate can be expressed as:

$$\frac{d\xi}{dt} = -\frac{3D_H\left[C_H\left(\frac{g}{\beta}\right) - C_H\left(\frac{\beta}{\alpha}\right)\right]}{v_m r_0^2}\frac{(1-\xi)^{\frac{2}{3}}}{1-(1-\xi)^{\frac{1}{3}}} \tag{2.24}$$

Integrating both sides gives

$$\xi = 1 - \left(1 - \sqrt{\frac{2D_H\left[C_H\left(\frac{g}{\beta}\right) - C_H\left(\frac{\beta}{\alpha}\right)\right]}{r_0^2 v_m} t}\right)^3 \tag{2.25}$$

If diffusion is the RDS and other processes are faster and approach equilibrium, one can substitute the hydrogen concentration and equilibrium constant, and the relationship among the diffusion coefficient, activation energy, and temperature into Eq. (2.25) to obtain

$$\xi = 1 - \left(1 - \sqrt{\frac{2D_H^0 \exp\left(\frac{E}{RT}\right) K_{pa}^{\frac{1}{2}} K_{ca}^{\frac{1}{2}} K_{sp}\left(P_{H_2}^{\frac{1}{2}} - P_{H_2}^{eq\frac{1}{2}}\right)}{r_0^2 v_m} t}\right)^3 \tag{2.26}$$

where K_{pa} is the equilibrium constant for the physical adsorption process, K_{ca} is the equilibrium constant of the chemical process, K_{sp} is the equilibrium constant for the surface permeation process, and D_H^0 is the diffusion coefficient constant.

The previous derivation does not consider the volume changes between products and reactants. However, if the volume change is non-negligible, the volume ratio z of the products and reactants should be taken into account [16], and the relation between reacted fraction (ξ) and time (t) is

$$\int_0^{\xi} \frac{[z - (z-1)(1-\xi)]^{\frac{1}{3}} - (1-\xi)^{\frac{1}{3}}}{3(1-\xi)^{\frac{2}{3}}} d\xi = \frac{D_H^0 \exp\left(\frac{E}{RT}\right) K_{pa}^{\frac{1}{2}} K_{ca}^{\frac{1}{2}} K_{sp}\left(P_{H_2}^{\frac{1}{2}} - P_{H_2}^{eq\frac{1}{2}}\right)}{r_0^2 v_m} t \tag{2.27}$$

In addition to the spherical model, the Chou model also provides specific equations for reactants with other different shapes [17], e.g., for cylindrical samples:

$$\xi = 1 - \left[1 - \frac{1}{r_0}\sqrt{\frac{2D_H\left(C_H\left(\frac{g}{\beta}\right) - C_H\left(\frac{\beta}{\alpha}\right)\right)}{v_m} t}\right]^2 \left[1 - \frac{2}{h_0}\sqrt{\frac{2D_H\left(C_H\left(\frac{g}{\beta}\right) - C_H\left(\frac{\beta}{\alpha}\right)\right)}{v_m} t}\right] \tag{2.28}$$

For quadratic samples:

$$\xi = 1 - \left[1 - \frac{1}{L_0}\sqrt{\frac{2D_H\left(C_H\left(\frac{g}{\beta}\right) - C_H\left(\frac{\beta}{\alpha}\right)\right)}{v_m}t}\right]^2 \left[1 - \frac{2}{H_0}\sqrt{\frac{2D_H\left(C_H\left(\frac{g}{\beta}\right) - C_H\left(\frac{\beta}{\alpha}\right)\right)}{v_m}t}\right] \tag{2.29}$$

where h_0 is the length of the cylinder, L_0 and H_0 are the width and height of the sample, respectively.

Although the equation form of the Chou model is similar to that of the Jander model for spherical powder samples, the Chou model quantitatively describes the effects of temperature, gaseous hydrogen pressure, and powder particle size on the reaction rate. The model can be applied to a wider range of applications with higher accuracy when different reactant shapes and product/reactor volume ratios are considered. In particular, Chou model proposed the concept of "characteristic reaction time", which equals to $1/k_d$ and can be expressed as:

$$t_c = \frac{r_0^2 v_m}{D_H^0 \exp\left(\frac{E}{RT}\right) K_{pa}^{\frac{1}{2}} K_{ca}^{\frac{1}{2}} K_{sp}\left(P_{\mathrm{H}_2}^{\frac{1}{2}} - P_{\mathrm{H}_2}^{eq\frac{1}{2}}\right)} \tag{2.30}$$

When $t = t_c$ and $\xi = 1$, t_c can be regarded as the critical time at which the reaction is completed. t_c can also be used to compare the speed of the reaction. A larger t_c value indicates that the reaction needs more time to be completed, i.e., the slower the reaction rate is.

2. Nucleation and growth controlled kinetic models

(1) Classical JMAK model

The nucleation rate $I(\tau)$ of new phases can be described either by a constant number of nuclei or a constant nucleation rate [8, 18]:

$$I(\tau) = N_0\delta(\tau - 0) \tag{2.31}$$

$$I(\tau) = I_0 \exp\left(-\frac{E_n}{RT}\right) \tag{2.32}$$

where τ is time, N_0 is the number of nuclei formed, $\delta(\tau–0)$ is the Dirac function, I_0 is the intrinsic nucleation rate constant, and E_n is the nucleation activation energy. The growth process of the new phase can be described by Eq. (2.33) [8, 9, 18]:

$$V(\tau) = \left[G_0 \int\limits_{\tau}^{t} \exp\left(-\frac{E_g}{RT}\right) d\eta \right]^{\frac{d}{m}} \tag{2.33}$$

where G_0 is the intrinsic growth rate, E_g is the growth activation energy, m is the growth mode parameter, and d/m is the growth factor.

By assuming random nucleation and isotropic growth, the interfacial movement of the new phase can be written as follows:

$$f(x) = 1 - \exp(x) \tag{2.34}$$

Thus, the reaction fraction (ξ) derived from a constant number of nuclei and a constant rate of nucleation can be written as follows, respectively:

$$\xi = 1 - \exp\left\{ -N_0 \left[G_0 \int\limits_{0}^{t} \exp\left(-\frac{E_g}{RT}\right) d\eta \right]^{\frac{d}{m}} \right\} \xi \tag{2.35}$$

$$= 1 - \exp\left\{ - \int\limits_{0}^{t} I_0 \exp\left(-\frac{E_n}{RT}\right) \left[G_0 \int\limits_{\tau}^{t} \exp\left(-\frac{E_g}{RT}\right) d\eta \right]^{\frac{d}{m}} d\tau \right\} \tag{2.36}$$

Under the isothermal condition, Eqs. (2.35) and (2.36) can be written as follows:

$$\xi = 1 - \exp\left\{ -N_0 G_0^{\frac{d}{m}} \exp\left(-\frac{E_g \frac{d}{m}}{RT}\right) t^{\frac{d}{m}} \right\} \xi \tag{2.37}$$

$$= 1 - \exp\left\{ -I_0 G_0^{\frac{d}{m}} \frac{1}{1 + \frac{d}{m}} \exp\left(\frac{-E_n - E_g \frac{d}{m}}{RT}\right) t^{1+\frac{d}{m}} \right\} \tag{2.38}$$

Both Eqs. (2.37) and (2.38) can be simplified as:

$$\xi = 1 - \exp\left(-k_{ng} t^n\right) \tag{2.39}$$

where n is the Avrami index, generally $n \geq 0.5$. Equation (2.39) is the equation of the classical JMAK model, which has been widely used in the process of hydrogen adsorption and desorption of metal hydrides. The nucleation and growth process are generally determined by the magnitude of n. Distinct n values indicate the different RDS. The relationship between n value and RDS is summarized in Table 2.2.

(2) JMAK model with self-catalytic nucleation (NI-JMAK with nucleation index combination)

Table 2.2 The corresponding RDS of different Avrami indexes

RDS	Growth dimension	n values	
		Constant nuclei number	Constant nucleation rate
Diffusion	1	1/2	3/2
	2	1	2
	3	3/2	5/2
Interface movement	1	1	2
	2	2	3
	3	3	4

The NI-JMAK model is developed based on the classical JMAK model, in which the self-catalytic contribution is considered [10–12]. The new phase nucleation rate adopts the assumption of continuous nucleation.

$$I(\tau) = \frac{\partial}{\partial \tau}\left[\int_0^{\tau} I_0 \exp\left(-\frac{E_n}{RT}\right) d\eta\right]^c \tag{2.40}$$

where c is the nucleation index representing the self-catalytic effect of nucleation.

The original NI-JMAK model is given by:

$$\xi = 1 - \exp\left\{-\int_0^t \frac{\partial}{\partial \tau}\left[\int_0^{\tau} I_0 \exp\left(-\frac{E_n}{RT}\right) d\eta\right]^c \left[G_0 \int_{\tau}^{t} \exp\left(-\frac{E_g}{RT}\right) d\eta\right]^{\frac{d}{m}} d\tau\right\} \tag{2.41}$$

By introducing the first Euler integral function and first-order approximation of temperature integral, the equation can be expressed as follows under isothermal condition:

$$\xi = 1 - \exp\left[-k_{ng,0}^{c+\frac{d}{m}} Beta\left(c, \frac{d}{m} + 1\right) \exp\left(\frac{-cE_n - \frac{d}{m}E_g}{RT}\right) t^{c+\frac{d}{m}}\right] \tag{2.42}$$

When ΔE_n/RT > > 0 and under non-isothermal condition, the NI-JMAK model can be expressed as:

$$\xi = 1 - \exp\left[-k_{ng,0}^{c+\frac{d}{m}} \frac{Beta\left(\frac{cE_n}{E_g}, \frac{d}{m} + 1\right)}{E_g^{\frac{d}{m}+1} E_n^{c-1}} \exp\left(-\frac{\frac{E_g d}{m} + cE_n}{RT}\right)\left(\frac{RT^2}{\beta}\right)^{c+\frac{d}{m}}\right] \tag{2.43}$$

When $\Delta E_n/\mathrm{RT} \approx 0$ and under non-isothermal condition, the NI-JMAK model can be expressed as:

$$\xi = 1 - \exp\left[-k_0^{c+\frac{d}{m}} \frac{R^{\frac{d}{m}}}{cE_g^{\frac{d}{m}} \beta^{c+\frac{d}{m}}} \exp\left(-\frac{E_g \frac{d}{m}}{RT} \right) T^{2\frac{d}{m}} (T - T)_0^c \right] \quad (2.44)$$

The NI-JMAK model gives five independent variables to describe the nucleation and growth process under isothermal condition. The complexity of this model prevents its application in the hydrogen absorption and desorption reactions of metal hydrides.

3. Chemical reaction controlled kinetic model

If the chemical reaction rate is the RDS, the chemical reaction occurs at the metal/hydride interface and the reaction is usually described by the following equation:

$$[\mathrm{H}](\alpha/\beta) + \mathrm{M}(\alpha) \leftrightarrow \mathrm{MH}(\alpha/\beta) \quad (2.45)$$

where $[\mathrm{H}](\alpha/\beta)$ is the hydrogen atoms at the interface of the α and β phases and $\mathrm{MH}(\alpha/\beta)$ is the hydride formed at the interface between the α and β phases.

The interface continuously moves towards the bulk of the reactant. To date, the Contracting Volume (CV) model and the Chou model have been proposed to describe the processes of chemical reaction at the interface.

(1) CV model

The CV model assumes that the hydrogen absorption and desorption rates are controlled by interfacial processes, and the rate of interfacial motion can be described as [19]:

$$\frac{\partial r}{\partial t} = -k_{\mathrm{int}} \quad (2.46)$$

where k_{int} is the reaction rate constant for interfacial control.

The combination of Eqs. (2.46) and (2.11) and subsequent integration gives the CV model:

$$1 - (1 - \xi)^{\frac{1}{d}} = \frac{k_{\mathrm{int}}}{r_{0_r}} = k_r t \quad (2.47)$$

where d is the geometric dimension, k_r is the rate constant for the interfacial chemical reaction.

(2) Chou model

The chemical reaction rate in the Chou model can be expressed as:

$$V_r = k_r^{ab}\left[C_H(\beta) - C_H^{eq}\left(\frac{\alpha}{\beta}\right) \right] \quad (2.48)$$

or

$$V_r = k_r^{ab} K_{pa}^{\frac{1}{2}} K_{ca}^{\frac{1}{2}} K_{sp} K_d \left[P_{\mathrm{H_2}}^{\frac{1}{2}} - P_{\mathrm{H_2}}^{eq\frac{1}{2}} \right] \tag{2.49}$$

The relationship between the migration rate of interface and the reaction rate is shown in Eq. (2.23). Combining Eqs. (2.11), (2.22), (2.23), and (2.49), the following equation is obtained:

$$1 - (1 - \xi)^{\frac{1}{d}} = \frac{k_r^0}{r_0} \exp\left(-\frac{E}{RT} \right) \left(P^{1/2} - P_{eq}^{1/2} \right) t \tag{2.50}$$

where k_r^0 is the equilibrium constant of the chemical reaction.

The kinetic equations of both the CV model and the Chou model show a high level of consistency. Unlike the CV model, the Chou model gives a physical meaning for the rate constant k_r as a function of temperature, particle size, pressure, and equilibrium pressure.

The above-mentioned kinetic models and corresponding integral form equations and parameters are summarized in Table 2.3. The reaction fraction-time curves of the models are shown in Fig. 2.4. The numbers following the models indicate the values of the parameters in the models. For example, in the VC model, 0.8 and 1.25 are utilized as z values, and in the C-JMAK model, 1, 2, and 3 are used as n values. An obvious difference can be identified for each curve derived from corresponding models, as shown in Fig. 2.4. These results indicate that the most reasonable model needs to be selected for the isothermal kinetic data analysis in the practical applications. It is worth noting that the curves of the VC model are very close to those of the G-B model, which indicates that the volume change is not the most important factor. Therefore, only a small number of studies used the VC model, and most of the researchers concentrated on the kinetic model in which the diffusion, nucleation, and growth steps are the RDS.

2.3 Factors that Affect the Kinetics of Hydrogen Absorption and Desorption Reactions in Mg-Based Hydrogen Storage Alloys

2.3.1 Effect of Temperature

The temperature mainly affects the reaction rate constant. As the temperature increases, the reaction rates of various steps (physical adsorption, chemical adsorption, surface penetration, hydrogen diffusion, β phase growth, and interfacial chemical reaction) are enhanced. The relationship between the rate constant and temperature is governed by the van't Hoff function. When the temperature increases by 10 K, the reaction rate increases by 2 ~ 4 times.

Table 2.3 Summary table of kinetic theory models of hydrogen absorption and discharge reactions in hydrogen storage alloys

RDS	Model name	Equation form	Applicable conditions	Parametrics
Diffusion	Jander	$[1-(1-\xi)^{\frac{1}{d}}]^2 = k_d t$	Constant temperature	d—dimensionality, taking 3, 2 and 1 k_d—reaction rate constant
	G-B model	$(1-\xi)\ln(1-\xi)+\xi = k_d t$	Constant temperature 2D fibrous	k_d—reaction rate constant
		$1-\frac{2}{3}\xi-(1-\xi)^{\frac{2}{3}} = k_d t$	Constant temperature three-dimensional spherical particles	k_d—reaction rate constant
	V-C model	$\frac{z-[1+(z-1)\xi]^{\frac{2}{3}}-(z-1)(1-\xi)^{\frac{2}{3}}}{z-1} = k_d t$	Constant temperature Three-dimensional spherical	z-volume expansion ratio of the alloy before and after hydrogen absorption and release; k_d—reaction rate constant
	Chou model	$\xi = 1-\left(1-\sqrt{\frac{2K_0 D_H^0 \exp\left(\frac{E}{RT}\right)\left(P_{H_2}^{\frac{1}{2}}-P_{H_2}^{eq\frac{1}{2}}\right)}{r_0^2 v_m}t}\right)^3$	Constant temperature Three-dimensional spherical	P_{H_2}- partial pressure of hydrogen r_0—particle radius E—activation energy
		$\xi = 1-\left[1-\frac{1}{r_0}\sqrt{\frac{2D_H\left(C_H\left(\frac{g}{\beta}\right)-C_H\left(\frac{\beta}{\alpha}\right)\right)}{v_m}t}\right]^2 \left[1-\frac{2}{h_0}\sqrt{\frac{2D_H\left(C_H\left(\frac{g}{\beta}\right)-C_H\left(\frac{\beta}{\alpha}\right)\right)}{v_m}t}\right]$	constant temperature Column samples	r_0-radius of the column sample h_0—length of the column sample

(continued)

Table 2.3 (continued)

RDS	Model name	Equation form	Applicable conditions	Parametrics
		$\xi = 1 - \left[1 - \frac{1}{L_0}\sqrt{\frac{2D_H\left(C_H\left(\frac{g}{\beta}\right) - C_H\left(\frac{\beta}{\alpha}\right)\right)}{v_m}t}\right]^2$ $\left[1 - \frac{2}{H_0}\sqrt{\frac{2D_H\left(C_H\left(\frac{g}{\beta}\right) - C_H\left(\frac{\beta}{\alpha}\right)\right)}{v_m}t}\right]$	Constant temperature Square Sample	L_0—length of the square sample h_0—height of the column sample
Nucleation	C-JMAK model	$\xi = 1 - \exp\left\{-N_0 G_0^{\frac{d}{m}} \exp\left(-\frac{E_g \frac{d}{m}}{RT}\right) t^{\frac{d}{m}}\right\}$	Constant temperature Number of constant nuclei	E_g—growth activation energy m—Growth model parameters d/m—growth factor
		$\xi = 1 - \exp\left\{-I_0 G_0^{\frac{d}{m}} \frac{1}{1+\frac{d}{m}} \exp\left(\frac{-E_n - E_g \frac{d}{m}}{RT}\right) t^{1+\frac{d}{m}}\right\}$	Constant temperature Constant nucleation rate	E_n—morphological nuclear activation energy E_g-growth activation energy m—Growth model parameters d/m—growth factor
Nucleation	NI-JMAK model	$\xi =$ $1 - \exp\left[-k_{ng,0}^{c+\frac{d}{m}} Beta\left(c, \frac{d}{m} + 1\right) \exp\left(\frac{-cE_n - \frac{d}{m}E_g}{RT}\right) t^{c+\frac{d}{m}}\right]$	Constant temperature	E_n—morphological nuclear activation energy E_g—growth activation energy m—Growth model parameters d/m—growth factor

(continued)

Table 2.3 (continued)

RDS	Model name	Equation form	Applicable conditions	Parametrics
		$\xi = 1 - \exp\left[-k_{ng,0}^{c+\frac{d}{m}} \frac{Beta\left(\frac{cE_n}{E_g}, \frac{d}{m}+1\right)}{E_g^{\frac{d}{m}+1} E_n^{c-1}} \exp\left(-\frac{\frac{E_g d}{m}+cE_n}{RT}\right)\left(\frac{RT^2}{\beta}\right)^{c+\frac{d}{m}}\right]$	Non-isothermal $E_n/RT \geq 0$	E_n—morphological nuclear activation energy E_g—growth activation energy m—Growth model parameters d/m—growth factor
		$\xi = 1 - \exp\left[-k_0^{c+\frac{d}{m}} \frac{R^{\frac{d}{m}}}{cE_g^{\frac{d}{m}} \beta^{c+\frac{d}{m}}} \exp\left(-\frac{E_g \frac{d}{m}}{RT}\right) T^{2\frac{d}{m}} (T-T)_0^c\right]$	Non-isothermal $E_n/RT \approx 0$	E_n—morphological nuclear activation energy E_g—growth activation energy m—Growth model parameters d/m—growth factor
Chemical reaction	CV model	$1-(1-\xi)^{\frac{1}{d}} = \frac{k_{int}}{r_{0_r}}$	Constant temperature	d—geometric dimension value K_r—Rate constants for interfacial chemical reactions
	Chou model	$1-(1-\xi)^{\frac{1}{d}} = \frac{k_r^0}{r_0} \exp\left(-\frac{E}{RT}\right)\left(P^{1/2} - P_{eq}^{1/2}\right)t$	Constant temperature	D—geometric dimension value K_r—Rate constants for interfacial chemical reactions

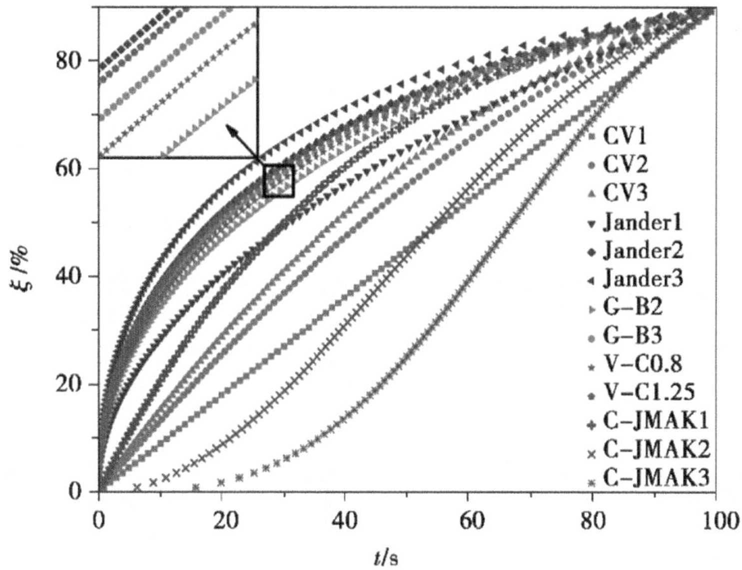

Fig. 2.4 Reaction fraction-time curves for different kinetic models (The values after the models indicate the values of the parameters in the corresponding models; the values after the CV, Jander, and G-B models are the values of d, after the V-C model are the values of z, and after the C-JMAK model are the values of n.)

$$\frac{k_{T+10}}{k_T} = 2 \sim 4 \tag{2.51}$$

where k_T is the rate constant at the temperature T (K) and k_{T+10} is the rate constant at the temperature $T + 10$ (K).

Based on the van't Hoff function, one can roughly estimate the effect of temperature on hydrogen absorption and desorption rates. However, to accurately analyze the effect of temperature, it is necessary to calculate the reaction rate constant or characteristic time at a specific temperature using the previously discussed kinetic models.

To calculate the reaction rate constant, it is necessary to choose a proper kinetic model. By fitting the ξ (fraction)–t (time) curves via different kinetic models, one can compare the correlation coefficients and computational errors of the fitted results. The model showing the smallest error is considered as the most appropriate one. Generally, the kinetic model is expressed in the form of $f(\xi) = kt$, where $f(\xi)$ has different forms according to different theoretical models. By evaluating the fitting errors of different theoretical models, the RDS can be determined. In the $f(\xi)$–t curve, the slope is the reaction rate constant k. The rate constants at different temperatures follow the Arrhenius equation:

$$k = A\exp\left(\frac{-E_a}{RT}\right) \tag{2.52}$$

Or

$$\ln k = -\frac{E_a}{RT} + \ln A \tag{2.53}$$

where E_a is the apparent activation energy of hydrogen absorption or desorption (J/mol), R is the gas constant, T is the reaction temperature, and A is the pre-exponential factor. The apparent activation energies of the reactions can be estimated from the graph of ln$k$ versus 1/T.

In the following discussion, by taking pure Mg, Mg-Er, and Mg-$LaNi_5$ as examples, we illustrate the methods used to estimate the reaction rate constant k or the characteristic time t_c. Figure 2.5a and b show the hydrogen absorption curves of pure Mg and Mg-Er alloy powders, respectively [20]. The pure Mg powder can absorb 4.2 wt.% of hydrogen in 5 min at 573 K, while only 3.1 wt.% of hydrogen is absorbed within 2 h when the temperature is lowered to 473 K. This result suggests that the hydrogen absorption rate is significantly suppressed at 473 K. The Mg-Er alloy powders can absorb 6.0 wt.% of hydrogen in 5 min at 573 K. Even at 473 K, it can absorb 3.67 wt.% of hydrogen within 2 h. These values are higher than those of pure Mg powders. In both cases, the hydrogen storage capacity increases with the rising temperature within the given reaction time. The hydrogen storage capacity of Mg-Er alloy is higher than that of pure Mg powder. To quantitatively compare the hydrogen absorption rate of pure Mg and Mg-Er alloy at different temperatures, it is necessary to calculate the reaction rate constants at each temperature using different kinetic models.

The kinetics of pure Mg and Mg-Er alloy powders are well described by the JMAK model when comparing the calculations of the kinetic models discussed in Sect. 2.2.2. The JMAK equation can be transformed to

$$\ln[-\ln(1-\xi)] = \ln k + n \ln t \tag{2.54}$$

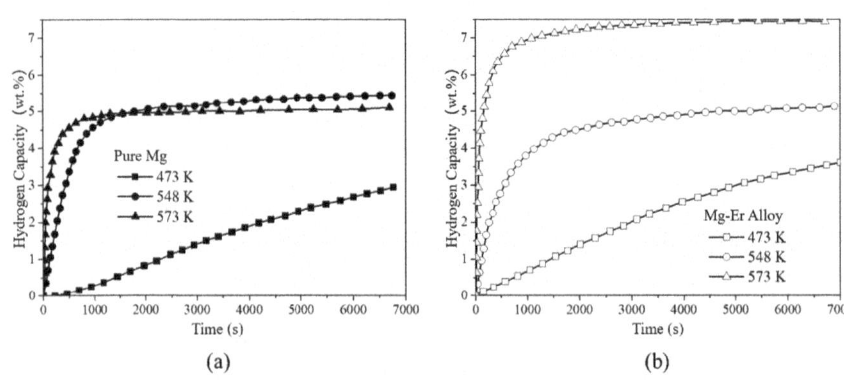

Fig. 2.5 Hydrogen absorption curves for Mg and Mg-Er alloys [20]

Figure 2.6a and b show the ln[–ln(1–ξ)] plot as a function of ln$t$. The slope obtained from the fitting is the Avrami exponent and the intercept with the coordinate axes is ln$k$. The Avrami exponent of pure Mg decreases from 1.74 to 0.6 as the temperature increases from 473 to 573 K. Meanwhile, ln$k$ increases from −14.38 to −2.99, indicating the enhanced reaction rates. The Avrami index of Mg-Er alloy is slightly smaller than that of pure Mg, and its ln$k$ is increases from −9.44 to −2.51 with temperature increasing from 473 to 573 K, which is higher than that of pure Mg. Figure 2.6c shows the ln$k$ plot as the function of 1000/T, and from the fitting line, we can obtain the reaction activation energy of the material. The apparent activation energy of hydrogen absorption of pure Mg is as high as 243.92 kJ/mol H_2, while that of Mg-Er alloy is 143.48 kJ/mol H_2, which is much lower than that of pure Mg, indicating that the alloying elements can effectively reduce the reaction activation energy

The hydrogen absorption kinetics of Mg-$LaNi_5$ alloy at different temperatures is used as an example to illustrate the effect of temperature on hydrogen absorption in Mg alloys with multiple phases. Figure 2.7 shows the hydrogen absorption fraction ξ as a function of reaction time t [21]. The Mg-30wt.% $LaNi_5$ alloy contains several phases, such as Mg, Mg_2Ni and $Mg_{12}La$. With the increase in temperature,

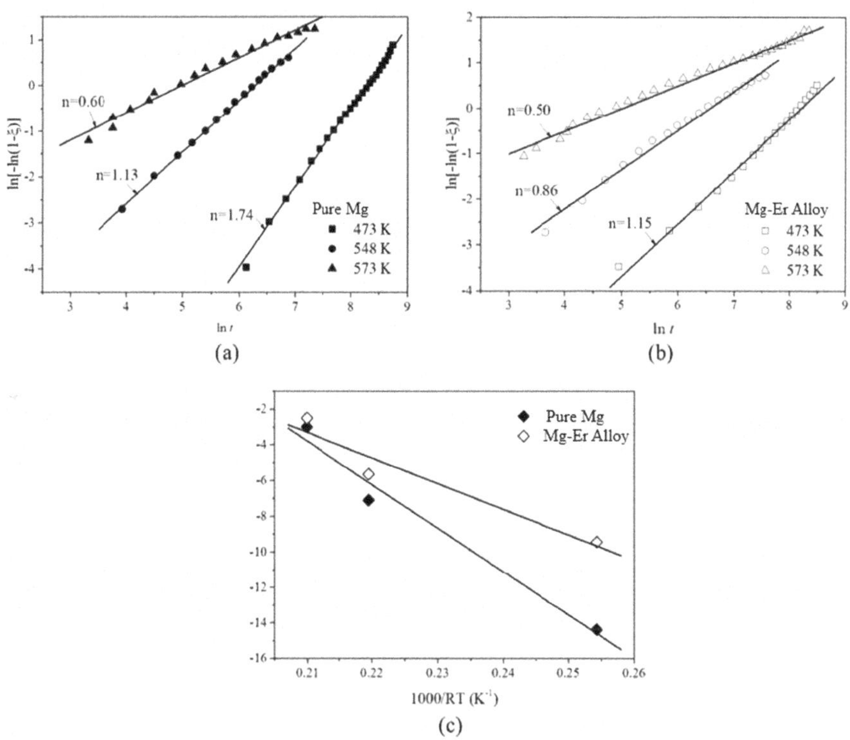

Fig. 2.6 Linear plot of ln[–ln(1–ξ)] versus ln$t$

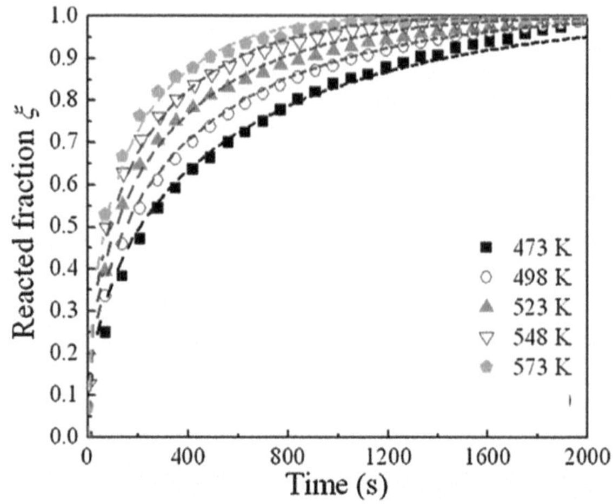

Fig. 2.7 Hydrogen absorption reaction fraction ξ versus time t for Mg-30wt.% $LaNi_5$ composite in the temperature range of 473–573 K[21]

the reaction fraction systematically increases, which indicates that the reaction rate is accelerated. To quantitatively compare the reaction rates at different temperatures, every model needs to be used for fitting the hydriding kinetic curves. In addition to the previously described calculation by $f(\xi)$–t linear relationship, the ξ-$f(t)$ relationship can also be directly obtained using the calculation software. Currently, there is growing consensus that the diffusion rate equation of the Chou model can better fit the experimental results, and the reaction fraction and time at different temperatures are consistent with the following equation.

$$\xi = 1 - \left[1 - \sqrt{\frac{t}{t_c}}\right]^3 \tag{2.55}$$

The corresponding fitted correlation coefficients r^2 were 0.9853 (473 K), 0.9985 (498 K), 0.9917 (523 K), 0.9945 (548 K), and 0.9905 (573 K), respectively. The residual errors are all less than 4.0%. The characteristic time t_c decreased with increasing temperature from 4914 s at 473 K to 1564 s at 573 K. Since the characteristic time represents the required time for the reaction to be completed, the calculated results indicate that the reaction rates increase with higher temperatures.

It is noted that the Arrhenius equation has already been internalized in the Chou model when it is used to resolve the relationship between the diffusion coefficients, reaction rate constants, and temperatures by fitting experimental data measured at different temperatures as shown in Eq. (2.56).

$$\xi = 1 - \left[1 - \exp\left(\frac{-E_a}{2RT}\right)\sqrt{\frac{t}{B_T}}\right]^3 \tag{2.56}$$

Thus, the activation energy of the reaction can be calculated by simultaneously fitting kinetic experimental data from various temperatures. It is also noted that Eq. (2.56) is transformed from Eq. (2.26), where the B_T is expressed as follows:

$$B_T = \frac{r_0^2 v_m}{2D_H^0 K_{pa}^{\frac{1}{2}} K_{ca}^{\frac{1}{2}} K_{sp}\left(P_{\mathrm{H}_2}^{\frac{1}{2}} - P_{\mathrm{H}_2}^{eq\frac{1}{2}}\right)} \tag{2.57}$$

The value of B_{T} remains constant for a given material and fixed reaction conditions. Additionally, by calculating the activation energies and equation coefficients B_{T} at any two temperatures, it is possible to predict the reaction kinetics at other temperatures. For example, utilizing obtained activation energies (25.2 ± 0.1) kJ/mol H_2 and B_{T} (8.25), the reaction fraction-time relationship at other temperatures can be written as follows:

$$\xi = 1 - \left[1 - \exp\left(\frac{-25200}{2 \cdot 8.314 \cdot T}\right)\sqrt{\frac{t}{8.25}}\right]^3 \tag{2.58}$$

The reaction fraction curves predicted by the Chou model are shown in Fig. 2.7. The predictions reasonably reproduced the experimental results with prediction errors of only 1.5% (498 K), 2.4% (523 K), and 1.2% (548 K). These results demonstrated the high accuracy of the method.

2.3.2 *Effect of Hydrogen Pressure*

From the Gibbs free energy, it can be seen that the pressure difference between a given hydrogen pressure and the equilibrium hydrogen pressure is the driving force for the hydrogen adsorption and desorption reaction. For the kinetic processes, the increased hydrogen pressure difference implies the increase of the driving force and the rates of hydrogen adsorption and desorption reactions. As can be seen from Table 2.3, the relationship between reaction fraction and hydrogen pressure, which is ignored by other models, is considered in the Chou model. When other factors such as temperature are fixed and only the hydrogen pressure is changed, the reaction fraction is expressed as follows (assuming diffusion is the RDS):

$$\xi = 1 - \left(1 - \sqrt{\frac{\left(P_{\mathrm{H}_2}^{\frac{1}{2}} - P_{\mathrm{H}_2}^{eq\frac{1}{2}}\right)}{B_P} t}\right)^3 \tag{2.59}$$

If the chemical reaction is the RDS, the equation can be expressed as follows:

$$1 - (1 - \xi)^{\frac{1}{d}} = \frac{\left(P^{\frac{1}{2}} - P_{eq}^{\frac{1}{2}}\right)}{B_P} t \tag{2.60}$$

where B_{p} is a temperature-dependent but pressure-independent constant. Figure 2.8 shows the hydrogen absorption curves of Mg_2Ni alloy under 0.275–1.133 MPa

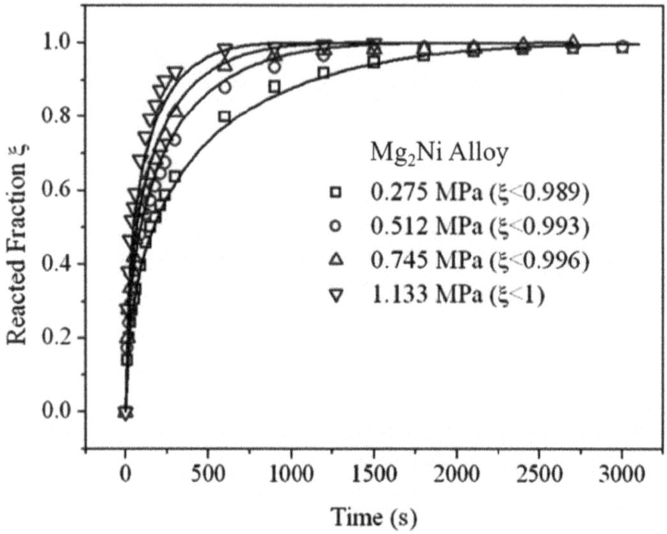

Fig. 2.8 Hydriding kinetics of Mg_2Ni alloy at different hydrogen pressures

hydrogen pressure. With increasing hydrogen pressure, the reaction fraction becomes larger, indicating that the increased hydrogen pressure leads to a higher reaction rate. The characteristic times (t_c) of the reaction at different hydrogen pressures were estimated to be 3650 s (0.275 MPa), 2452 s (0.512 MPa), 1762s (0.745 MPa), and 895 s (1.133 MPa). These values were calculated using the Chou model, showing that the characteristic time systematically decreasing with increasing hydrogen pressure, which indicates that the hydrogen absorption reaction accelerates under higher pressure conditions. It is noted that the equilibrium hydrogen pressure $P_{H_2}^{eq}$ and B_P can be obtained by fitting the kinetic curves at any two hydrogen pressures using Eq. (2.59) or Eq. (2.60). The kinetic curves at other hydrogen pressures can be predicted using the known equations. For example, the hydrogen absorption curve of Mg_2Ni (Fig. 2.8) is fitted by Eq. (2.59), giving $P_{H_2}^{eq} = 0.09511$ MPa and $B_p = 852.07$. The reaction fractions at other hydrogen pressures agree well with the predicted curves.

$$\xi = 1 - \left(1 - \sqrt{\frac{\left(\sqrt{P_{H_2}} - \sqrt{0.09511}\right)}{852.07} t}\right)^3 \tag{2.61}$$

2.3.3 *Effect of Particle Size*

Under constant temperature and pressure conditions, the rate of hydrogen adsorption and desorption can be accelerated by decreasing particle size. The smaller the particle size, the higher the surface area and reaction cross-section, which leads to reduce diffusion distance. The kinetic models discussed in Sect. 2.2.2 provide the relationship between the reaction fraction and the particle size. In the kinetic models of physical adsorption, chemical adsorption, surface permeation and interfacial chemical reaction, the reaction fraction is proportional to $1/r_0$. For the hydrogen diffusion process, the reaction fraction is proportional to $1/{r_0}^2$. However, for the nucleation and growth process, the influence of the particle size on reaction fraction does not have a clear relation, and must be evaluated by experimental results. The actual particle size of the powder usually follows a certain distribution pattern, and the particle size given by the model calculation should be the average particle size or the effective particle size of the samples.

Nanosizing is an effective method to increase the rate of hydrogen absorption and desorption in Mg-based hydrogen storage alloys. Nanostructures improve the density of various defects such as grain boundary/phase boundary, dislocation and stacking fault. These increase the contact area between Mg and hydrogen and reduce the diffusion distance of hydrogen atoms, thus favoring the hydrogen absorption and desorption rates [22]. In addition, as the hydrogen content in the nanoparticles increases, the energy barrier of hydride formation decreases, which reduces the apparent activation energy of hydrogen absorption and desorption reactions of Mg-based hydrogen storage materials. Ball milling is most commonly used to prepare hydrogen storage nanomaterials. The particle size of $Mg_{50}Co_{50}$ alloy prepared by ball milling is 1~2 μm, and the grain size is on the order of a few nanometers [23–25]. This $Mg_{50}Co_{50}$ alloy has a body-centered cubic structure and exhibits a higher hydrogen storage capacity and faster hydrogen absorption and desorption rates than conventional Mg-Co alloys.

Ti composite Mg nanocrystals, prepared by dehydriding Ti-MgH_2 nanocrystals, exhibit improved hydrogen absorption rates even at low temperatures [25]. As shown in Fig. 2.9, Ti-catalyzed Mg can absorb 6 wt.% hydrogen within 1 h at 530 K, while it takes 3 h at 623 K for the conventional Mg. Hydrogen absorption can be further enhanced by nanosizing. It was reported that Mg nanocrystals can absorb 5.2 wt.% H_2 under 426 K in 18 h. The Jander model fits well with the hydrogen absorption curves of MgH_2 nanocrystalline and Ti-catalyzed MgH_2. It shows that the diffusion of three-dimensional hydrogen atoms is the RDS of the reaction. The rate constant of Ti-catalyzed MgH_2 nanocrystals is as high as 3.06 s^{-1} at 573 K, which is 40 times higher than that of conventional MgH_2. The inset of Fig. 2.9 shows the linear relationships of Ti and conventional MgH_2 samples of the reaction fraction function as the function of time, which are fitted by the Jander model.

2.3.4 Effects of Catalysts

In addition to reducing the particle size, the combination of Mg-based materials and catalysts can also enhance the hydrogen absorption and desorption kinetics of Mg-based alloys. Recently, catalyst addition has been intensively investigated,

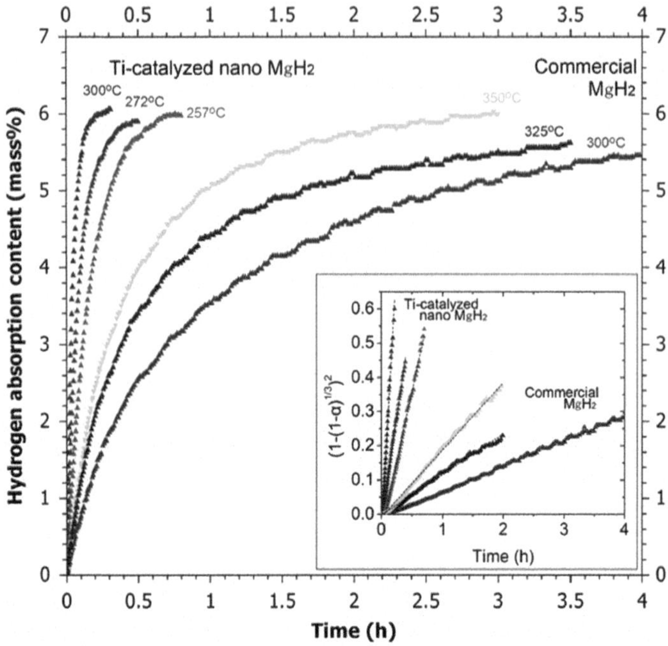

Fig. 2.9 Hydrogen absorption kinetics curves of MgH_2 nanocrystalline samples catalyzed by Ti and conventional MgH_2 samples at 1 MPa hydrogen pressure

including metal oxides, transition metals, rare earth elements, fluoride, and halide. Generally, four mechanisms have been proposed for improving kinetic properties with catalysts, including hydrogen pump, overflow, channel, and electron transfer effects. The hydrogen pump effect is that the doped transition metals preferentially absorb and desorb hydrogens by forming transition metal hydrides, improving the reaction kinetics by weakening the stability of hydrides. The overflow effect means that hydrogen is preferentially adsorbed and dissociated on the catalyst, and then hydrogen atoms are transferred from the catalyst surface to the reactant surface to promote the reaction. The channel effect means that the catalyst acts as a fast hydrogen diffusion channel during hydrogen absorption and desorption processes. The electron transfer mechanism accompanies charge modifications of catalysts, forming a chemical interaction with hydrogens, which can promote the chemical absorption and desorption of hydrogen. The addition of catalysts does not change the thermodynamic properties of the reaction, such as hydrogen storage capacity and enthalpy change, but significantly reduces the apparent activation energy of the reaction.

Metal oxides are commonly used as catalysts for Mg-based hydrogen storage alloys. For example, Nb_2O_5-catalyzed Mg can absorb 6.9 wt.% H_2 within 60 s. It can also release hydrogen completely within 140 s with a hydrogen desorption rate of 0.011 wt.%/s [26]. It has been suggested that the electron exchange between polyvalent metals and hydrogen molecules accelerates the gas–solid reaction process.

CeO_2-catalyzed Mg also exhibits an improved hydrogen absorption rate compared to Mg. The as-prepared Mg can absorb 3.46 wt.% H_2 within 5 min at 593 K, while 5 wt.% CeO_2-catalyzed Mg can absorb 3.95 wt.% H_2 under the same condition [27]. In addition, 3.6 wt.% H_2 can be desorbed at 593 K within 30 min, which is much higher than that of pure MgH_2 (1.0 wt.% H_2). In recent years, the catalytic effects of transition metal sulfides such as MoS and FeS on MgH_2 have also been reported. The MgH_2-5MoS_2 complex can start releasing hydrogen at 532 K and can release 4.0 wt.% hydrogen within 20 min at 553 K [28].

A number of studies have reported the effect of the addition of transition metals (like Ni, Nb, Ti, and Fe) and rare earth elements (like La, Ce, Pr, Nd, and Sm). Notably, Ti and Ni among the transition metals have a significant impact on enhancing the kinetic properties of MgH_2, lowering hydrogen desorption temperature and activation energy. The fine powder of Mg-TM (Ti, Fe, Ni)-La ternary alloy exhibits outstanding hydrogen absorption kinetics. It also shows a lower hydrogen desorption temperature compared to binary Mg-TM and Mg-RE complexes [20, 28–31]. Disordered NbH_x greatly enhances the hydrogen absorption and desorption kinetics of MgH_2, releasing 7.0 wt.% H_2 within 9 min at 573 K and absorbing 4.0 wt.% H_2 at 373 K. Multicomponent Mg-TM-RE alloys typically comprise various phases. For example, $Mg_{80}Ce_{18}Ni_2$ alloys contain 57 wt.% of CeMg, 29 wt.% of Ce_2Mg_{17}, 7 wt.% of CeMg and 5 wt.% of $CeMgNi_4$ [32, 33]. Since Ni and REH_x are commonly recognized as active species, their uniform distribution in hydrogen storage alloys is critical to maximize catalytic activity. REH_2-Mg-Mg_2Ni nano-composite materials can be obtained in situ by hydrogenation, which shows excellent hydrogen absorption and desorption kinetics and cyclic stability [34, 35]. Similarly, $Nd_4Mg_{80}Ni_8$ is a single-phase alloy. It also decomposes to NdH_2, α-Mg, and Mg_2Ni during the hydrogen absorption and desorption processes. NdH_2 nanoparticles are dispersed on the Mg-based alloy matrix due to the strong affinity between Nd and H atoms. The homogeneous NdH_2-Mg-Mg_2Ni nanocomposite exhibits high hydrogen storage capacity and excellent hydrogen absorption and desorption kinetics due to its well-distributed fine structure. The characteristic time of the hydrogen absorption reaction (Chou model with RDS of diffusion) decreases from 153.5 min at 373 K to 1.6 min at 573 K, indicating a significant increase in reaction rate. The calculated activation energy of the hydrogen absorption reaction is 82.3 kJ/mol H_2, which is lower than that of pure Mg (160–240 kJ/mol H_2), indicating that the reaction barrier was reduced by the addition of rare earth and Ni. The hydrogen desorption reaction data were well-fitted using the Chou model (RDS: chemical reaction). The calculated characteristic times are 8.8 min (564 K), 3.0 min (588 K), and 1.5 min (620 K), respectively, indicating an increase in hydrogen desorption rate at higher temperatures. The hydrogen desorption activation energy is 97.5 kJ/mol H_2, which is much lower than that of pure MgH_2 and other magnesium alloys Fig. 2.10.

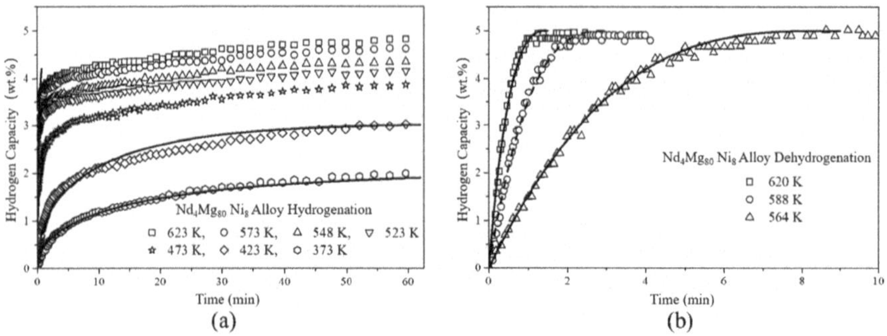

Fig. 2.10 Hydrogen absorption and hydrogen desorption kinetic curves of $Nd_4Mg_{80}Ni_8$ alloy at different temperatures

References

1. SUPPER W, GROLL M, U M. Reaction kinetics in metal hydride reaction beds with improved heat and mass transfer [J]. Journal of the Less Common Metals, 1984, 104(2): 279–286.
2. OSUMI Y. Properties and Applications of Metal Hydrides [M]. Beijing: Chemical Industry Press, 1990.
3. JANDER W. Reactions in the solid state at high temperatures [J]. Zeitschrift füranorganische und allgemeineChemie, 1927, 163: 1–30.
4. GINSTLING A, BROUNSHTEIN B. Concerning the diffusion kinetics of reactions in spherical particles [J]. Journal of Applied Chemistry USSR, 1950, 23(12): 1327–1338.
5. CARTER R E. Kinetic Model for Solid-State Reactions [J]. The Journal of Chemical Physics, 1961, 34(6): 2010–2015.
6. CHOU KC, XU K. A new model for hydriding and dehydriding reactions in intermetallics [J]. Intermetallics, 2007, 15(5–6): 767–777.
7. AVRAMI M. Kinetics of phase change. I General theory [J]. The Journal of Chemical Physics, 1939, 7.
8. AVRAMI M. Kinetics of phase change. II transformation-time relations for random distribution of nuclei [J]. The Journal of Chemical Physics, 1940, 8(2): 212–224.
9. AVRAMI M. Granulation, Phase change, and microstructure kinetics of phase change. III [J]. The Journal of Chemical Physics, 1941, 9(2): 177–184.
10. PANG Y, LI Q. A review on kinetic models and corresponding analysis methods for hydrogen storage materials [J]. International Journal of Hydrogen Energy, 2016, 41(40): 18072–18087.
11. PANG Y, LI Q. Insight into the kinetic mechanism of the first-step hydrogen desorption of Mg $(AlH_4)_2$ [J]. Scripta Materialia, 2017, 130: 223–228.
12. PANG Y, SUN D, GU Q, et al. Comprehensive determination of kinetic parameters in solid-state phase transitions: An extended Jonhson-Mehl-Avrami-Kolomogorov model with analytical solutions [J]. Crystal Growth & Design, 2016, 16(4): 2404–2415.
13. BOOTH F. A note on the theory of surface diffusion reactions [J]. Transactions of the Faraday Society, 1948, 44(0): 796–801.
14. CRANK J. The mathematics of diffusion [M]. London: Oxford university press, 1979.
15. VALENSI G. Kinetics of the oxidation of metallic spherules and powders [J]. Comptes Rendus, 1936, Comptes Rendus(4): 309–312.
16. CHOU K-C, LUO Q, LI Q, et al. Influence of the density of oxide on oxidation kinetics [J]. Intermetallics, 2014, 47: 17–22.
17. CHOU K-C, HOU X-M. Kinetics of high-temperature oxidation of inorganic nonmetallic materials [J]. Journal of the American Ceramic Society, 2009, 92(3): 585–594.

18. KEMPEN A T W, SOMMER F, MITTEMEIJER E J. Determination and interpretation of isothermal and non-isothermal transformation kinetics; the effective activation energies in terms of nucleation and growth [J]. Journal of Materials Science, 2002, 37(7): 1321–1332.
19. CARSTENSEN J T. Stability of solids and solid dosage forms [J]. Journal of Pharmaceutical Sciences, 1974, 63(1): 1–14.
20. ZOU J, ZENG X, YING Y, et al. Study on the hydrogen storage properties of core-shell structured Mg-RE (RE = Nd, Gd, Er) nano-composites synthesized through arc plasma method [J]. International Journal of Hydrogen Energy, 2013, 38(5): 2337–2346.
21. PAN Y-B, WU Y-F, LI Q. Modeling and analyzing the hydriding kinetics of Mg-$LaNi_5$ composites by Chou model [J]. International Journal of Hydrogen Energy, 2011, 36(20): 12892–12901.
22. LI Q, LU Y, LUO Q, et al. Thermodynamics and kinetics of hydriding and dehydriding reactions in Mg-based hydrogen storage materials [J]. Journal of Magnesium and Alloys, 2021, 9(6): 1922–1941.
23. SHAO H, MATSUDA J, LI H-W, et al. Phase and morphology evolution study of ball milled Mg-Co hydrogen storage alloys [J]. International Journal of Hydrogen Energy, 2013, 38(17): 7070–7076.
24. SHAO H, ASANO K, ENOKI H, et al. Fabrication, hydrogen storage properties and mechanistic study of nanostructured $Mg_{50}Co_{50}$ body-centered cubic alloy[J].Scripta Materialia [J]. Scripta Materialia, 2009, 60(9): 818–821.
25. SHAO H, XIN G, ZHENG J, et al. Nanotechnology in Mg-based materials for hydrogen storage [J]. Nano Energy, 2012, 1(4): 590–601.
26. BARKHORDARIAN G, KLASSEN T, BORMANN R. Fast hydrogen sorption kinetics of nanocrystalline Mg using Nb_2O_5 as catalyst [J]. Scripta Materialia, 2003, 49(3): 213–217.
27. MUSTAFA N S, ISMAIL M. Hydrogen sorption improvement of MgH_2 catalyzed by CeO_2 nanopowder [J]. Journal of Alloys and Compounds, 2017, 695: 2532–2538.
28. WANG L, HU Y, LIN J, et al. The hydrogen storage performance and catalytic mechanism of the MgH_2-MoS_2 composite [J]. Journal of Magnesium and Alloys, 2022, 11(7): 2530–2540.
29. LI Q, CHOU K C, LIN Q, et al. Hydriding kinetics of the $LaNiMg_{17}$-H system prepared by hydriding combustion synthesis [J]. Journal of Alloys and Compounds, 2004, 373(1–2): 122–126.
30. LI Q, LIN Q, JIANG L J, et al. On the characterization of $La_{1.5}Mg_{17}Ni_{0.5}$composite materials prepared by hydriding combustion synthesis [J]. Journal of Alloys and Compounds, 2004, 368(1–2): 101–105.
31. PANG X, RAN L, CHEN Y A, et al. Enhancing hydrogen storage performance via optimizing Y and Ni element in magnesium alloy [J]. Journal of Magnesium and Alloys, 2022, 10(3): 821–835.
32. OUYANG L Z, YANG X S, ZHU M, et al. Insights into the composition exploration of novel hydrogen storage alloys: evaluation of the Mg-Ni-Nd-H phase diagram [J]. Journal of Physical Chemistry C, 2014, 118(15): 7808–7820.
33. LIU J W, ZOU C C, WANG H, et al. Facilitating de/hydrogenation by long-period stacking ordered structure in Mg based alloys [J]. International Journal of Hydrogen Energy, 2013, 38(25): 10438–10445.
34. LI Q, LUO Q, GU QF. Insights into the composition exploration of novel hydrogen storage alloys: evaluation of the Mg-Ni-Nd-H phase diagram [J]. Journal of Materials Chemistry A, 2017, 5(8): 3848–3864.
35. LI Q, LI Y, LIU B, et al. The cycling stability of the in situ formed Mg-based nanocomposite catalyzed by YH_2 [J]. Journal of Materials Chemistry A, 2017, 5(33): 17532–17543.

Chapter 3
Preparation of Mg-Based Hydrogen Storage Materials

In recent years, the research scope of Mg-based hydrogen storage materials has expanded significantly, leading to great progress in their preparation methods. Different preparation techniques significantly impact the microstructure, phase composition, and surface morphologies of Mg-based hydrogen storage materials, which closely affects the hydrogen storage properties. Traditionally, Mg-based hydrogen storage alloys are prepared using the melting method, which can mass-produce the bulk hydrogen storage materials. To further enhance the hydriding/dehydriding kinetic properties of Mg-based alloys, the characteristics of nanocrystalline or amorphous materials, e.g., high diffusion coefficients and activity, are helpful. Moreover, the different Mg-based hydrogen storage materials can also be combined to form Mg-based composites to improve the comprehensive performance. For example, catalysts were added to Mg-based hydrogen storage alloy and composited between MgH_2 and porous materials. The diversity and abundance of Mg-based composite hydrogen storage materials and the complexity of structural modification offer the potential to develop high-performance hydrogen storage materials. It is considered that alloyed Mg-based materials can utilize the effect of the secondary phases and improve hydrogen storage properties with structural optimization. The synergistic effect of catalysts and nanosizing can improve the material's hydrogen storage performances by utilizing their "hydrogen pumping", "stress" and "nano-sizing" effects.

The synthesis methods of Mg-based hydrogen storage materials are varied based on different materials. Based on the morphology and composition of the materials, the synthesis methods are categorized into three types in this chapter: the processing method of Mg-based hydrogen storage alloys, Mg-based nanomaterials, and Mg-based composites, followed by the introduction of powdering and hydrogenation methods. The preparation of Mg-based hydrogen storage alloys involves melting and powder metallurgy, which can be used to prepare commonly used systems, such as Mg-TM (TM: transition metals) and Mg-RE (RE: rare-earth) alloys. The preparation methods of Mg-based nanomaterials include ball milling, deposition, and

Q. Li and F. Pan, *Magnesium-based Hydrogen Storage Materials*,
https://doi.org/10.1007/978-981-96-5531-1_3

liquid-phase reduction, etc. The preparation method part of Mg-based composites mainly introduces the incorporation of modified non-hydrogen storage materials. This chapter focuses on the preparation methods of Mg-based hydrogen storage materials and Mg-based hydrides. We will summarize these methods in detail, covering their definition, principles, advantages, disadvantages, applicable systems, and process parameters, etc.

The future developments in Mg-based hydrogen storage materials:

① Structural optimization of micro-alloyed Mg-based materials with novel preparation methods, which can introduce the alloying elements as the second phase in Mg, acting as catalytic components and improving hydrogen storage performances.

② Mg-based composites are prepared through specialized processes, the merit of the "synergistic effect", including nano-interfacial, hydrogen pumping, and stress effects, further improving their hydrogen storage performance.

3.1 Preparation of Mg-Based Hydrogen Storage Alloys

3.1.1 Melting Method

The melting method involves heating and melting the raw materials with a specific composition ratio, followed by cooling the metal liquid into alloy blocks. This method is the traditional method for preparing Mg-based hydrogen storage alloys and the most frequently utilized melting methods are vacuum induction melting, arc melting, and resistance melting, etc. These methods have the advantages of mass production capability and low cost.

(1) Vacuum induction melting method

The induction melting method is an approach that heats metal materials through electromagnetic induction under a vacuum environment. In vacuum induction melting, raw materials are placed into a crucible surrounded by an electromagnetic coil. The current in the induction coil induces an electromagnetic induction potential, generating heat to melt metals and realizing the smelting.

Induction heating is one of the non-contact heating methods, the energy transmit is based on electro-magnetic induction. The metal materials in the crucible are heated by the generated vortexing magnetic field, and gradually, the metals are melted. The advantages of this method including uniform heating, rapid heating rates, good controllability, easily to achieve automatic control, stable product quality, a favorable working environment, and a relatively small footprint. Figure 3.1 illustrates the schematic of the induction heating principle [1, 2].

Vacuum induction melting is one of the most efficient, rapid, and environmentally friendly heating methods. It is widely utilized to prepare Mg-based hydrogen storage alloys. Unlike traditional methods, induction melting can melt high- and

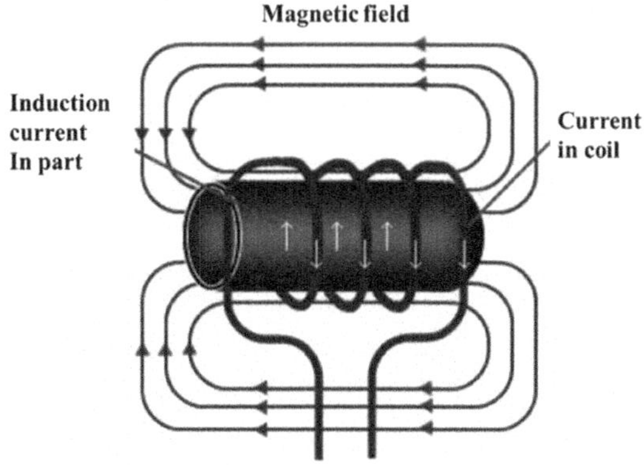

Fig. 3.1 Schematic illustration of the induction heating principle

low-melting-point metals. During the melting process, the electromagnetic force exerted on liquid metal materials enables the automatic stirring of liquid, ensuring uniform composition of the alloys. Additionally, operating under vacuum conditions can improve the quality of the products by reducing the gas impurities in the metals [3]. However, despite the advantages of mass production, low-cost, and extensively used in industry, vacuum induction melting still suffers from high energy consumption and difficulty in controlling alloy microstructure. Moreover, high-purity (99.9% or higher) raw materials are needed to minimize the impact from impurity. It is also necessary to introduce inert gas to prevent combustion of Mg. Liquid alloy can be formed into ingots via casting or rapid solidification processing, obtaining bulk metal or powders. Finally, the metal hydride products are obtained through activation processes. The vacuum induction furnace is operated at a lower voltage than the medium frequency induction furnace, typically lower than 750 V, to prevent the insulation breakdown caused by gas discharge and over-high voltage. When the vacuum chamber pressure reaches 0.67 Pa, the power supply can be introduced to heat metals. During the melting process of Mg-based hydrogen storage alloys, an extra 5% of Mg is added to compensate the evaporation losses of Mg. High temperatures can ensure that the elements are melted uniformly. Usually, the melted alloy is kept at ca. 1173 K and electromagnetic stirred for 5 min. The ingots can be obtained by filling the liquid metal into the molds and solidifying for 2 h or more.

Reilly et al. [4, 5] synthesized Mg_2Ni alloy by melting Mg and Ni, as depicted in Fig. 3.2, Mg_2Ni is sensitive to hydrogen, forming Mg_2NiH_4 under 3 MPa hydrogen pressure at 598 K. The decomposition temperature of Mg_2NiH_4 is 547 K under the pressure of 0.1 MPa H_2, which is significantly lower than that of MgH_2. However, the Mg_2Ni alloy prepared by the melting method is difficult to active, requiring several cycles of hydrogen absorption and desorption process. Once the sample is activated, the alloy can readily absorb hydrogen at the temperature and pressure higher than 473 K and 2 MPa. Other researchers [6] also employed the induction melting method to prepare Mg_2Ni alloys. It was reported that after three cycles of activation, it can absorb 2.67 wt.% H_2 at 573 K, equivalent to 74% of its theoretical value.

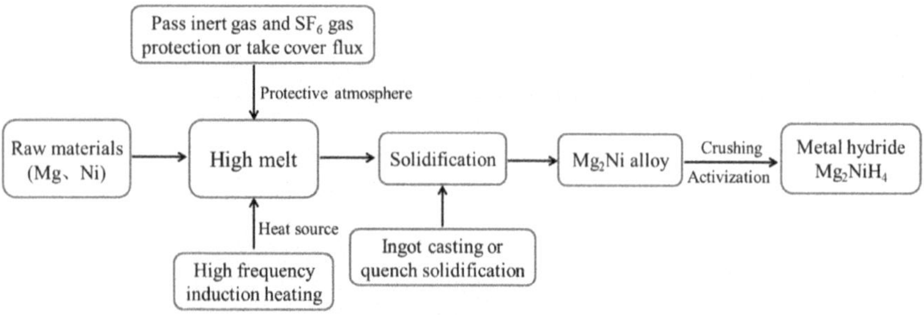

Fig. 3.2 Flow chart for the preparation of Mg_2Ni alloys by vacuum induction melting method

(2) Arc-melting method

The arc-melting method utilizes the electric arc between two metal electrodes, or between the metal electrode and the smelted material, to melt the metals. It can thermally melt metals, such as titanium (Ti), zirconium (Zr), molybdenum (Mo), niobium (Nb), tantalum (Ta), tungsten (W), and other refractory metals and alloys. This technique is particularly crucial to prepare Mg and refractory metal-based hydrogen storage alloys. Vacuum arc melting is one of the widely employed techniques, which utilizes metals as the self-consuming electrodes under vacuum conditions (10^{-2}~10^{-1} Pa) and a direct current arc as a heat source to gradually melt the self-consuming electrodes, allowing the molten metal to drip into a water-cooled copper mold. The liquid transforms to ingot after solidification. During this progress, the molten droplets are refined in high vacuum 5000 K arc region. The molten metal gathers in the copper mold to form a molten metal pool and further under-went the vacuum refining and then sequence solidified. A typical vacuum electric arc furnace is composed of a furnace, a power supply system, a vacuum system, a water-cooling system, and an automatic control system, as shown in Fig. 3.3. Vacuum arc remelting enables the production of large-size and ton-class metal ingots. Nowadays, the largest metal ingot produced with vacuum arc furnace is ca. 50~70 tons with diameters exceeding 2 m. Moreover, the continuous casting of the molten metal in the water-cooled copper crystallizer efficiently eliminates defects, such as shrinkage, segregation, and central porosity. Meanwhile, it also let the impurities move upwards to the top to form slag, which is a protective layer of the liquid metal. The disadvantages of vacuum arc remelting include two points: (i) the manufacture of the consumable electrode is an extra process compared with other melting methods; (ii) columnar crystals dominated the crystal structures in the resulting ingots, from the bottom to the top, the columnar crystals are gradually changed, the upper one is much coarser, and the angel with radial directions decreases. The coarse column-like grains are not fit for forging and extrusion processing [7].

Due to the high temperature of the arc furnace as well as the low melting point and high volatility of Mg, the utilization of arc-melting mainly includes the preparation of refractory alloy additives and catalysts. Amongst them, Ti-based [8, 9] and Zr-based [10, 11] alloys are usually selected as catalysts for MgH_2. For example, $Ti_{50}Zr_{50}$ alloy

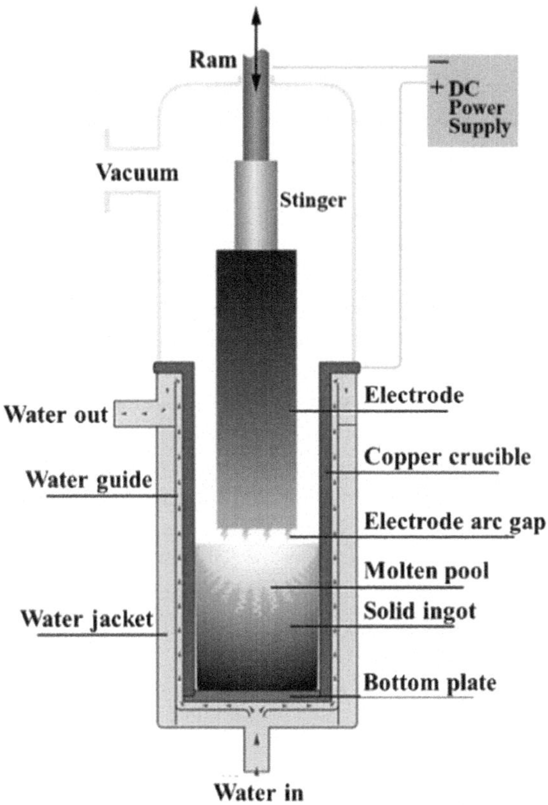

Fig. 3.3 Structure schematic of vacuum electric arc furnace

was synthesized using arc melting. The Mg-Ti-Zr hydrogen storage materials were then prepared by melting $Ti_{50}Zr_{50}$ and Mg [8]. The resulting Mg-Ti-Zr nanocomposite exhibits improved hydrogen absorption and desorption cycling performance. What needs to be emphasized is that Ti and Zr can transform to metal hydrides during the cycling process, effectively preventing the particle growth of Mg and the resulting Mg-based hydrogen storage materials exhibit rapid hydrogen absorption at moderate temperatures, it absorbs 4.3 wt.% H_2 within 5 min at 473 K and desorbs 2.8 wt.% H_2 within 5 min at 523 K. To improve the hydrogen desorption thermodynamics of MgH_2, $TiCr_{1.2}Fe_{0.6}$ alloy is first prepared via vacuum arc-melting and utilized as a catalyst, and MgH_2-5% $TiCr_{1.2}Fe_{0.6}$ was prepared by ball-milling $TiCr_{1.2}Fe_{0.6}$ and MgH_2 [11]. Due to grain refinement and the catalytic effect of $TiCr_{1.2}Fe_{0.6}$, the dehydrogenation temperature of MgH_2 is reduced from 699 to 508 K.

Arc melting employs electric energy to create an electric arc between the electrode and the material to melt the materials. The melting temperature is typically adjusted by changing arc voltage and other parameters, but it is still challenging. For the melting of Mg-based hydrogen storage alloys, the voltage and current need to be carefully controlled to obtain the desired microstructure. Nowadays, large cross-section electrodes are commonly utilized, along with short arc operations. The arc length can be as small as 30 mm for the small diameter ingots (10 mm) with arc voltage ranging from 15 to 40 V.

(3) Resistance furnace melting method

The resistance furnace melting method is heating a crucible with a resistance wire and the raw materials are placed in the crucible. Subsequently, block alloys can be obtained using gravity casting of the molten metal. Due to its relatively low heat energy input, this method is more suitable for alloys with lower melting points, such as Mg- and Al-based alloys. Thus, the resistance furnace melting method is widely applied in the preparation process of Mg-based hydrogen storage materials [12, 13]. Resistance heating offers lower heat flux density than induction arc heating methods, enabling precise temperature control and the various compositions of Mg-based hydrogen storage alloys. However, due to the low melting point and the high vapor pressure of Mg, ensuring the compositional accuracy of Mg with higher-melting-point alloys is still a challenge. To address this issue, researchers developed gas-protected (the mixture of SF_6 and CO_2) resistance furnace melting and flux (RJ-2) covered resistance melting methods for preparing Mg-based alloys. Mg-Ni-La ternary alloys have been successfully prepared using the flux-covered resistance melting method [14], as shown in Fig. 3.4.

A certain amount of additional Mg surplus is needed when melting Mg with other volatile elements in a resistance furnace. A covering flux is added on the top of the dried crucible, and Mg and Mg-based intermediate alloy are added into the furnace

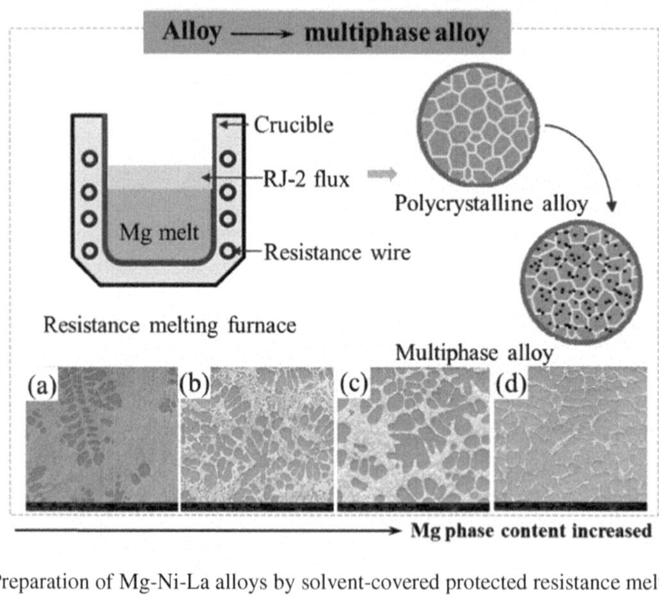

Fig. 3.4 Preparation of Mg-Ni-La alloys by solvent-covered protected resistance melting method [14]

when the temperature reaches ca. 873 K. Subsequently, the temperature is raised to 1073~1153 K with a dwell time of 0.5~1 h. During the heating and temperature-holding processes, samples are intermittently stirred using high-purity graphite rods to obtain a more homogeneous chemical constitution. Finally, the samples are cooled in the furnace or poured into a mold.

(4) Melt Spinning

Melt Spinning (MS), also known as rapid cooling, is a technique used for rapidly cooling metals or alloys. It is often employed in preparing thin metal ribbons with specific structures. A typical melt-spinning process involves spraying molten metal or alloy onto the surface of a high-speed rotating roller (water- or liquid-nitrogen-cooled) under the protection of Ar atmosphere. The molten materials are solidified rapidly after they contact the surface of the cooling roller. Typically, the rotation speed of the cooling roller is kept between 10 and 60 m/s to prevent the formation of droplets or broken ribbons. The continuous rotation of the cooling rollers facilitates the removal of the solidified alloy ribbons and provides a fresh surface for receiving the hot molten metal fluid. This method enables continuous production of metal ribbons [15]. The schematic illustration of melt spinning is shown in Fig. 3.5. Due to the rapid unfold speed of the alloy ribbon on the cooling roller surface, the system can achieve an extremely high cooling rate due to the contact with the highly thermally conductive cooler, enabling the formation of amorphous/nanocrystalline thin ribbons.

The melt-spinning process is affected by many factors and the quality and size of the as-prepared alloy ribbon are determined by the operation and parameters of the systems. The main factors are as follows:

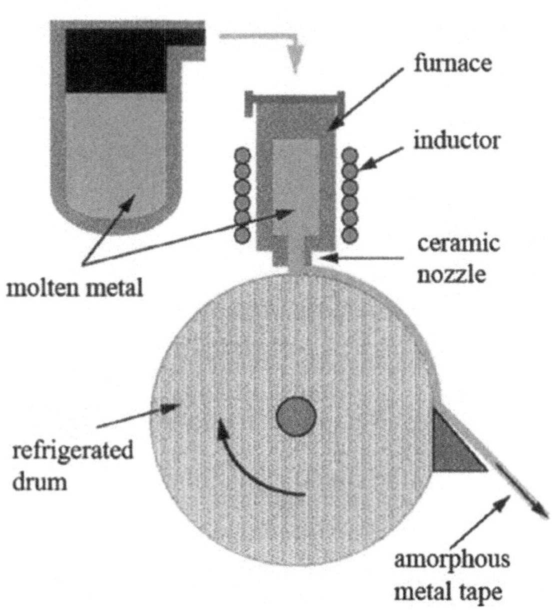

Fig. 3.5 Schematic diagram of the melt-spinning process principle

① The distance between the nozzle and cooling roller.
② The shapes of nozzles.
③ The flow rate of melted alloy on the rotating rollers is closely related to the speed of the rollers. It primarily affects the width and thickness of the ribbon.
④ The faster roller speed leads to the thinner alloy ribbon.
⑤ The cooling temperature of the roller mainly affects the microstructure of the ribbon. Different alloys result in different microstructures of the ribbon at different temperatures [16].

Melt spinning technology offers several advantages over other methods for preparing Mg-based hydrogen storage materials, including a simple production process, high production efficiency (10~60 m/s), and minimal environmental pollution. However, it also suffers from the low viscosity of molten material, high heat input requirements, and the need for frequent equipment maintenance. Nevertheless, melt spinning technology has proved to be one of the most effective methods for preparing Mg-based amorphous and nanocrystalline alloys with various compositions. Therefore, this method is highly suitable for mass production, to date, in the field of Mg-based materials, this method is mainly used for the production of Mg-TM (TM = Ni, Cu, Co, Fe, etc.) and Mg-Ni-RE (RE = La, Y, Nd, etc.) alloy. During the preparation of Mg-based alloys, the rotational velocity of the cooling rollers typically served as the melt spinning rate, ranging from 10 to 40 m/s.

Figure 3.6 [17] compares the microstructure between the as-cast Mg-10Ni alloy and the Mg-10Ni alloy prepared using the melt-spinning method. In the as-cast Mg-10Ni alloy (Fig. 3.6a, b), the eutectic structure consists of a fine Mg_2Ni phase (white) and a dendritic α-Mg phase (dark grey). The numerous interfaces caused by the eutectic structure suppress the diffusion of hydrogen atoms during the hydrogen absorption and desorption reactions in Mg-Ni alloys. However, the grain size of the α-Mg phases is ca. a few tens of micrometers, limiting the hydrogen diffusion in alloys, since the large grain size cannot provide enough transport channels (grain/phase boundaries) for H atoms. TEM images (Fig. 3.6c–e) show the morphologies of Mg-10Ni prepared by melt spinning with a linear velocity of 35 m/s. we can see that nano-scale Mg_2Ni (black) isometric grains are uniformly distributed in the refined white α-Mg matrix. The increased α-Mg grain boundaries and dispersed distribution of nanocrystalline Mg_2Ni intermetallic compounds are expected to enhance the hydrogen absorption and desorption properties of Mg-10Ni alloy.

Furthermore, the amorphous Mg-Ni-RE nanoalloys (RE = rare earth element) can be prepared using melt spinning, achieving improved hydrogen storage performances. Various rotation speeds of the cooling rollers yield alloys with diverse microstructures and phases, including amorphous, partially crystallized nanoalloys, and fully crystallized nanoalloys. In $Mg_{22}Y_3Ni_9Cu$ alloy, the proportion of amorphous microstructure notably increases with higher spinning speeds. Importantly, employing melt spinning at different speeds can adjust the thermodynamic and kinetic properties of hydrogen absorption and desorption in Mg-based hydrogen storage alloys [18]. The thermodynamic and kinetic properties of $Mg_{24}YNi_9Cu$ alloys prepared under different melt spinning speeds are provided in Fig. 3.7a, with

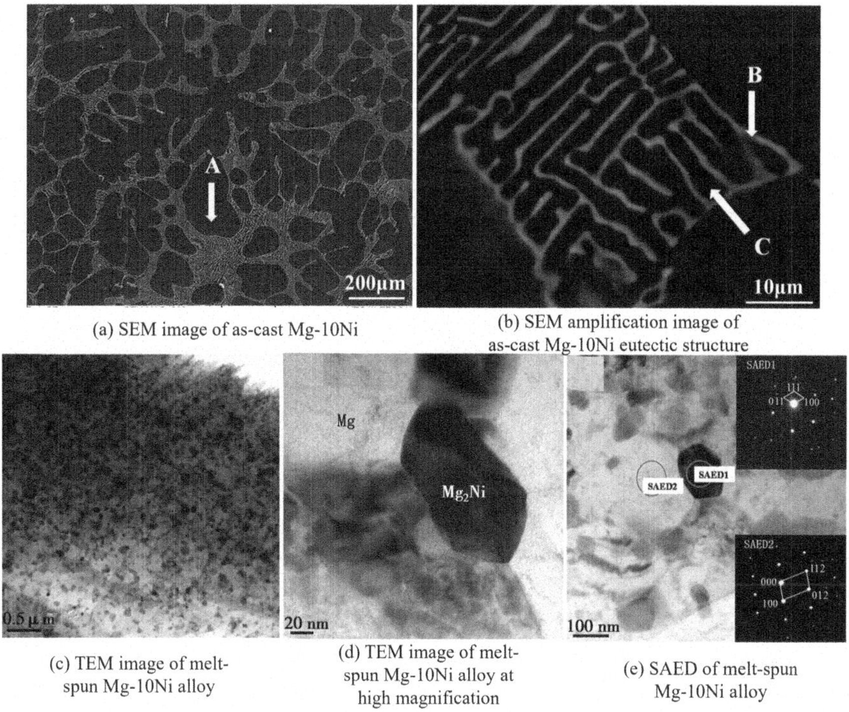

Fig. 3.6 The microstructure of Mg-10Ni alloy [17]

the increase of the rotation speed, the thermal stability of MgH_2 was decreased, resulting in suppressed ΔH_{ab} and ΔH_{de}. The hydrogen desorption temperature is also lowered from 518.6 to 501.1 K. These findings demonstrate the influence of melt spinning speed on the thermodynamic and kinetic properties of $Mg_{24}YNi_9Cu$ alloy. The samples prepared with rotation speeds of 15 and 20 m/s exhibit enhanced hydrogen absorption capacity and kinetics compared to as-cast alloys; besides, the activation energy of the hydrogen desorption reaction (E_k) decreases with the increasing spinning speed.

Table 3.1 summarizes the relationship between process parameters and hydrogen storage properties of various Mg-based hydrogen storage alloys prepared using melt-spinning methods. It can be seen that this method is highly effective in producing amorphous/nanocrystalline Mg-based hydrogen storage alloys without decreasing hydrogen storage capacity. However, during the hydrogen absorption and desorption process, the amorphous phase in alloys suffers from crystallization, to this end, uniformly introducing nanoscale phases of alloying elements or their compounds in the Mg matrix are required, so as to form a stable multiple-phase nanostructure. Thin ribbons produced by melt spinning technique exhibit excellent reactivity with hydrogen. The presence of nanosized phases facilitates the chemisorption of H_2 molecules at the surface of the alloys and enhances the hydrogen diffusion

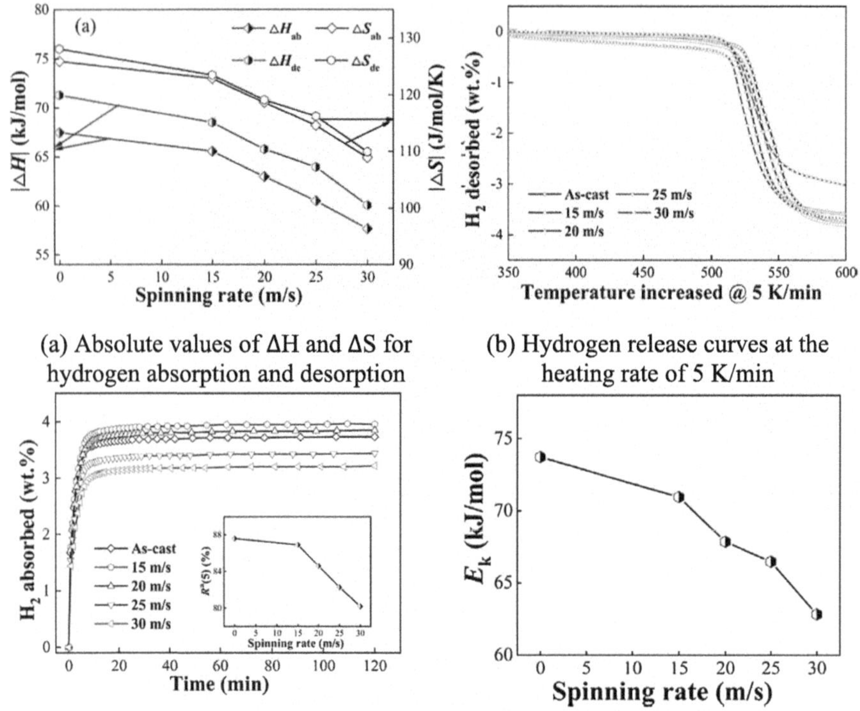

(a) Absolute values of ΔH and ΔS for hydrogen absorption and desorption

(b) Hydrogen release curves at the heating rate of 5 K/min

(c) Isothermal hydrogen absorption kinetic curves

(d) Hydrogen release activation energy E_k

Fig. 3.7 Thermodynamic and kinetic properties of $Mg_{24}YNi_9Cu$ alloys prepared by melt spinning at different rotation speeds [18]

into the bulk. Consequently, melt spinning emerges as a promising technique for preparing amorphous/nanocrystalline Mg-based hydrogen storage materials, which can optimize their thermodynamic and kinetic properties.

3.1.2 Powder Metallurgy

Powder metallurgy is a processing method that heating metal powders or pellets below the melting point temperature of the main components to achieve the material or product with desired strength or characteristics through particle bonding and other physicochemical effects. During this process, the surface curvature increases with the decrease in particle size of ultrafine powder, leading to high surface energy. Meanwhile, the pressure caused by the surface inward tension rises inside the ultrafine powders. Consequently, the physical properties of fine powders are modified, leading

Table 3.1 Process parameters and hydrogen storage properties of Mg-based hydrogen storage alloys prepared by melt spinning method

Mg-based hydrogen storage materials	Rotation speed/ ($m\text{-}s^{-1}$)	Sample size	Hydrogen storage capacity/ wt.%	Final particle (P)/grain (G) size	Hydrogen absorption kinetics	Hydrogen desorption kinetics	References
Mg-10Ni	35	Width: 10 mm; Thickness: 150 μm	5.75	P: 31.6 μm	5 min/ 3.91 wt.%/523 K/2.5 MPa	–	[17]
Mg-10Ni-2Mm	20.9	Width: 3 mm	4.20	G: nanoscale	7 min/3 wt.%/573 K/2.8 MPa	From 493 K	[19]
$Mg_{12}YNi$	20	Width: 3 mm; Thickness: 30~50 μm	4.74	G: 10 μm	20 min/ 4.74 wt.%/573 K/3 MPa	10 min/4 wt.%/573 K	[20]
$Mg_{80}Ni_{10}Y_{10}$	40	Width: 10 mm; Thickness: 35 μm	5.30	G: 5–20 nm	10 min/ 4.8 wt.%/ 523 K/2 MPa	30 min/5 wt.%/523 K	[21]

to a lower melting point and reduced sintering temperatures. To prepare bulk Mg-based hydrogen storage alloys via powder metallurgy, it is necessary to mix two or more metal powders with appropriate ratios and pelletize them through pressing. Then, the samples are calcined up to the melting points under a hydrogen, argon, or vacuum environment. Compared with the melting method, powder metallurgy possesses the merit of lower reaction temperature. In addition, the issue of high Mg vapor pressure can be solved by the powder metallurgy method, enabling the precise control of the compositions of Mg-based alloys.

Powder metallurgy techniques have been further optimized. Based on traditional methods, more advanced technologies, e.g., discharge plasma sintering [22–24], laser sintering [25–27], microwave sintering [28–30] and strong magnetic field-assisted sintering [31–33], have been developed. All these methods have been employed to prepare Mg-based hydrogen-storage alloys in the past years.

(1) Laser sintering method

The heating methods used in the sintering process significantly impact the properties of prepared alloys. Recently, laser and microwave heating techniques were utilized in the sintering process. The laser sintering technologies are classified into several types. Amongst them, selective laser sintering (SLS) is the primary approach for preparing Mg-based hydrogen storage composites. In SLS, a high-energy laser beam is used as an irradiation source, and powder material partially melts into a liquid or semi-solid phase after irradiation, while some powder particles remain in a solid phase. Subsequently, solidification and particle rearrangement lead to the densification of

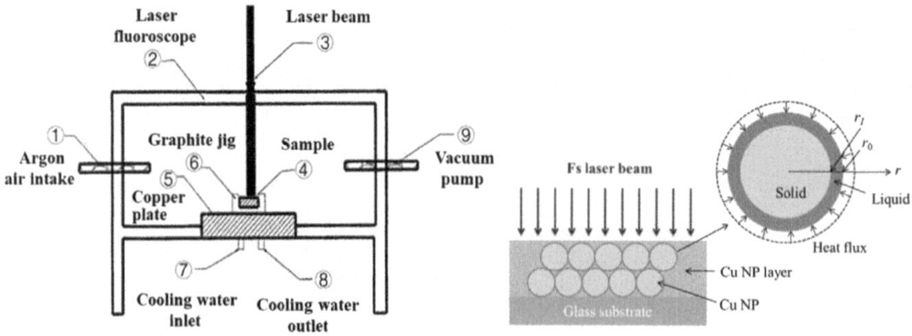

Fig. 3.8 Laser sintering chamber and laser sintering schematic for Mg-based hydrogen storage alloys and their composites [36]

powders. This technology offers unique advantages for sintering those materials that are difficult to sinter when using conventional sintering furnaces. The concentrated beam and limited penetration depth make it suitable for sintering small-area or thin-film products, facilitating the sintering of powder or film with different compositions [34]. The process and parameters for preparing Mg-based hydrogen storage alloys and their composites via laser sintering are summarized as follows [35]: metal or alloy powder is weighed and uniformly mixed according to the designed chemical ratio. The mixture is then pressed into a pellet through pressing. Subsequently, the pellet is sintered using a CO_2 laser with a power ranging from 500 to 1600 W with a scanning speed of 100–120 mm/min. The spot diameter of the laser beam is as large as 3~5 mm. The scanning cycle is repeated three times and the sintering time is ca. 3~10 s. Figure 3.8a depicts a schematic illustration of the laser sintering chamber for Mg-based hydrogen storage alloys and their composites. The laser sintering process for Mg-based alloys is similar to the sintering of copper-nitrogen dispersed layers under femtosecond laser pulse irradiation [36] (Fig. 3.8b).

Laser sintering is a process with rapid heating and cooling. The rapid cooling and phase transformation of samples efficiently induce the refinement of grain and microstructure, along with the formation of crystal defects and microcracks. These can further enhance the hydrogen storage properties of the alloy. Mg–Ni hydrogen storage alloys prepared using laser sintering technology consist of various phases, including Mg_2Ni, $MgNi_2$, pure Mg, pure Ni, and MgO. With the increase of laser power, the weight loss of the Mg element increases [26, 27, 37]. Notably, at 1200 W, significant weight loss was observed. Mg_2Ni alloys prepared by laser sintering can achieve a maximum hydrogen storage capacity of 3.44 wt.%. Figure 3.9 shows the Mg-20 wt.% $LaNi_5$ composites prepared by laser sintering. The composite exhibits a reticulated distribution of Mg_2Ni + Mg eutectics (Arrow A). White clusters of $LaMg_{12}$ can be observed within the reticulated structure (Arrow C) and the black area is the α-Mg (Arrow B). Despite the morphological structure is similar to that prepared by the traditional sintering method, the grain size of Mg-20 wt.% $LaNi_5$ is smaller when laser sintering is utilized, favoring enhanced hydrogen absorption properties.

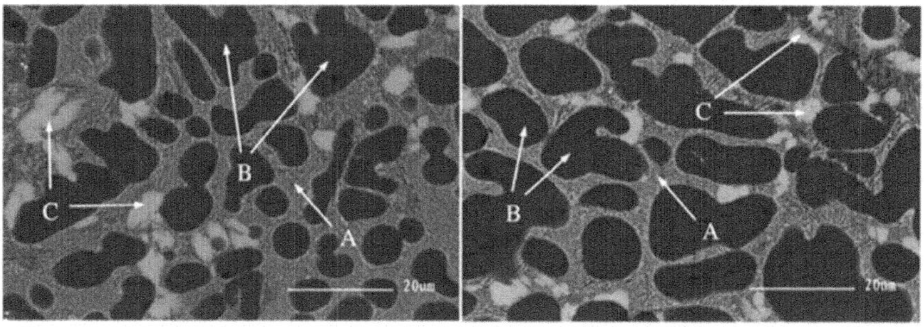

Fig. 3.9 Backscattered SEM images of laser-sintered Mg-20 wt.% $LaNi_5$ composites at different power levels [25]

Furthermore, the eutectic Mg_2Ni and $LaMg_{12}$ phases exhibit a beneficial catalytic effect on hydrogen absorption and desorption. The hydrogen storage capacity reached 5 wt.% after 5 activation cycles [25].

Due to the instantaneous action of the high-energy laser beam on the material, the liquid or semi-solid materials have one or several phases formed. It enables the rapid fabrication of Mg-based hydrogen storage alloys and their composites with excellent metallurgical bonding and performance. Typically, the sintering processes can be finished within a few dozen seconds. Various Mg-based hydrogen storage alloys with different composites have been successfully prepared using laser sintering, including Mg-Ni binary alloys, Mg-Ni-RE ternary alloys, and their composites. In addition, laser sintering possesses the advantages of low Mg weight loss, optimized microstructure, reduced grain size, and increased defect density (vacancies and cracks). It enables the preparation of a "core/shell" structure, in which the primary and secondary phases exhibit high hydrogen capacity and excellent catalytic properties, respectively, enhancing the hydrogen storage performance and electrochemical properties of Mg-based hydrogen storage composites. The design of laser sintering chamber devices is simple and cost-effective, which enables sample preparation under an inert atmosphere. In addition, it is characterized by laser beam penetration fast and rapid cooling, high yield, low energy consumption, simple process, excellent product quality, and low gas consumption natures during the preparation of Mg-based composite hydrogen storage materials [35].

(2) Microwave sintering method

Microwave heating is a process in which materials interact with microwaves, absorbing electromagnetic energy and converting it into heat. The heat transfer mechanism is different from conventional ones, such as conduction, radiation, and convection. In conventional sintering methods, the material's surface is initially heated and the heat is then conducted to the bulk, resulting in a temperature gradient. In contrast, microwaves generate heat inside of the material first and conduct heat energy to the entire samples, as illustrated in Fig. 3.10 [38]. This heating mechanism effectively enhances diffusion processes and reduces energy consumption. Due to the rapid heating nature, microwave heating significantly decreases processing time. The temperature needed for material processing is also lowered, enhancing physical and mechanical properties [39]. A schematic diagram of the microwave tube furnace is shown in Fig. 3.11 [40].

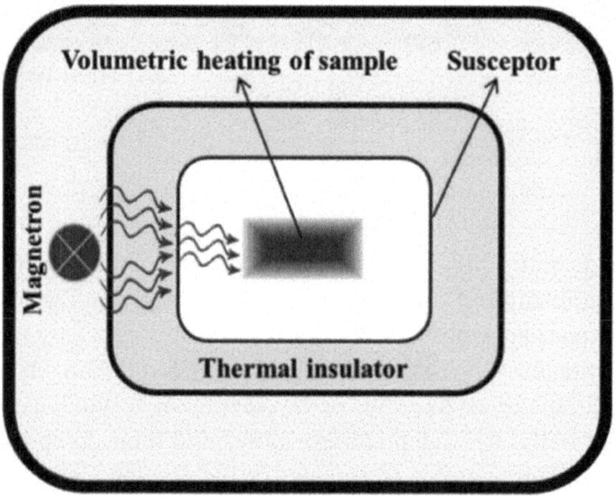

Fig. 3.10 Schematic diagram of a microwave sintering furnace [38]

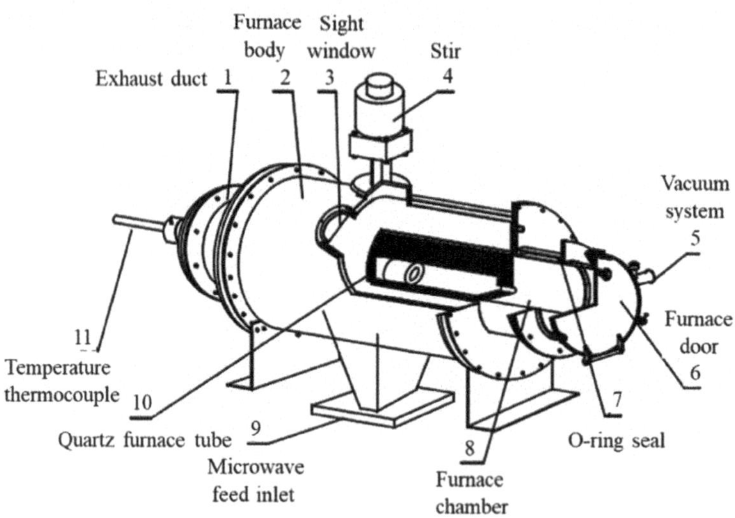

Fig. 3.11 Schematic diagram of a microwave tube oven [40]

The principle of microwave sintering technology relies on the phenomenon of dielectric loss of materials in electromagnetic fields. The reason is that electromagnetic waves can penetrate materials, and generate heat in bulk, realizing the desired sintering temperature, therefore, microwave sintering is a crucial sintering and densification technique. During the sintering process, dielectric polarization phenomena appear in the microwave electromagnetic field. Based on Faraday's law, the generated induction currents are in the same direction as the electric field, leading to energy dissipation within the material. This kind of energy loss is derived from dipole and interface polarization, and dielectric dissipation of the material arises through current absorption. Under an adiabatic condition, the latent heat in the heating process is negligible. The heating rate of the material per unit volume under the electromagnetic field of microwaves can be expressed as follows:

$$\frac{dT}{dt} = \frac{2\pi f \varepsilon_0 \varepsilon' E^2}{c_p \rho} \tag{3.1}$$

where ρ is the density of a material, c_p is the heat capacity, ε_0 is the dielectric constant, ε' is the dielectric loss, f is the operating frequency, and E is the electric field strength.

Microwave heating has the advantages of being fast, uniform, efficient, and promoting chemical reactions. Using this technique to prepare Mg-based hydrogen storage alloys, we can address the issues, such as low melting point and volatility of Mg. It was demonstrated that $Mg_{17}Al_{12}$ prepared using a small industrial microwave sintering furnace exhibited optimal hydrogen storage performance [28]. Under the optimized preparation conditions (4 kW and 30 min), the synthesized $Mg_{17}Al_{12}$ can absorb 3.83 wt.% H_2 at 623 K, and desorb 76.50% of the absorbed H_2. However, if the sintering time is too short, the $Mg_{17}Al_{12}$ cannot be fully activated, leading to poor hydrogen absorption and desorption performance. When the sintering time is extended, the grains begin to grow, the specific surface area of the powder decreases, and the alloy's activity diminishes. As a result, after hydrogen absorption, the MgH_2 phase decreases while the Mg_2Al_3 phase increases, leading to a reduction in hydrogen desorption capacity.

Li et al. [29] from Shanghai University successfully utilized microwave sintering to prepare ternary hydrogen storage alloys by the mixtures of Mg, La, and Ni powders, such as 80Mg-6.48La-13.52Ni, 70Mg-9.72La-20.28Ni and 60Mg-12.96La-27.04Ni. The chemical composition was designed using the Mg-La-Ni phase diagram. Among them, the 70Mg-9.72La-20.28Ni alloy exhibited notable hydrogen adsorption and desorption properties. It absorbs 4.1 wt.% H_2 within 600 s and desorb 3.9 wt.% H_2 within 1500 s at 573 K. The activation energy of the hydrogen desorption reaction was estimated as 113.5 kJ/mol H_2. The optimal microstructure and appropriate ratio of Mg, Mg_2Ni, and LaH_x phases lead to superior hydrogen storage performance even at 523 K (Fig. 3.12). In addition, graphite-doped Mg-Ni-Ce alloy was prepared by combining mechanical ball milling and microwave sintering methods [30]. The introduced graphite is a lubricant agent during ball milling and improves heat efficiency during microwave sintering. In addition, graphite also catalyzes hydrogen

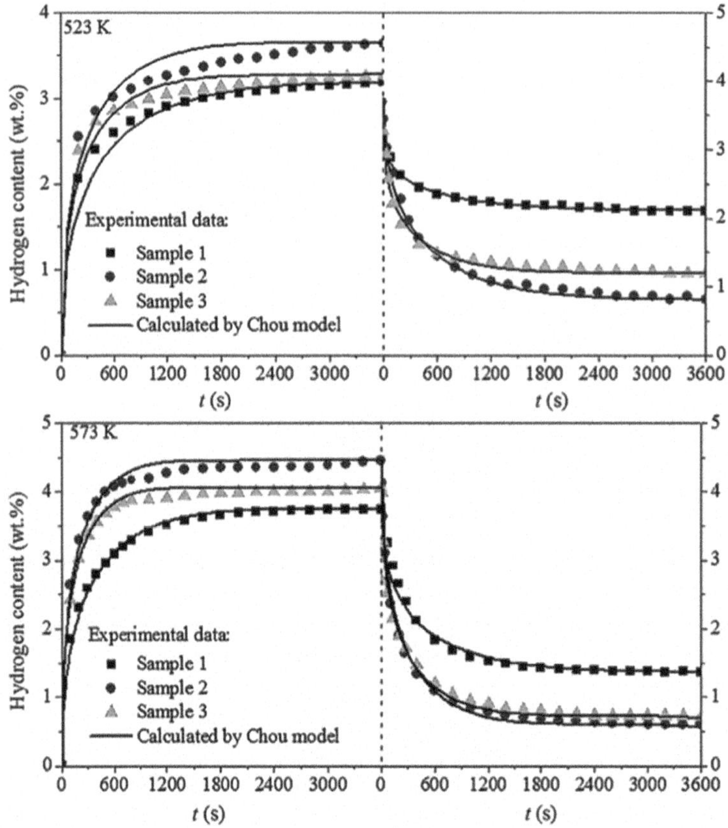

Fig. 3.12 Kinetic curves of hydrogen absorption and desorption in Mg-La-Ni ternary alloy (80Mg-6.48La-13.52Ni for Specimen 1; 70Mg-9.72La-20.28Ni for Specimen 2; 60Mg-12.96La-27.04Ni for Specimen 3) [29]

adsorption and desorption reactions, enhancing the hydrogen storage performance of Mg-based composites. The 5 wt.% graphite-$Mg_{17}Ni_{1.5}Ce_{0.5}$ alloy absorbs 5.34 wt.% H_2 and desorb 5.30 wt.% H_2 at 573 K, the activation energy is as low as 54.5 kJ/mol H_2 and desorbs hydrogen even at 511 K. When using microwave sintering to prepare Mg-based hydrogen storage alloys, it is typically required to mix and pelletize powders. Subsequently, the pellets are placed in a vacuumed chamber. The sintering parameters include a frequency of 2.45 GHz and a power range of 3~4 kW, which leads to a heating rate of 13~25 K/min, and the sintering time is ca. 20~30 min.

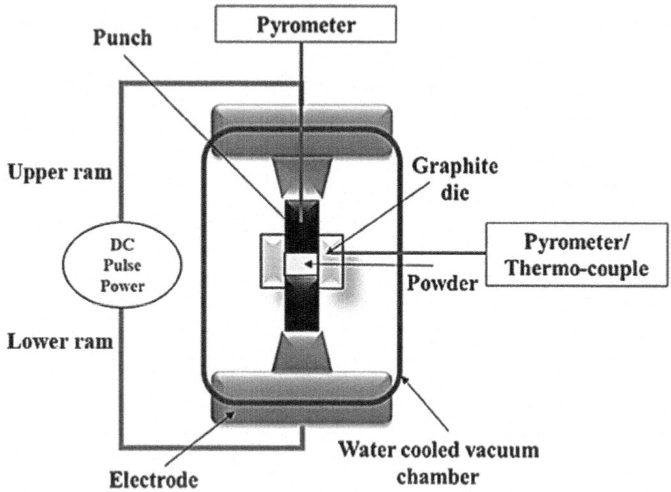

Fig. 3.13 Schematic diagram of the discharge plasma sintering system [41]

(3) Spark Plasma Sintering

Spark Plasma Sintering (SPS) has received considerable attention as a rapid solidification technique for powders, and the SPS system is composed of three main components (Fig. 3.13) [41].

① Devices and sintering mold for generating uniaxial pressure.
② Pulsed current generators are employed for generating plasma to activate materials.
③ Resistance heating equipment.

The working principle of SPS is generally as follows: First, a direct current pulse voltage generated by a specialized power supply induces discharge plasma within the voids of the powder. This discharge plasma purifies the metal particle surface and enhances the sintering activity. When the pulse voltage surpasses a specific threshold, the insulating layer between the powder particles is broken down and discharges. The discharge heats the particles rapidly by the spontaneous generating of heat, which resulting the material's evaporation and surface melting. A pulsed current heating with uniaxial pressure promotes element diffusion in bulk and grain boundaries, accelerating the sintering densification process. The operation temperature of SPS is generally lower than conventional methods by 473–773 K and the sintering time of SPS is ca. 5~10 min. SPS is featured in rapid heating rates, short processing times, and high sintering efficiency. Furthermore, its unique plasma activation and rapid sintering effects inhibit grain growth and efficiently preserve the original particle microstructure. These distinctive features of SPS facilitate the preparation of Mg-based hydrogen storage composites [42, 43].

Due to the significant difference in melting temperature between Mg and other metals, preparing Mg-based composites using traditional melting processes is still challenging, however, it is achievable with long-time mechanical ball milling and high-temperature sintering. Moreover, avoiding oxidation or contamination during the ball milling process is challenging. To overcome these limitations, the SPS method was proposed. For example, Mg-50 vol.% $V_{7.4}Zr_{7.4}Ti_{7.4}Ni$ was synthesized via SPS technology. The alloy can absorb 3.05 wt.% H_2 at 303 K, and desorb 2.55 wt.% H_2 at 573 K [22], significantly better than pure Mg. The decomposition temperature of MgH_2 in the composite is 591 K, 117 K lower than that of pure Mg. The hydrogen absorption capacity of the Mg-50 vol.% $V_{7.4}Zr_{7.4}Ti_{7.4}Ni$ prepared via SPS is comparable to that of Mg-40 wt.% $Ti_{0.28}Cr_{0.50}V_{0.22}$ prepared via mechanical ball milling for 40 h. SPS methods achieved a shorter reaction time and a lower hydrogen desorption temperature during Mg-based alloy preparations. Similarly, Mg-50%$V_{7.4}Zr_{7.4}Ti_{7.4}Ni_{7.4}$, Mg-30%$V_{38}Zr_{25}Ti_{15}Ni_2$, and Mg-30%$ZrMn_2$ were also prepared using SPS [23]. These alloys enable hydrogen absorption and desorption at 573 K, and the hydrogen storage capacities reached 4.4 wt.%, 5.88 wt.%, and 6.34 wt.%, respectively. In all materials prepared with SPS, a thin layer of nanocrystalline Mg was identified, which plays a crucial role in enhancing the hydrogen storage performance of Mg-based composites [24].

SPS technology is employed widely to prepare Mg-based hydrogen storage materials, particularly alloys or composites that are challenging to produce using traditional methods. The SPS method is applied to the systems containing elements with distinct melting temperatures from Mg. Typically, the SPS method mixes powders with the designed composition, then sintering the mixture in a vacuum at 823 K and a pressure of 30 MPa for 10 min. The resulting alloy is typically in the form of a ϕ 20 $\times$ 5 mm cylinder. It is noted that a subsequent powdering process is needed before it is utilized as a hydrogen storage material.

(4) Strong magnetic field-assisted sintering method

In addition to laser and microwave sintering techniques, external fields, e.g., magnetic field, can also provide energy to facilitate phase transitions, serving as supplemental tools in the sintering preparation of alloys. The magnetic field has been utilized to prepare a range of Mg-based hydrogen storage alloys, including Mg-Fe-H [31] and Mg-La-Ni [32]. Magnetic Assisted Combustion Synthesis (MACS), also known as "Strong Magnetic Field Assisted Sintering", is a novel approach utilizing hydrogen combustion synthesis technology. In the MACS method, the energy is provided by the strong magnetic field and the heat source, which enables powder particles to interact, initiating gas–solid self-propagating combustion reactions. The magnetic field accelerates the reaction process and reduces the reaction temperature, preventing excessive particle growth and Mg evaporation. The equipment used in the MACS is depicted in Fig. 3.14 [33]. The process involves mixing and compressing the sample powder. The samples are pelletized to the shape with $\phi15 \times (2\sim3)$ mm and then

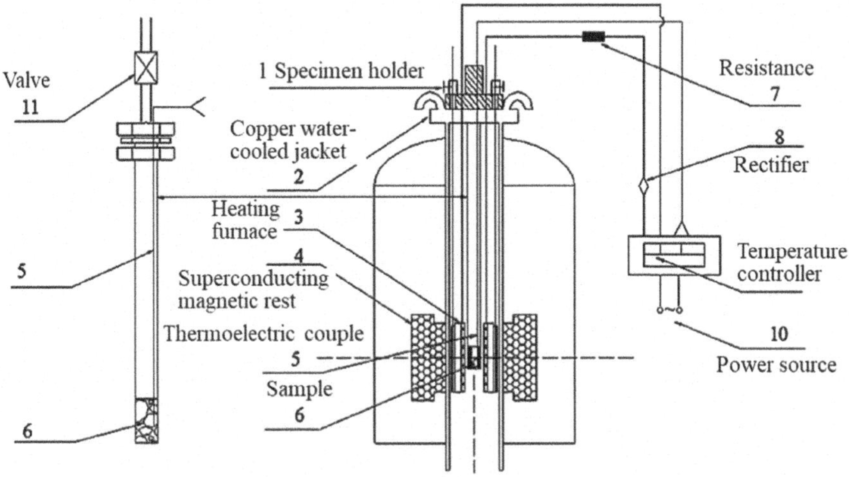

Fig. 3.14 Schematic diagram of a strong magnetic field-assisted sintering unit [33]

placed in a high-pressure reactor filled with 0.5 MPa argon gas. Then, the high-pressure reactor is subjected to a stable and strong magnetic field (0~8 T). The material is then heated using a tube furnace with a ramping rate of 20 K/min, and finally, sintering is carried out at 1073 K for 4 h.

It was revealed that applying a strong magnetic field impacts various characteristics of Mg-Fe composites, including structural and morphological features, phase composition, crystal size, and elemental distribution [31]. The synthesis of Mg_2FeH_6 from Mg and Fe powders via MACS significantly enhances their hydrogen stora0ge properties. This improvement can be attributed to the modification of phase composition and the reduced crystal size. Notably, the 2Mg-Fe composite prepared under a magnetic field of 5 T can desorb 6.73 wt.% H_2 within 1000 s at 593 K and 0.1 MPa H_2, as shown in Fig. 3.15.

The La_2MgNi_9 alloy prepared using this method crystallized in a $PuNi_3$-type structure. In the alloy, (La, Mg)Ni_3 is the primary phase, accompanied by a trace amount of impurity phase of $LaNi_5$. Interestingly, the content of $LaNi_5$ fluctuates as a function of the magnetic field and shows a minimum content at 4 T [32]. Alloy powders prepared under 1 T are characterized by porous morphology, fine particles, and rough surfaces. After hydrogenation, the alloy decomposed into La_2Ni_7, $MgNi_2$ and $LaNi_3$ phases. Figure 3.16 illustrates the Pressure-Composition-Temperature (PCT) curves of La_2MgNi_9 at room temperature. Notably, the LMN-1 alloy, obtained under a 1 T magnetic field, exhibits the smallest hysteresis coefficient (0.480) and the highest hydrogen desorption amount (1.307 wt.%) with the highest hydrogen desorption amount and rates.

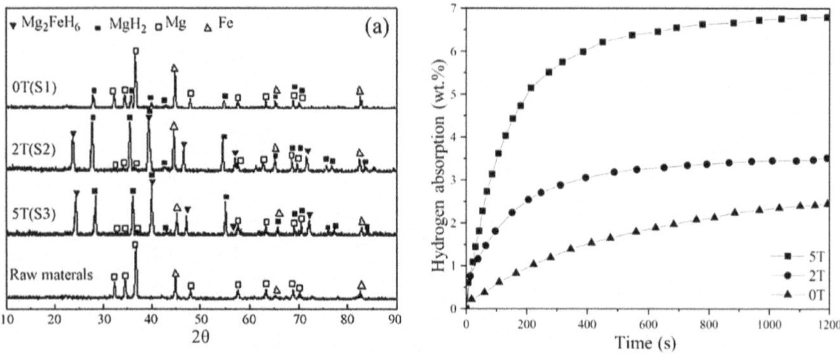

(a) X-ray diffraction patterns (b) Hydrogen desorption kinetic curves

Fig. 3.15 Phase constituents and hydrogen desorption properties of Mg-Fe-H phases prepared under different intensities of magnetic field [31]

Fig. 3.16 P-C-T curve of hydrogen absorption (A) and desorption (D) at 300 K for La_2MgNi_9 alloy prepared by strong magnetic field assisted sintering method [32]

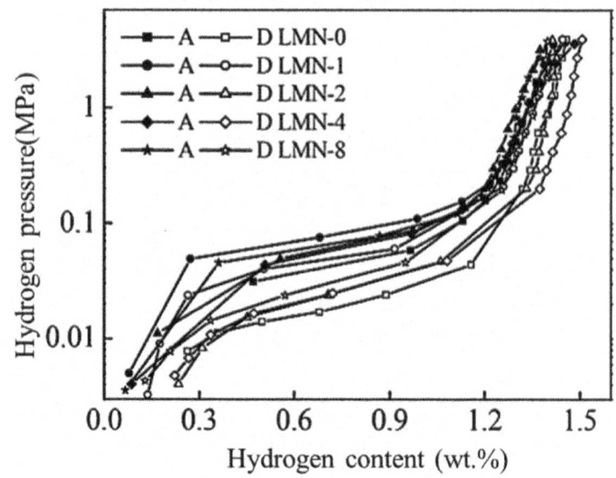

(5) Self-propagating high-temperature combustion synthesis method

Combustion Synthesis (CS) is also known as self-propagating high-temperature combustion synthesis. This method utilizes the reaction heat to realize the high-temperature condition, which enables chemical reactions between different materials. The materials synthesized with this method possess the particular advantageous of enhanced hydrogen storage capacity, furthermore, the activation treatment is not required, which ensures precise composition control, these merits also minimize the synthesis time and energy consumption.

The Hydriding Combustion Synthesis was proposed by Yagi and coworkers at Tohoku University in Japan in 1995. This method was first employed to prepare Mg-Ni alloy (Mg_2Ni) [44]. Subsequently, the same group introduced a novel approach utilizing the hydriding combustion synthesis method for synthesizing hydrogenated

Mg–Ni alloys (Mg_2NiH_4) in 1997. The Mg and Ni mixed powders are placed under high temperature and high hydrogen pressure conditions and directly synthesized Mg_2NiH_4 in a single step. This method uses the heat from the exothermic reaction between metal powder and hydrogen to facilitate the overall reaction, forming a self-exothermic solid-phase reaction. The reaction process mainly consists of the following three steps.

$$Mg + H_2 = MgH_2, \quad \Delta H^\circ = -74.5 \text{ kJ/mol} \tag{3.2}$$

$$Mg + 2Ni = Mg_2Ni, \quad \Delta H^\circ = -372 \text{ kJ/mol} \tag{3.3}$$

$$Mg_2Ni + 2H_2 = Mg_2NiH_4, \quad \Delta H^\circ = -64.4 \text{ kJ/mol} \tag{3.4}$$

The second step (3.3) is the reaction to synthesize the Mg_2Ni phase, which is completed instantaneously at a specific temperature and generates a large amount of reaction heat. The third hydrogenation step (3.4) occurs during the cooling process, resulting in the final hydrogenation of Mg-Ni alloy (Mg_2NiH_4).

Figure 3.17 shows the hydrogenation combustion synthesis equipment used for the preparation of Mg_2NiH_4 [44]. The reactor comprises a Ni-Cr-Fe alloy tube with an inner length and diameter of 520 and 70 mm, respectively. Through a sample holder, three pairs of thermocouples are linked to three orifices below the stainless-steel sample chamber. The thermocouples are not only used to measure temperature but also to fix the sample chamber, ensuring its placement at the center of the alloy tube.

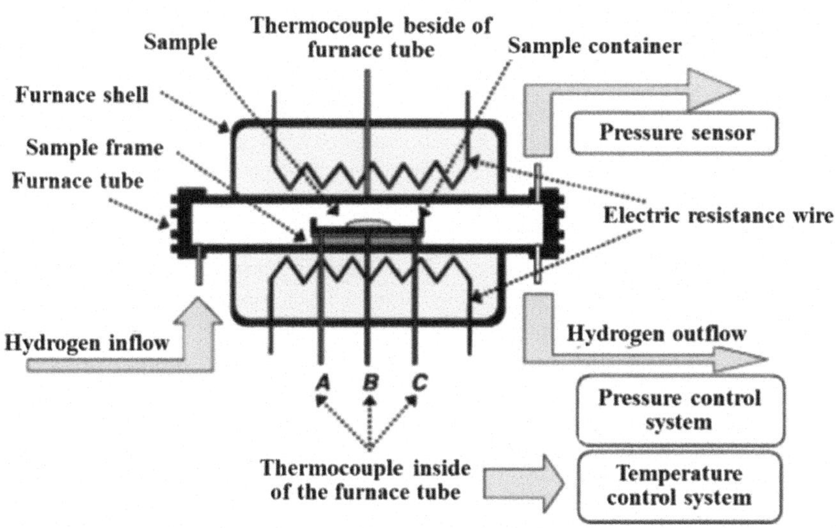

Fig. 3.17 Schematic diagram of the HCS reactor [44]

The middle one of the thermocouples mentioned above, labelled as B, was employed to regulate the reaction temperature with an accuracy of $\pm$ 1 K, controlling the temperature of the alloy tubes. The Mg powders (99.9% purity, particle size < 180 μm) and nickel powders (99.9% purity, particle size 2~3 μm) are selected as the starting materials. After uniform mixing the powders by sonication in acetone with an atomic ratio of 2:1, the mixture was dried and calcined at 873K under 4 MPa H_2. Various hydrogen storage alloy systems, including MgH_2, Mg_2NiH_4, Mg-Ni-Cu, Mg-Ti-V-Mn, and Mg-Ni-La, have been synthesized using the HCS method. Furthermore, this method can also be used for the commercial production [1, 2].

The HCS (Self-Propagating High-Temperature Synthesis) method utilizes the reaction enthalpy of exothermic reactions, resulting in a short synthesis time (typically a few hours). The synthesis reaction temperature is significantly lower than the melting point of Mg, thus reducing Mg volatilization and allowing direct preparation of the corresponding hydrogen storage alloy with a precise stoichiometric ratio. The resulting products exhibit high activity, and their hydrogen storage capacity approaches theoretical value before activation. Overall, the HCS method is featured by energy efficiency, time efficiency, high purity, and high activity natures, making it a novel and promising approach for preparing Mg-based hydrogen storage alloys. For instance, Mg_2NiH_4 prepared by the HCS method stores 3.6 wt.% H_2, significantly higher than the samples prepared by the conventional sintering method (IM). The hydrogen absorption rate and capacity of the HCS samples surpassed that of the IM samples (4 cycles) without any activation. The ad-/de-sorption curves at different temperatures confirm the improved hydrogen storage performance of HCS samples. Notably, in the low-temperature region (<423 K), the hydrogen storage capacity of the HCS samples was doubled that of IM samples. Moreover, PCT tests revealed that the equilibrium pressure of the HCS sample was consistently lower than that of the IM sample, indicating higher hydrogen storage capacity at lower hydrogen pressure for the HCS samples [45] (Fig. 3.18).

Li et al. [46] from Nanjing University of Technology improved the HCS method and proposed a laboratory-scale approach for producing Mg_2NiH_4. The maximum pressure and temperature of the HCS method are 6.0 MPa and 1073 K, respectively, with a homogeneous temperature zone of 20 cm. Unlike traditional methods, no pelletizing process is needed, and the mixed powder can be used as the starting material. The preparation process is simplified as follows: The sample underwent sintering at 873 K for 30 min, and after 50 min cooling process, the temperature is decreased to 680 K and kept for 30 min. Finally, the temperature was then cooled down to room temperature naturally in the furnace.

The HCS technique is frequently combined with the ball milling process to enhance the hydrogen storage properties of materials. Figure 3.19 shows the preparation process of $Mg_{90}Al_{10}$ samples using the HCS method. Subsequently, 0.6 g of $Mg_{90}Al_{10}$ was ball-milled with 5 mol.% of Ni and 0.012 g of graphite. The powder-to-ball ratio was kept at 1:30. After 400 r/min ball-milling for 10 h, $Mg_{90}Al_{10}$-5 nano Ni-MM sample was successfully obtained [47]. To investigate the impact of ball-milling process on the hydrogen absorption and desorption properties of the HCS

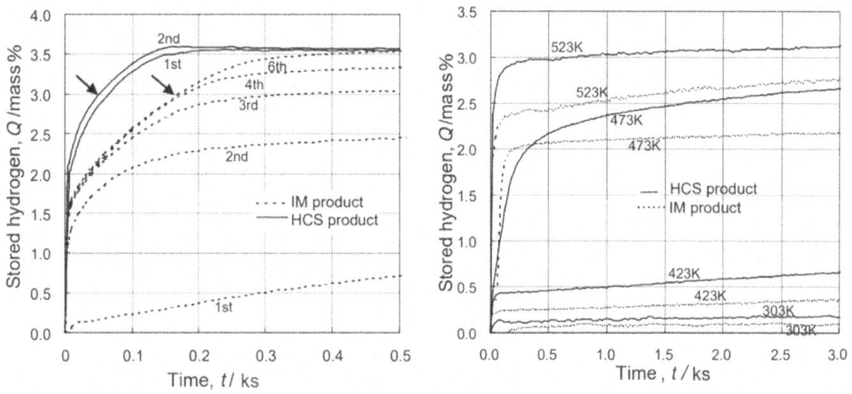

(a) Hydrogenation curves of HCS and IM samples at 603 K

(b) Hydrogenation curves of HCS and IM samples at 303, 423, 473, and 523 K.

Fig. 3.18 Comparison of hydrogenation curves for HCS and IM samples [45]

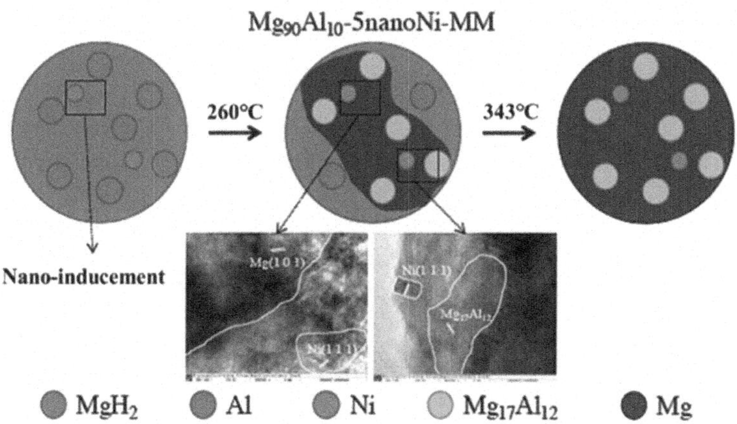

Fig. 3.19 Schematic illustration of the preparation process of $Mg_{90}Al_{10}$-5 nano Ni-MM sample [47]

samples, the Mg_2NiH_4 sample prepared by HCS method was ball-milled at 200 r/min for 40 h (0.1 MPa Ar, powder-to-ball ratio = 1:30), the hydrogen storage properties before and after the ball-milling process were compared (Fig. 3.20a), the hydriding rate and hydrogen storage capacity were improved after the ball milling process, in which, 2.76 wt.% H_2 was absorbed at 313 K (2.78 wt.% at 373 K) within 100 s. The starting desorption temperature of the ball-milled sample is ca. 370 K, while the sample before the ball-milling process starts the hydrogen desorption at ca. 570 K (Fig. 3.20b), the results demonstrate that the ball-milling process can significantly improve the thermodynamic properties of hydrogen storage materials [48].

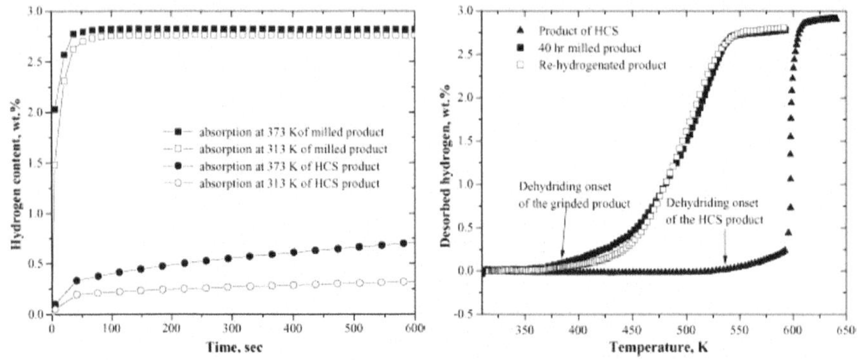

(a) Hydrogen absorption curves of the HCS samples before and after the ball milling process at 313 and 373K

(b) Hydrogen desorption curves in the heating process before and after the ball milling process of the HCS samples

Fig. 3.20 Comparison of hydrogen uptake and release kinetics of the HCS samples before and after ball milling [48]

(6) Replacement-diffusion method

The process of preparing alloys through the diffusion reaction between elements is known as the diffusion method, including the reduction-diffusion method, the coprecipitation-reduction method, the replacement-diffusion method, and the ball-milling-diffusion method. However, considering that Mg is the active metal, it is challenging to prepare Mg-based hydrogen storage alloys using the reduction-diffusion and coprecipitation-reduction methods. Instead, the replacement-diffusion method is more reasonable to be used to prepare Mg-based alloys.

The Replacement-Diffusion Method (RDM) [49], which utilizes the reactivity of Mg, is an effective approach for preparing Mg-based hydrogen storage alloys. The detailed procedures are as follows: dissolving Mg in anhydrous $NiCl_2$ or CuBr, along with dried dimethylphthalamine or acetonitrile, after 2~3 h vigorous stirring, Nickel or copper atoms are deposited onto the Mg atoms via the displacement reaction, subsequently, the vacuum-dried products are placed in a high-temperature furnace and subjected to the pyrolysis in argon atmosphere at 773~853 K for 2~3 h. Based upon the thermal diffusion, greyish-black powder of Mg_2Ni or Mg_2Cu was obtained [49, 50]. Figure 3.21 depicts the preparation process of Mg_2Ni or Mg_2Cu composites and the specific reactions for the preparation of Mg_2Ni are as follows.

$$NiCl_2 + 3Mg \rightarrow 2MgNi + MgCl_2 \tag{3.5}$$

$$2Mg + 2MgNi \xrightarrow{\Delta} 2Mg_2Ni \tag{3.6}$$

$$Mg_2Ni + 3H_2 \rightarrow MgH_2 + MgNiH_4 \tag{3.7}$$

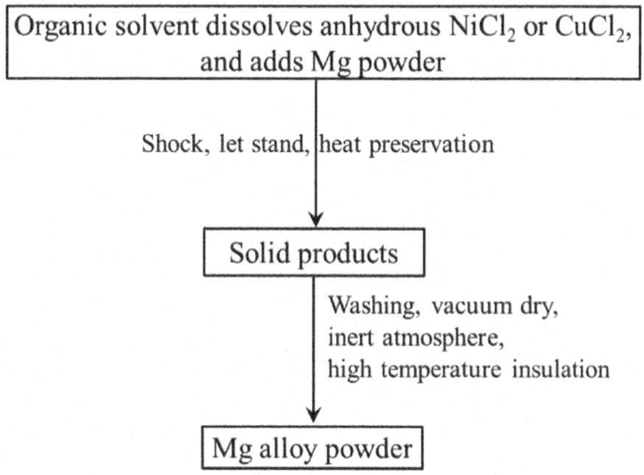

Fig. 3.21 Replacement-diffusion process flow chart [50]

The $Mg_2Ni_{0.75}Cu_{0.25}$ sample obtained via the replacement-diffusion method exhibited superior hydrogen absorption and desorption properties versus the samples prepared by the smelting method. Notably, these samples do not need to be pulverized and can be completely activated in a single hydrogen absorption and desorption cycle. The P-C-T plateaus observed at 623 K, 598 K, and 573 K were smooth, with a small hysteresis loop (<0.1 MPa), absorbing 3.2 wt.% H_2 [51]. Additionally, $Mg_2Ni_{0.75}Fe_{0.25}$ (molar ratio 2:1) was prepared using the replacement-diffusion method. The enthalpy change of hydrogen desorption was 65.2 kJ/mol H_2, which was lower than that of pure Mg. Furthermore, Ni and Fe catalyze the reaction between Mg and hydrogen, and the hydrogen storage capacity reaches approximately 3.3 wt.% for the optimal composition [52].

In all, the replacement-diffusion method is simple and can effectively suppress the Mg, which results in uniformly composed alloy powders, and the crushing process is unnecessary. The surface of alloy hosts numerous cracks, contributing to a high specific surface area that enhances activity and facilitates hydrogen absorption. Consequently, the hydrogen absorption and desorption rates are accelerated. It also reduces the thermal decomposition temperature of the hydride. For instance, the hydrogen desorption temperature of Mg_2NiH_4 is ca. 518 K, indicating excellent hydrogen storage properties. Moreover, a range of multifunctional Mg-based alloys, such as $Mg_2Ni_{1-x}Cu_x$ and $Mg_2Ni_{1-x}Pt_x$, can be synthesized using this method [53]. The replacement-diffusion method significantly enhances the hydrogen absorption and desorption rates. However, it is primarily suitable for preparing highly reactive metals like Mg.

3.2 The Preparation of Mg-Based Hydrogen Storage Nanomaterials

3.2.1 Ball Milling

Ball milling is widely used to prepare Mg-based nanomaterials for hydrogen storage. During the ball milling process, the formation of ultrafine powder leads to a large specific surface area and surface energy, increasing the tendency of particle aggregation and reducing surface energy. A suitable amount of additive of surfactant can effectively modify the surface properties of the powder, reducing surface energy and preventing powder oxidation during milling. This method is one of the most commonly employed techniques for preparing Mg nanoparticles for hydrogen storage.

(1) High-energy ball milling

A high-energy ball milling is a method to prepare nanoparticles utilizing the rotation or vibration of the ball mill, which causes the hard impact, grind, and mix to the raw materials. The high-energy ball mill typically consists of several parts, including the feeding, discharging, rotating, and transmission parts (reducer, small transmission gear, motor, and electric control system), etc. The hollow shaft of the high-energy ball milling is made of cast steel. The rotary gear is produced by casting hobbing processing. The cylinder body is embedded with a wear-resistant plate, providing excellent anti-abrasion ability. Overall, the high-energy ball mill method is smooth and reliable.

The high-energy ball milling operates within a temperature range of 278 to 323 K, with a voltage of (220 $\pm$ 10) V and a frequency of 50 Hz. The cylinder speed is set to 32 r/min. This method combines impact, grinding, and vibration, the grinding balls experience high-speed rotary movement, which impacts the sample obtaining the refined particles. An efficient high-energy ball mill delivers substantial mechanical energy to the sample powder within a short duration, enabling mechanical alloying or mechanical activation of the material, or even result in the formation of nanocrystalline or amorphous material without material oxidation and pollution. Additionally, high-energy ball milling can increase the specific surface area and introduce numerous defects and lattice distortions. To date, the high-energy ball milling method is widely used in synthesizing Mg-based hydrogen storage nano-materials and associated composite hydrogen storage materials.

In the high-energy ball milling process, the optimized milling parameters vary greatly based on the different purposes and materials. Usually, rotation speed, rotation time, ball-to-powder ratio, filling factor, and milling atmosphere are the main technical parameters. In particular, ball milling time and atmosphere are the two most essential factors. For example, ball milling time significantly affects the morphology, dispersion and desorption performance of the Ni-catalyzed MgH_2. The distribution of Ni is highly localized after 1 h of ball milling, however, with increased ball milling time, the size of the Ni particles decreases, and their distribution is more homogeneous. This facilitates the transformation of Ni to the metallic hydride of Mg_2NiH_4 [54]. Moreover, in $Mg_{87}Ni_{10}Al_3$ alloys prepared by ball milling under H_2 atmosphere, hydrogen-assisted ball milling triggers the conversion of Mg into nanocrystalline of MgH_2, and the content of MgH_2 increases with the increased milling time. As the hydriding progresses, MgH_2 will react with Ni to form Mg_2NiH_4. Thermal analysis reveals that the hydrogen desorption starts at 473 K after a short ball milling process. Additionally, the hydrogen desorption temperature of nanocrystalline MgH_2 prepared by ball milling under an H_2 atmosphere is significantly lower than that of pure MgH_2. With prolonged ball milling time, the stability of the hydride slightly improves, resulting in faster hydrogen absorption and desorption kinetics at 573 K [55].

In high-energy ball milling, metals, and intermetallic compounds undergo a transformation that modifies their long-range ordered structure. The phenomenon is also known as mechanical disordering. This process leads to the rearrangement of the structure of alloys, typically requiring a long period of ball milling, sometimes, mechanical amorphization will happen with certain alloy compositions and milling durations. As a result, Mg-based hydrogen storage materials prepared via high-energy ball milling usually have an amorphous or nano-crystallization structure. These structures effectively optimize the hydrogen storage properties of alloys. As shown in Fig. 3.22a [56], the XRD patterns of Mg-50Ni alloys after the 30 and 50 h ball milling process present amorphous/nanocrystalline co-exist and complete amorphous states. The former structure is illustrated in Fig. 3.22b, the nanocrystalline particles exist in the amorphous basal body. This structure combined the excellent mass transfer properties of nanocrystalline materials and the high catalytic activity of the amorphous phase. Consequently, Mg-50Ni amorphous and nanocrystalline composites exhibit increased hydrogen absorption/desorption kinetics and improved hydrogen storage capacity.

BCC (body-centered cubic) structured $Mg_{50}Co_{50}$ alloys [57] with grain diameter of a few nanometers were prepared by ball milling Mg and Co powders at 200 r/min between 0.5 and 400 h. The hydrogen absorption capacities of the alloys are ca. 2.67 wt.%, 2.42 wt.%, and 2.07 wt.% at 258, 303, and 373 K, respectively (Fig. 3.23). Notably, the 258 K hydrogen absorption temperature is the lowest value reported for MgCo composites. Additionally, amorphous and nanocrystalline $CeMg_{11}Ni + x$ wt.% Ni ($x = 100, 200$) composites were prepared via ball milling, increasing Ni content and milling time promotes the amorphization of the alloys, significantly enhancing their hydrogen absorption and desorption kinetics [58]. With extended milling time, the maximum hydrogen storage capacities of the alloys ($x = 100$ and 200) reached 5.949 wt.% and 6.157 wt.%, respectively, moreover, the hydrogen desorption rate is improved and the corresponding activation energies are decreased as well (Fig. 3.24).

The internal structure of a Mg-10Ni alloy was optimized through melt quenching. Subsequently, MWCNTs and titanium trifluoride (TiF_3) composite catalysts were introduced via short-term high-energy ball milling [59]. After a short time of high-energy ball milling, TiF_3 particles, and MWCNTs were dispersed on the

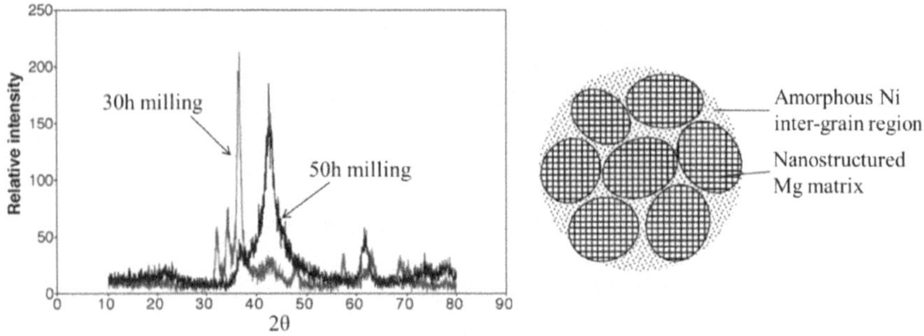

(a) XRD patterns of ball-milled Mg-50Ni alloys

(b) Structure schematic illustration of nanocrystalline/amorphous composites

Fig. 3.22 Microstructure of Mg-50Ni alloys subjected to high energy ball milling for different time [56]

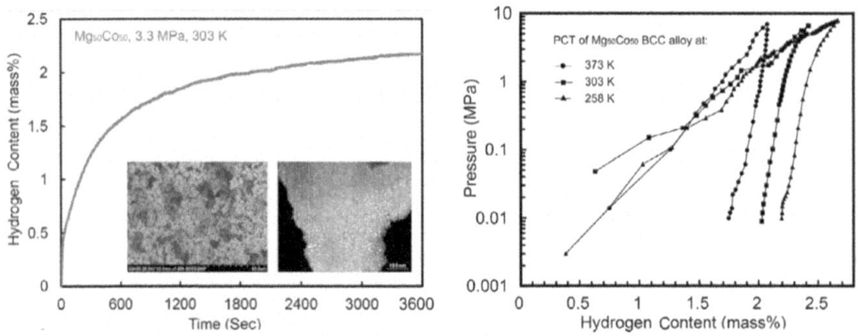

(a) Hydrogen absorption kinetics curve of $Mg_{50}Co_{50}$ alloy under 3.3 MPa hydrogen pressure and at 303 K (left is SEM image, and right is dark field TEM image in the insert)

(b) PCT curves of BCC $Mg_{50}Co_{50}$ alloy at different temperatures

Fig. 3.23 Hydrogen absorption kinetics and P-C-T curves of $Mg_{50}Co_{50}$ alloy [57]

surface of Mg-10Ni nanocrystalline particles (Fig. 3.25). Although the internal microstructure of the Mg-rich alloy remains unchanged after rapid quenching, the successful introduction of TiF_3 and MWCNTs results in improved surface activity and increased heterogeneous nucleation interfaces and additional diffusion channels. These enhancements collectively lead to excellent hydrogen absorption and desorption thermodynamics and kinetics.

Among the controllable parameters in ball milling, the primary factors influencing the process include milling speed, milling time, and the ratio of balls to powder.

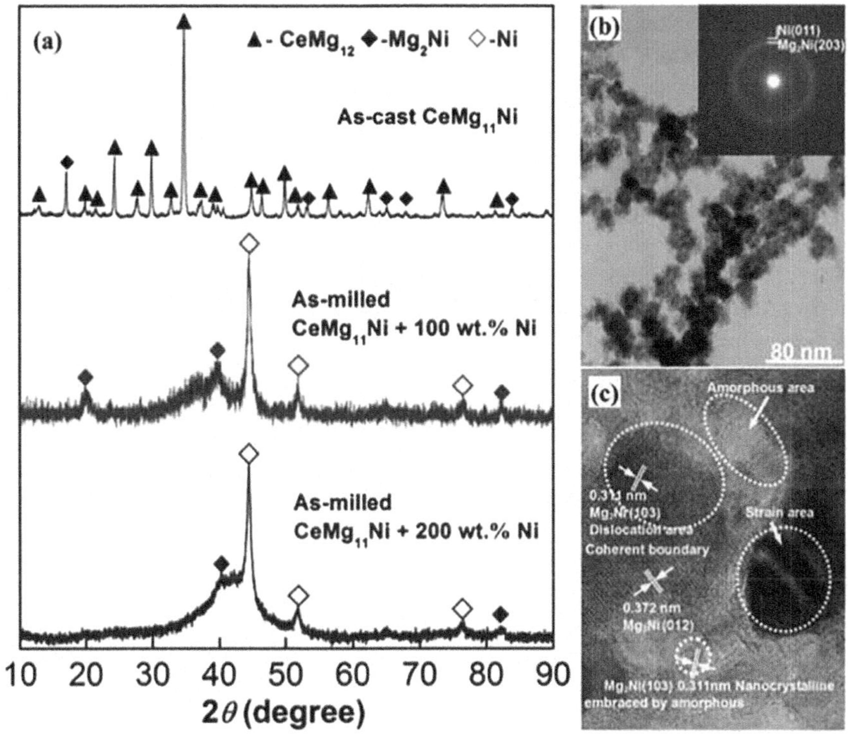

Fig. 3.24 Phase constitution and TEM images of $CeMg_{11}Ni + x$ wt.% Ni alloys after ball milling [58]

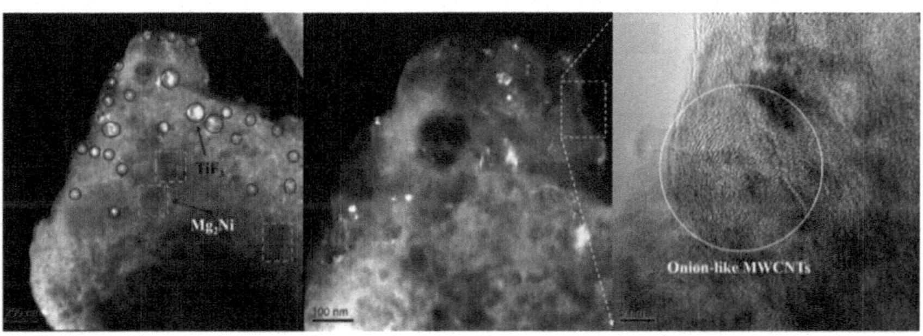

(a) Low-magnification dark field TEM image

(b) High-magnification dark field TEM image

(c) Locally enlarged bright field TEM image of (b)

Fig. 3.25 TEM microscopic morphology of Mg-10Ni alloy after surface catalytic modification by high-energy ball milling with the addition of MWCNTs and TiF_3 [59]

Typically, the grinding bottles and balls are made of Cr-Ni stainless steels. The ball-to-powder mass ratio ranges from 10:1 to 40:1, the rotational speeds vary from 100 to 850 r/min, and the milling time is typically selected between 0.5 and 400 h. The selected parameters are varied based on materials, for instance, alloys featuring a high Mg content tend to aggregate during long-time milling, and when milling time exceeds 30 h, the amorphous/nano-crystalline composite structure appears in the prepared Mg-based materials, therefore, short-term intermittent milling (typically around 1 h) are helpful to avoid undesired modifications.

To further improve the hydrogen storage properties of Mg-based materials, novel preparation methods were developed based on the ball milling method, e.g., ball milling process combined with hydrogen combustion and annealing method, etc. Hydrogen combustion synthesis (HCS) and mechanical ball milling (MM) are widely used for preparing Mg-based hydrogen storage alloys. HCS prepares highly active hydrides with simple procedures, while MM facilitates the synthesis of various nanocrystalline, metastable crystals and amorphous hydrogen storage materials with excellent hydrogen absorption capabilities. Li et al. [48] synthesized Mg_2Ni alloys through the combination of HCS and MM methods following the procedures of 1) Mg_2NiH_4 was synthesized via HCS, 2) 40 h MM (200 rpm, 0.1 MPa Ar) of the resulting Mg_2NiH_4. This approach significantly enhanced the hydrogen absorption and desorption properties of Mg_2Ni. The as-synthesized Mg_2Ni absorbs 2.76 wt.% H_2 within 100 s at 313 K, outperforming the un-milled samples. Moreover, the ball-milled sample desorbs hydrogen at 370 K, which is 190 K lower than that of the samples prepared using conventional HCS.

To improve ball-milling efficiency, physical fields were usually selected to assist high-energy ball-milling processes, e.g., ultrasound- and discharge-plasma-assisted methods are usually employed to enhance the efficiency of ball-milling. The acoustic vibration originating from the ultrasonic vaporizes the liquid phase and generates the instantaneous high temperature and pressure, resulting in powerful shock waves within the materials, facilitating their crush and promoting the generation of new phases. For instance, ultrasonic-assisted ball milling for 60 h can directly yield $Mn_xMg_{1-x}Fe_2O_4$ nanoparticles/clusters with particle size of ~30 nm [60].

In 2005, the Key Laboratory of Advanced Energy Storage Materials of Guangdong Province introduced cold-field plasma into the mechanical ball milling process, which led to the development of Dielectric Barrier Discharge Plasma (DBDP) assisted ball milling technology and corresponding equipment (Fig. 3.26) [61]. The combination of plasma field and mechanical ball milling can be realized by introducing a medium-blocking discharge structure into a gas-controlled ball mill. The synergistic effect of mechanical and plasma energy during the ball milling process not only significantly enhances the efficiency of mechanical alloying but also accelerates in-situ solid-phase and gas-solid-phase reactions, resulting in unique microstructures that greatly enhance the properties of the prepared materials. The principle of plasma ball milling is that a high-frequency and voltage current is input into the electrodes at both sides of a discharge ball mill tank with a dielectric barrier structure, by adjusting the discharge parameters of the plasma, the introduced gases, e.g., argon, nitrogen, oxygen, ammonia, etc., are excited to generate low-temperature discharge plasma.

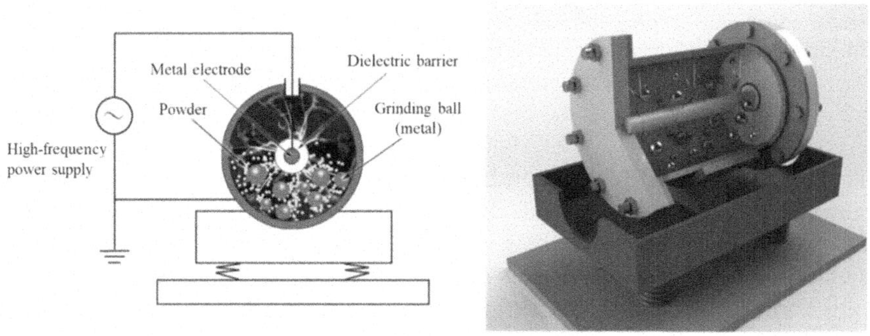

Fig. 3.26 Schematic illustration of the principle and instrument of dielectric barrier discharge plasma-assisted ball milling technology [61]

By varying the vibration frequency or rotation speed of the ball-milling process, the position of the powders, balls, and electrodes are changed, enabling plasma-assisted ball milling through corona discharge. The principle is shown in Fig. 3.26a. A DBDP power supply (25 kV, 15.5 kHz) is connected to the electrodes and the ball milling tank, resulting in uniform corona discharge [61, 62].

DBDP is one of the low-temperature plasma that can be realized at ambient pressure, and it is characterized by non-equilibrium properties such as high electron concentration and average energy. DBDP exhibits a broad operational pressure ranging from 0.1 to 10×10^{-5} Pa with continuously producing plasma, the plasma contains plenty of conductive ions and electrons in the macroscopically neutral gaseous medium. Since the plasma is a high-energy and high-activity atmosphere, it can serve as a heat source. Although the temperature of the electron generated by dielectric barrier discharge in the cold field plasma is extremely high, the overall macroscopic temperature of the atmosphere is acceptable (from room temperature to ca. 573 K). The dielectric barrier layer effectively prevents the escalation from micro-discharge, the risk of converting DBDP into spark or arc discharge, and the potential damage to the ball mill system from hot plasma. Due to the existence of the abundance of excited microscopic particles, plasma collisions with neutral particles or nano-powders supply thermal energy and lead to excitation, ionization, and light energies. This collision-induced impact of material surfaces or chemical activation within the gas phase and nano-powders can initiate the chemical processes, which are typically challenging under conventional conditions. Moreover, the cooling rate can reach approximately 10^5 K/s when the reactants leave the plasma. This rapid cooling process effectively "freezes" the treated powder in a particular state, favoring the production of nanoparticles.

DBDP-assisted high-energy ball milling ensures uniform exposure of milled powders to DBDP. The combination of thermal effect of DBDP with the impact of electrons and mechanical ball milling, activates the powders and accelerates powder refinement and alloying processes. To generate the DBDP, a high voltage (24~25 kV) alternating current at a frequency of 12~15.5 kHz is needed. Under the impact of

DBDP, a powder mixture and steel balls are loaded into a cylindrical stainless steel tank under high-purity argon (0.1 MPa) atmosphere with a ball-to-powder ratio of 30:1. The tank is subjected to vibration at an amplitude of 20 mm and a frequency of 25 Hz for 2 h to achieve the desired hydrogen storage material. Ouyang et al. [63] reported the preparation of Mg(In)-MgF_2 complexes through DBDP-assisted ball milling utilizing Mg and In powders and polytetrafluoroethylene (PTFE) as raw materials. After the composite underwent further processing via DBDP-assisted ball milling, the Mg(In)-MgF_2 composites were obtained, the whole preparation process required only 2 h. The polytetrafluoroethylene was *in-situ* generated on ultrafine MgF_2 particles (~300 nm), which serve as catalysts, leading to a low activation energy (127.7 kJ/mol H_2) and high hydrogen storage capacity of 5.16 wt.% at 609 K.

(2) Mechanical alloying

If metallic powders, or the mixtures of metallic and non-metallic powders, are ball-milled sufficiently, the solid phase transition will happen between the powders to form alloys. This process, also known as mechanical alloying (MA), was developed based on high-energy ball milling. MA involves the repeated process of cold welding, fracture, and re-welding of powder particles, which effectively reduces particle size, and increases specific surface area as well as introduces numerous defects and lattice distortions. In the early 1970s, the mechanical alloying method developed by Benjamin was used to prepare oxide-dispersed and nickel-/iron- reinforced high-temperature alloys for aerospace and high-temperature environments. Currently, this technique is considered as a crucial technology for synthesizing stable and metastable advanced materials with broad applications [64]. The mechanical alloying procedures can be categorized into four steps:

① The strike of the balls in the ball-milling tank imparts energy to the powders, causing localized temperature increases and promoting cold welding and homogeneous of the local composition.
② The continuous cold welding and fracturing facilitate the diffusion of powder particles and the formation of the solid solution.
③ The reduction in particle size of the powders promotes localized homogenization spread to the whole sample.
④ Powder particle distortion happens, leading to the formation of a metastable structure [2]. The specific process is shown in Fig. 3.27.

Nowadays, two main approaches are proposed for mechanical alloying. The first one is based on atomic diffusion. During the ball milling process, powder particles experience collisions with high-energy balls, leading to extrusion, severe plastic deformation, fracture, and cold welding. Consequently, the powders undergo continuous refinement with the exposure of more unreacted surfaces. This stepwise refinement results in the crystalline structure being a layered form, significantly enhancing the contact area for atomic reactions. Meanwhile, the diffusion distance is reduced, therefore, the diffusion coefficient is increased until the end of the refinement reaction to form alloys, e.g., Al-Zn, Al-Cu, Mg alloys, etc. The second approach is

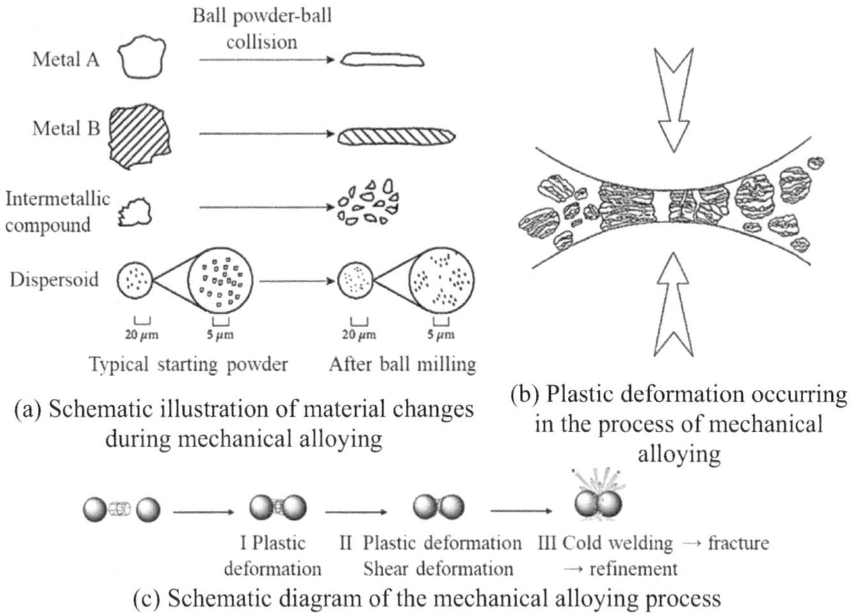

Fig. 3.27 Changes in microstructure and interatomic bonding of materials during mechanical alloying process

the explosion reaction. After the powders are ball-milled for a while, the alloying reaction occurs within a short time. A large amount of heat is released to form the alloy. This mechanism is often named as the explosion reaction, high-temperature self-propagating reaction (SHS), combustion synthesis reaction, or self-driven reaction. Similar phenomena have been observed in the mechanical alloying of $Ni_{50}Al_{50}$ powders, Mo-Si, Ti-C, and NiAl/TiC alloys [65].

The commonly used ball millers include stirring-type, planetary-type, and vibratory-type, as illustrated in Fig. 3.28. In the stirring-type high-energy ball mill, grinding media are crushed into alloy powders under the action of strike, friction, and shear. The planetary ball mill features several symmetrically equipped rotating discs and a rotation of the ball mill tank. Under the effect of inertial force, the powder samples in the tank undergo high-frequency strike and friction, resulting in rapid grinding into fine alloy powder. Vibratory ball milling utilizes high-frequency vibration of the grinding tank and balls to quickly crush the grinding media into alloy powder through high-frequency strike, friction, and shear effects. Among the methods mentioned above, vibratory ball milling has the highest energy; while stirring ball milling has the lowest energy. Planetary ball milling is characterized by high efficiency and high energy, making it commonly used in the preparation of Mg-based hydrogen storage materials.

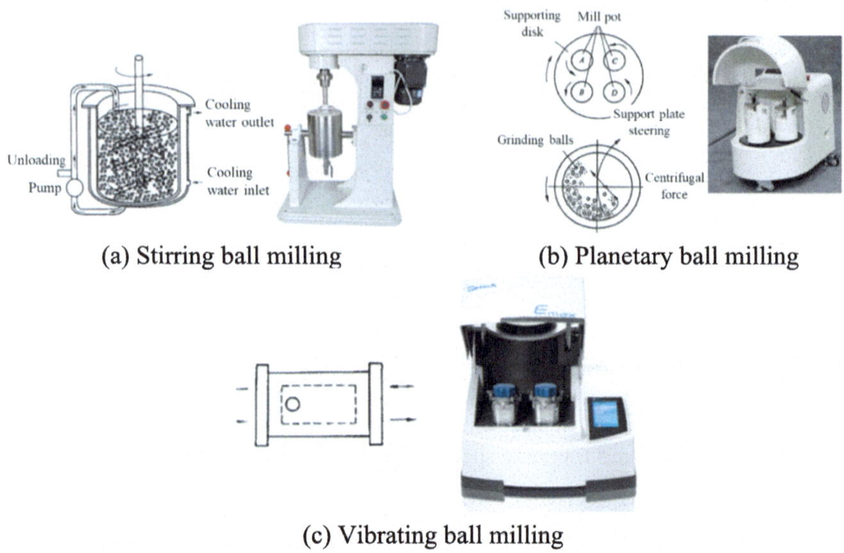

Fig. 3.28 Ball movement in common ball milling and ball milling tanks

The preparation of Mg-based hydrogen storage alloys or composites by mechanical alloying differs from sintering and smelting methods and has the following features:

① It is possible to prepare alloys with elements having distinct melting points or densities, such as Mg-Ni, Mg-Ti, Mg-Co, and Mg-Nb, etc. This process is carried out at room temperature and is not limited by melting point or relative density of elements.

② Ball milling destroys the oxide layer on the surface of Mg-based alloys, providing a fresh surface exhibiting high chemical activity, which helps to enhance their activation properties.

③ The ball milling process reduces particle and grain size, increasing specific surface area and introducing plenty of defects and lattice strains.

④ The process is simple. Ultrafine powder of hydrogen storage alloys can be obtained after the ball milling process, the further step of pulverization is not needed before utilization.

In 1987, Ivanov et al. made great progress in utilizing mechanical alloying to synthesize Mg_2Ni alloys [66]. However, due to the poor kinetic and thermodynamic properties of Mg and Mg_2Ni, the resulting samples cannot meet the requirements of the practical application. Thus, their research focus transfers to heteroatom-doping towards Mg_2Ni with other elements to realize multifunctional or amorphous hydrogen storage alloys. Consequently, mechanical alloying has become widely employed in preparing Mg-based hydrogen storage alloys or composites, successfully synthesized Mg-TM, Mg-RE, Mg-Tm-RE, and catalyst-doped Mg-based hydrogen storage composites [67].

The high-energy ball milling process in mechanical alloying can effectively prepare nanocrystalline/amorphous metastable Mg-based hydrogen storage materials. Mechanical alloying has been utilized to optimize the microstructure of Mg_2Ni alloys [68], aiming to enhance the thermodynamic properties of the alloys by preparing nanocrystalline or amorphous Mg_2Ni. Figure 3.29a shows that the primary phase of the as-cast alloy is the Mg_2Ni phase, along with a small amount of metallic Mg and $MgNi_2$ phases. The corresponding as-cast microstructures are illustrated in Fig. 3.29c, d. The overall microstructure is coarseness and lacks rapid transfer channels, hindering hydrogen absorption. With high-energy ball milling, the ordered structure of phases in the crystalline alloy diminishes, and the microstructure undergoes continuous refinement. After 24 h of high-energy ball milling, the alloy transforms to composites with amorphous and nanocrystalline structures, and further transforms to completely amorphous structure after 38 h of milling (Fig. 3.29b, e, f).

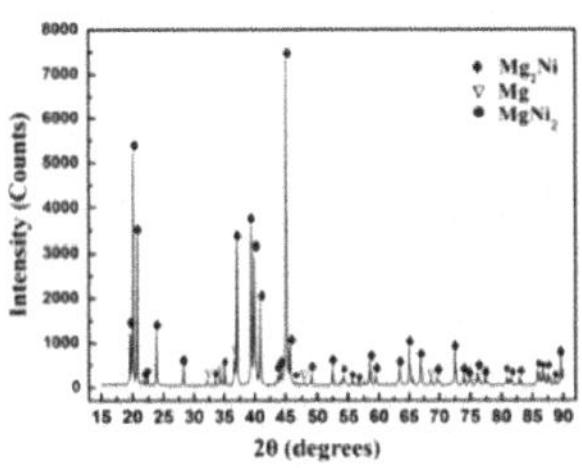

(a) XRD pattern of as-cast Mg_2Ni alloy

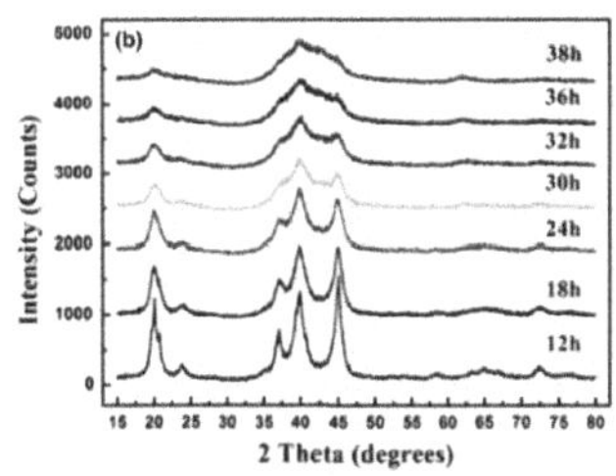

(b) XRD patterns of Mg_2Ni after ball milling

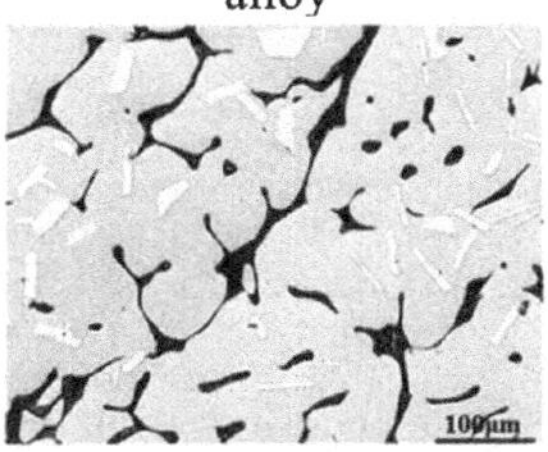

(c) SEM image of Mg_2Ni alloy in as-cast state

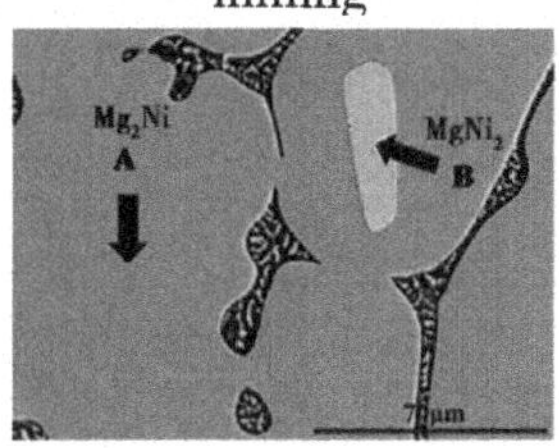

(d) Enlarged microstructure of as-cast Mg_2Ni alloy

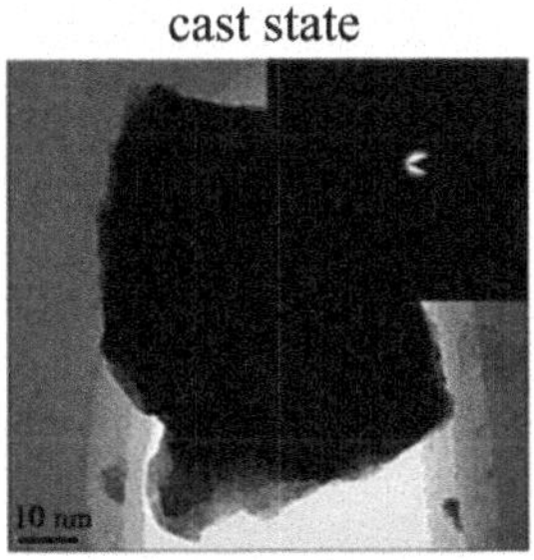

(e) Low magnification TEM image of high-energy ball-milled Mg_2Ni

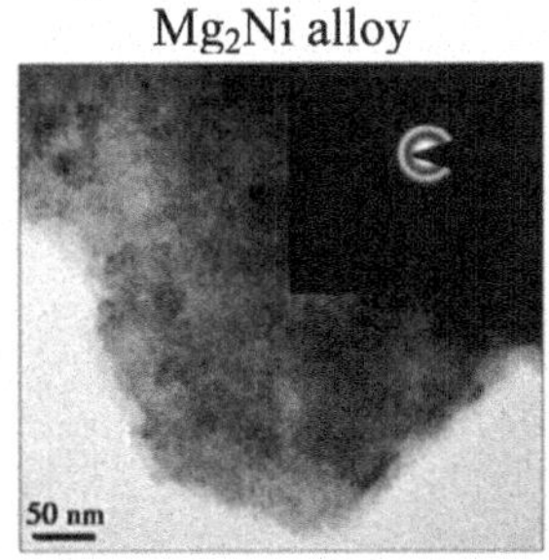

(f) High magnification TEM image of high-energy ball-milled Mg_2Ni

Fig. 3.29 Phase constitution and microstructure of as-cast Mg_2Ni alloys and after ball milling [68]

Table 3.2 displays the hydrogen storage properties of Mg-based hydrogen storage materials prepared via ball milling. Despite the numerous advantages discussed above, the ball milling method also suffers from the following drawbacks: impurities can be easily introduced during long-time ball milling processes, oxidation of the powders, and increased production costs. Additionally, inhomogeneities potentially exist in the particle and microstructures of the prepared materials. The process is also energy- and time-consuming, posing challenges for industrial-scale production and application.

Table 3.2 Hydrogen storage properties of different Mg-based hydrogen storage materials prepared by ball milling method

Mg-based hydrogen storage materials	Maximum hydrogen storage capacity/wt.%	Thermodynamic properties (T/K, P/MPa)		Kinetic properties/min (wt.% H_2)	References
Mg_2Ni	3.5	$T_{abs} = 573$	$P_{abs} = 3$	$t_{abs} = 30$	[69, 70]
Mg_2Cu	2.7	$T_{abs} = 573$	$P_{abs} = 3$	–	[67]
$Mg_{90}Al_{10}$	6	$T_{abs} = 617$	$P_{abs} = 2$	$t_{abs} = 30$	[70]
Mg-10Ni-5Fe-5Ti	5.4	$T_{abs} = 573$ $T_{des} = 573$	$P_{abs} = 1.2$ $P_{des} = 0.1$	$t_{abs} = 5$ (5.31) $t_{des} = 60$ (5.18)	[71]
$Mg_{90}Y_{1.5}Ce_{1.5}Ni_7$	5.6	$T_{abs} = 473$ $T_{des} = 633$	$P_{abs} = 3$ $P_{des} = 0.05$	$t_{abs} = 2$ (5) $t_{des} = 3$ (5.4)	[72]
MgH_2-5% VTi	4.7	$T_{abs} = 573$ $T_{des} = 573$	$P_{abs} = 2$	$t_{abs} = 1.2$ (4) $t_{des} = 10$ (4.4)	[73]
MgH_2-5% FeTi-5% CNTs	5.4	$T_{abs} = 573$ $T_{des} = 573$	$P_{abs} = 2$	$t_{abs} = 1$ (5) $t_{des} = 6$ (4.9)	[73]
$La_7Ce_3Mg_{80}Ni_{10}$-5TiO_2	5.1	$T_{abs} = 473$ $T_{des} = 573$	$P_{abs} = 3$ $P_{des} = 0.0001$	$t_{abs} = 0.75$ (4) $t_{des} = 2.8$ (3)	[74]

3.2.2 Deposition

The deposition methods can primarily be classified into electrodeposition and gas-phase deposition. Both are effective techniques for synthesizing thin films or nanomaterials that facilitate reactions on the surface of a substrate. In recent years, the deposition method has been widely employed in preparing Mg-based hydrogen storage materials.

(1) Electrodeposition

Alloy electrodeposition is a process in which two or more metal ions are co-deposited at the cathode, resulting in uniform and refined coating. The prerequisite for the co-deposition is that at least one metal component in the alloy can be individually deposited from its salt solution, and the deposition potentials of each metal in the alloy need to be equal or closely matched. There are four commonly used methods to promote the alignment of deposition potentials of the metals.

① Lowering the concentration of positively charged metal ions and increasing the concentration of negatively charged metal ions.
② Employ suitable complexing agents to shift the equilibrium potential of metal ions towards the negative direction, enhancing anodic polarization and achieving closer alignment of metal ion precipitation potentials.
③ Choose appropriate additives that have minimal impact on the equilibrium potential of the metal but exhibit a significant blocking effect on metal electrodeposition. These additives may be adsorbed onto the cathode surface or forming complexes, selectively hindering the electrodeposition process of the metal components in the alloy.
④ Utilize the depolarizing effect induced by the more negative potential components during co-deposition. Due to the interaction between the metals in the system, the Gibbs free energy and the equilibrium potential of the system changes, which causes the positive shifts of metal components with the negative potentials during electrodeposition and reduces polarization.

Compared with bulk and powder alloys, thin-film Mg-based hydrogen storage alloys offer unique advantages and provide a relatively cutting-edge research field. The preparation of rare earth-based and Mg-based thin film alloys mainly relies on vapor phase deposition and electrodeposition methods, however, the electrodeposition method stands out due to its simple equipment, convenient operation process, and flexibility to adjust the thicknesses, morphologies, and crystallization magnitudes of the deposited films [75]. Despite the challenges in electrodeposition in aqueous solutions of Mg, which are caused by its negative electrode potential, Mg-Ni hydrogen storage alloys are successfully electrodeposited by employing suitable additives and complexing agents. The specific co-deposition parameters for Mg-Ni alloy are as follows: temperature between 323 and353 K, pH value between 1.0 and 4.0, and current density is ca. 6~10 A/dm^2 [76]. The optimal parameters for Mg-Ni co-deposition were identified at the current density of 8.0 A/dm^2, electrolyte pH of 2.0, and temperature of 333 K. However, the highest electrochemical hydrogen storage capacity of the Mg-Ni alloy is only 75.457 mA·h/g, which is much lower than that of alloys synthesized by mechanical alloying. Therefore, this result is far behind the requirements of practical applications and further improvement of the alloy deposition process is needed.

Considering that the active metals (e.g., Mg) and some high melting point metals (e.g., rare earth metals) cannot be deposited individually in the aqueous solution, non-aqueous electroplating is commonly used, e.g., organic solvent electroplating and molten salt, etc. For instance, by introducing sodium hypophosphite to an aqueous solution, Mg-Ni alloys can be deposited through the catalytic effect, achieving a maximum discharge capacity of up to 186 mA·h/g [77]. Similarly, Mg-Al alloys can be prepared by pre-electrodeposition of dimethyl magnesium in an organic electrolyte to obtain Mg, followed by electrodeposition [78]. The roughness of the electrodeposited compounds increases with increased pre-electrodeposition time, meanwhile, the tendency of dendrite formation decreases with increased Mg ion concentration in the electrolyte. In organic solvent plating, the operation temperature varies from 303 to 363 K with current densities from 5 to 60 A/dm^2, higher than those used in aqueous plating. The pH value is adjusted using hydrochloric acid or sodium hydroxide solution, however, organic solvent electroplating suffers from flammability, toxicity, and not withstand high temperatures, high thresholds are also required for the equipment. Additionally, the low solubility of metal salts in organic solvents leads to a low solution conductivity and a significant concentration polarization. Consequently, the resulting compounds exhibit low Mg content, high inclusions/impurities with porous structures and low dense morphology.

Molten salt plating is a method which is used to achieve a strongly bonded metal coating on a substrate by employing applied electric current within molten inorganic or organic salts. This method addresses the problem of difficult to deposit for some metals, furthermore, unlike aqueous solutions, molten salt plating does not lead to hydrogen evolution, avoiding issues like hydrogenation, low current efficiency, cracks in plated layers, and microporosity. Additionally, molten salts exhibit high stability in decomposition voltage, low gas solubility, and reduced side reactions, resulting in high current efficiency and rapid deposition. The deposited metal can diffuse to the substrate metal at an appropriate diffusion rate by adjusting the temperature and deposition speed, enhancing the bonding strength between the deposited layer and the substrate metal. Mg-Al-Ni alloys were prepared using low-temperature salt solution electrodeposition [79] and the key parameters, e.g., electrodeposition temperature and time, etc., were systematically investigated (Fig. 3.30). The microstructures of the prepared alloys were also investigated, and the results show that grain size decreases with increasing electrodeposition temperature. Higher temperature promotes alloy nucleation, decreasing Mg content and increasing aluminum and nickel contents, and vice versa. After 5 min of deposition, tiny spherical particles containing Ni and Al were formed, with extended deposition time (to 30 min), the particle size increased, and a small amount of Mg was detected. Mg and Al content increased with deposition time, reaching the maximum at ca. 40 min.

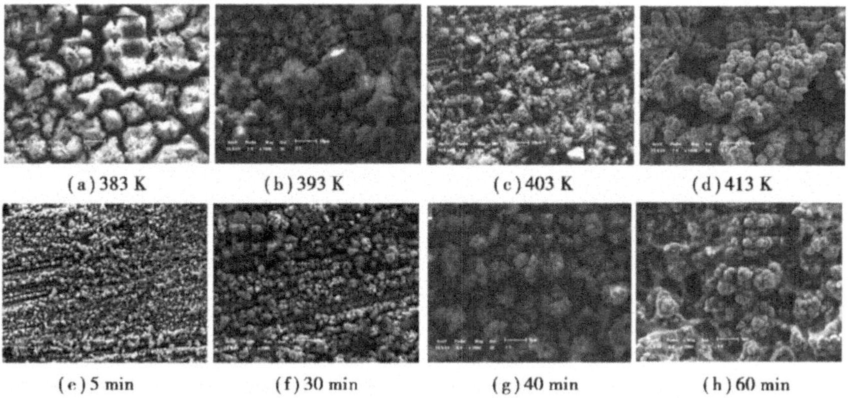

Fig. 3.30 Morphologies of electrodeposited Mg-Ni-Al alloys [79]

Mg-Ni alloys can be obtained via the electrodeposition in urea-acetamide-sodium bromide molten salts [80]. During the reduction and precipitation process at the cathode, no crystalline polarization can be observed in the alloy, suggesting that the nucleation growth dominates the deposition process. The optimal conditions were determined to be 40 min deposition at 383 K with Mg/Ni ratio of 2:1. Lower temperature promotes the Mg deposition, while a shorter deposition time benefit for nickel deposition, and the discharge capacity of Mg-Ni alloy varies based on different current densities (Fig. 3.31), reaching maximum (87.56 mA·h/g) at the current density of 15 mA/cm^2. This trend agrees well with the relationship between current density and Mg content, the possible reason is that Mg-Ni solid solution and amorphous phases are formed when the Mg-Ni atomic ratio increases in alloys, contributing to the high electrochemical hydrogen storage capacity.

Fig. 3.31 Discharge capacity of Mg-Ni alloy film deposited for 40 min at different current densities [80]

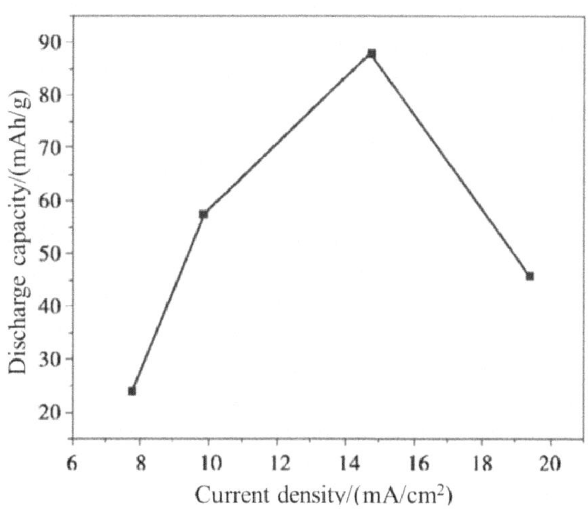

(2) Gas phase deposition

Gas phase deposition utilizes the gases or convert solid substances to gases phase, enabling physical or chemical reactions in the gas phase, the growth of hydrogen storage material powders will happen at the cooling down process. Based on different deposition principles, gas phase deposition can be categorized into physical vapor deposition (PVD), chemical vapor deposition (CVD), and magnetron sputtering (MS).

PVD is the method that using electric arc or plasma to heat the raw material to a gaseous state to obtain particulate products with a large temperature gradient during cooling. This method can produce fine powders with 5~100 nm particle diameters with the merits of high purity, uniform particle size, and crystal morphology. PVD is suitable for preparing fine powder of oxides, composite oxides, carbides, or metals [62]. During the PVD process, materials are evaporated by heating or ion bombardment of the target (plating material). The former involves vacuum and arc evaporation, etc., while the latter requires magnetron sputtering, coating and ion plating.

The plasma arc evaporation method is an effective way of producing Mg-based hydrogen storage nanoparticles. The principle is that a plasma arc is used as heat source, generating temperatures ranging from 10,000–30,000 K to melt and refine metal or alloy. The metal nanoparticles are obtained by the condensation of the gas-phase materials. In the plasma arc evaporation system, a plasma generator (plasma gun) is used instead of a graphite electrode, and a direct current arc transfer method with a conductive bottom anode is employed to establish a conductive circuit. Considering that Mg-based hydrogen storage alloys are oxidized easily, an inert or reducing atmosphere, such as Ar, H_2, is required in the furnace. Introducing H_2 into the Mg-based alloy nanoparticle preparation increases the rate of alloy particle generation. Compared with the ordinary arc, plasma arcs are featured with high arc temperatures and controlled atmospheres [7, 62]. The main structure of the plasma arc evaporator consists of a furnace, main power supply system, vacuum system, water-cooling system, etc., as illustrated in Fig. 3.32. The basic melting process includes.

① Sample preparation and vacuumize the chamber.
② Plasma generation with Argon as the ionizing gas.
③ Sample melting, alloying, and evaporation.
④ Solidification in water-cooled copper crucible.

Plasma arc evaporation is a method with nucleation/growth of particles in supersaturated vapors, the average particle size of metal nanopowders prepared using this method can be expressed as follows [81]:

$$D = \left(\frac{6C_0M}{\rho\pi N}\right)^{\frac{1}{3}} \tag{3.8}$$

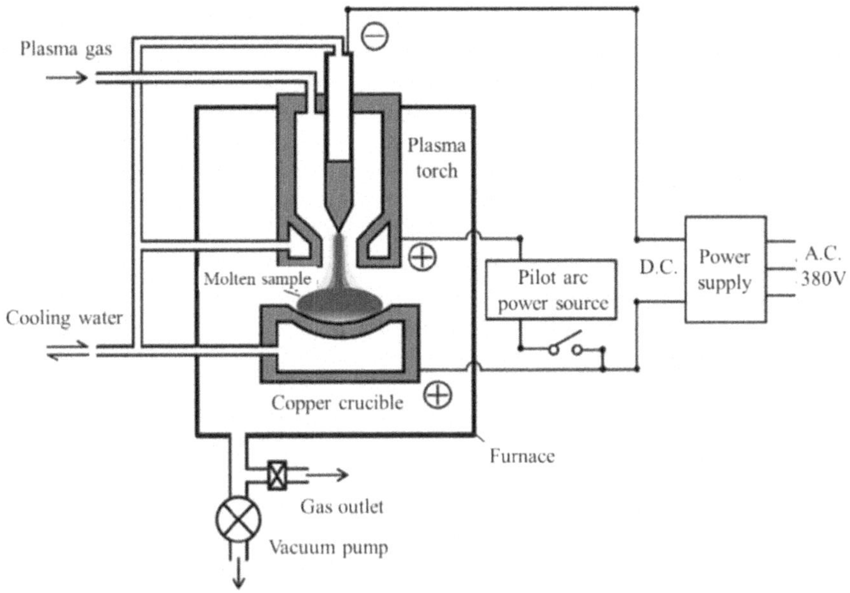

Fig. 3.32 Schematic diagram of the plasma arc instrument structure

where C_0 is the metal vapor density (mol/cm^3), N is the number of particles per unit volume (cm^3) of vapor, ρ is the density of particles (g/cm^3), and M is the molar mass (g/mol).

The average particle size is determined by C_0 and N, which are associated with the evaporation rate and the temperature distribution within the reactor. Since accurately controlling temperature distribution is challenging, precisely controlling the sample particle size via this method is problematic. However, the size of nanoparticles can be controlled under certain limitations with specific parameters. The variation of current in plasma arc evaporation significantly impacts the evaporation rates of materials. The critical current value depends on several factors: operation atmosphere composition, pressure, electrode-to-sample distance, sample composition, and tungsten electrode diameter. Therefore, the current density is adjusted according to the specific circumstances, typically ranging from 150 to 250 A. The tungsten electrode is positioned 1~2 mm above the sample, and atmospheric conditions are adjusted based on parameters required for nanoparticle preparation from different metals.

Mg-3.9 wt.% Ni_2Al_3 nanocomposites were synthesized using the Hydrogen Plasma-Metal Reaction (HPMR) method, which is one of the hydrogen plasma arc evaporation techniques [82]. The average particle size of obtained Mg was 90 nm, and Ni_2Al_3 nanoparticles were uniformly dispersed on the Mg surface. These nanoparticles were transformed into $Mg_2NiH_{0.3}$ and Al after hydrogen absorption. However, with hydrogen absorption and desorption cycles, $Mg_2NiH_{0.3}$ and Al are converted back to Ni_2Al_3 nanoparticles. This method enhances the hydrogen absorption rate and storage capacity of the Mg-Ni_2Al_3 nanocomposites. Specifically, 6.4 wt.% of hydrogen was absorbed in 10 min at 573 K, and 6.1 wt.% of hydrogen was desorbed within 10 min at 623 K. The apparent activation energies of the hydrogenation and dehydrogenation reactions were determined as 55.4 and 115.7 kJ/mol H_2, respectively (Fig. 3.33).

Mg, Ni, Cu, Co, Fe, and Al nanoparticles can also be prepared using the HPMR method. Unlike Mg particles, Ni, Cu, Co, Fe, and Al nanoparticles exhibit a granular structure with particle sizes of 30~50 nm (Fig. 3.34) [57]. The difference is the faster evaporation and generation rate of Mg, which results in the planar shape of Mg nanoparticles with a hexagonal structure and an average particle size of 300 nm. The hydrogen absorption kinetics of several Mg-based nanoparticles synthesized by the HPMR method are summarized in Fig. 3.35a, the pure Mg nanoparticles show rapid hydrogen absorption at 573 K, reaching a maximum value of 7.54 wt.% within 30 min at 573 K [83]. The hydrogen absorption content of the other Mg-M-based (M = Ni, Cu, Co) nanoparticles decreased differently. At 573 K, Mg_2Ni and $Mg_{85}Ni_{15}$ nanoparticles have relatively low hydrogen capacities, but their hydrogen adsorption

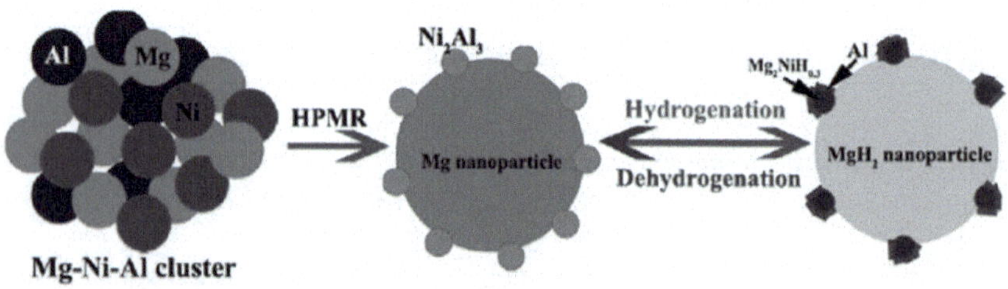

(a) Schematic illustration of the preparation and activation process of Mg-Ni_2Al_3 nanocomposites

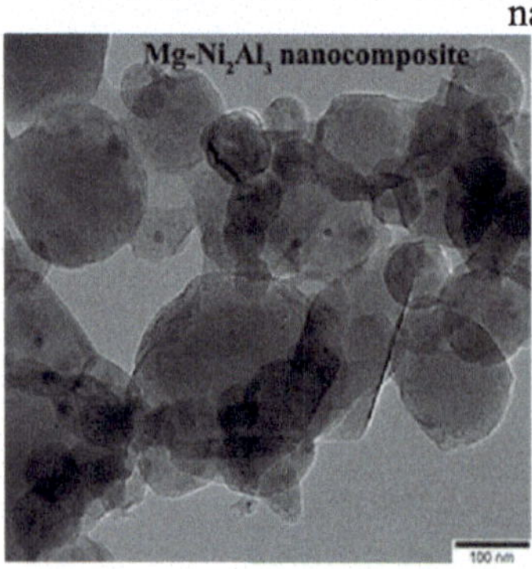

(b) TEM image of Mg-Ni_2Al_3 nanocomposites

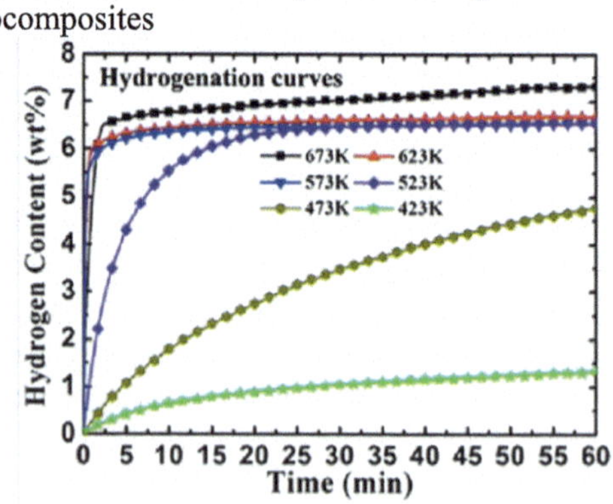

(c) Hydrogen absorption kinetics of Mg-Ni_2Al_3 nanocomposites under 4 MPa H_2

Fig. 3.33 Microstructure evolution in hydrogen absorption and desorption process and hydrogen absorption kinetics of Mg-Ni_2Al_3 nanocomposites [82]

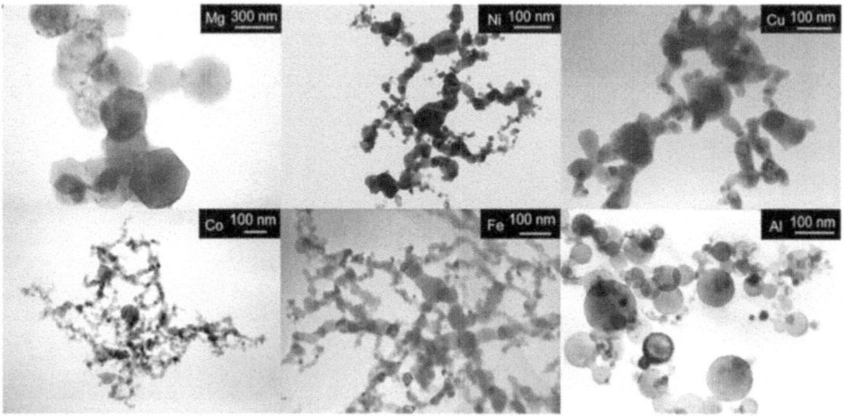

Fig. 3.34 TEM images of Mg, Ni, Cu, Co, Fe, and Al metal nanoparticles synthesized by hydrogen plasma metal reaction [57]

rates were much faster than those of Mg nanoparticles. After hydriding under 3.5 MPa H_2 and at 573 K for 13 min, the hydrogen capacity of $Mg_{85}Ni_{15}$ nanocomposites reaches 4.73 wt.% H_2. Furthermore, Mg_2Ni nanoparticles exhibited excellent cyclic stability (Fig. 3.35b), the size effect of Mg particles on the hydrogen absorption kinetics was explained by a simple schematic in Fig. 3.35c.

The merits of the HPMR process are simple, reliable, and easy to learn, enabling large-scale production. The formation mechanism of nanoparticles or thin films is relatively understandable, and most metal substances can be prepared by plasma arc evaporation. However, challenges exist, e.g., poor bonding strength between the coating layer and the substrate and difficulty in preparing nanoparticles or thin films of substances having high melting points and low vapor pressures. Such conditions also may cause the potential of impurity contamination due to the evaporation of substances contained in the crucible.

Magnetron sputtering (MS) is a commonly used "high-speed, low-temperature" sputtering method developed in the 1970s. The principle of MS is to create a cross electromagnetic field above the surface of the cathode target. When secondary electrons generated by sputtering are accelerated into high-energy electrons at the cathode, they do not move directly to the anode but oscillate under the electromagnetic field in an approximate pendulum motion. High-energy electrons constantly collide with gas molecules, and transferring their energy to the latter ones and ionizing them. Correspondingly, those high-energy electrons transform into low-energy electrons. These low-energy electrons eventually move along the line of magnetic force to the anode and are absorbed by the anode, avoiding the strong impact of high-energy electrons on the electrode, eliminating the damage caused by heating and electron irradiation and realizing the low-temperature characteristic of magnetron sputtering. Due to the existence of magnetic field, the complex movements of electrons increase their ionization rate, resulting in high-speed sputtering. The technical features of

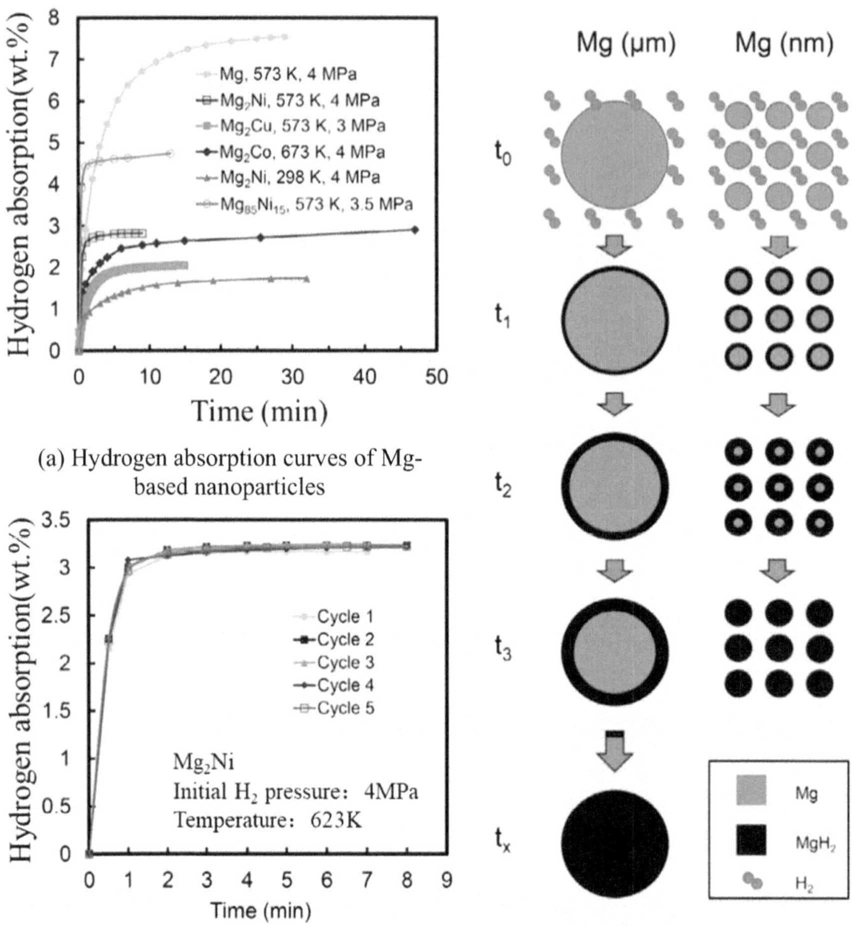

(a) Hydrogen absorption curves of Mg-based nanoparticles

(b) Hydrogen absorption cycles of Mg2Ni nanoparticles

(c) Schematic diagram of the effect of nanometer size on hydrogen absorption kinetics

Fig. 3.35 Schematic diagram of hydriding kinetics and hydrogenation process of Mg-based nanoparticles [83]

magnetron sputtering are characterized by generating a magnetic field perpendicular to the direction of the electric field at the cathode, generally achieving with permanent magnets [84, 85]. The schematic illustration of magnetron sputtering is shown in Fig. 3.36.

MS can be used to prepare a variety of functional thin films, including multi-component mixtures and compound films. It is characterized by low impedance of the sputtering plasma, resulting in high discharge currents ranging from 1 to 100 A at a voltage of approximately 500 V. The process also offers a film deposition rate ranging from 1 to 10 nm/s, ensuring a good consistency of film formation with

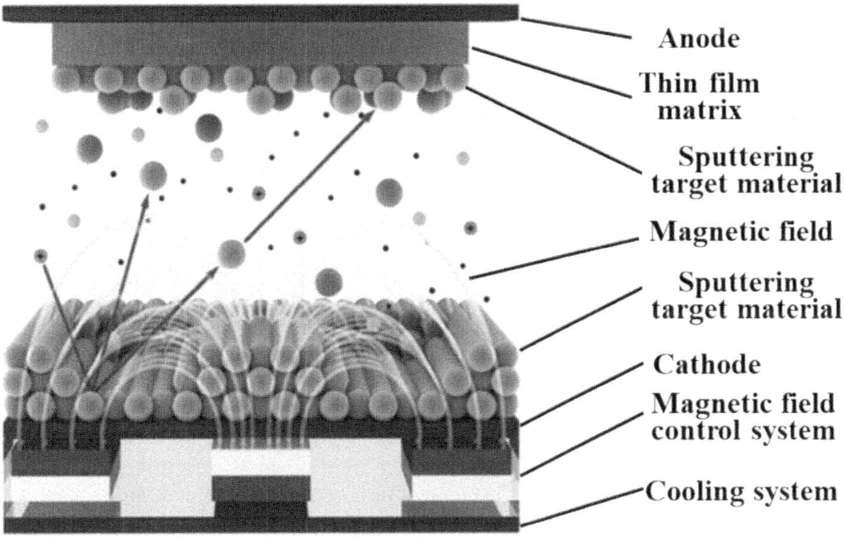

Fig. 3.36 Schematic diagram of the principle of magnetron sputtering [84]

different conditions of several meters long. Furthermore, the low substrate temperature rises results in more dense films with improved film/substrate consistency, as the sputtered particles have about tens of eV of energies. The parameters of magnetron sputtering can be adjusted based on the requirements of thin films, making it particularly suitable for large-area coating, ensuring more uniform deposition over large areas [85]. However, this method also has some disadvantages, e.g., MS suffers from a lower deposition rate and low target material utilization versus the evaporation method. The deposited substrate may experience a temperature rise due to plasma irradiation and other effects, sometimes the deposited film tends to be relatively inhomogeneous.

The preparation of Mg-based materials into thin films via MS can significantly enhance the hydrogen storage properties of the materials. For instance, Mg-Ni multilayer films were deposited on {001} plane of Si wafer through dual-target magnetron sputtering. The operation pressure was kept at 8×10^{-1} Pa during deposition, the power used for depositing Mg and Ni was 120 W and 90 W, respectively [86]. The microstructure of the Mg-Ni multilayer films typically comprises fine-crystalline Ni layers and <001> crystalline-oriented Mg layers. Hydrogen absorption measurement on Mg-Ni multilayer thin films revealed that different preferred orientations of Mg result in varied hydrogen storage performances. After the hydrogen absorption reaction at 533 K, the {002} diffraction peak of Mg disappears, while the peaks of Mg_2NiH_4 and MgH_2 appear and the {101} peak is retained.

It was observed that the {001}-oriented Mg film exhibited easier hydrogen absorption compared to the {101}-oriented Mg film at a specific temperature. This phenomenon could be explained by the energy changes associated with hydride nucleation and growth on different crystal planes. Therefore, to enhance the hydrogen absorption property of Mg-based films, adjusting the parameters in MS is necessary to selectively deposit {001}-oriented films to facilitate hydrogenation. The MS integration technology with computers has emerged as a promising research direction, the coating processes are precisely controlled using digitals, and by simulating the coating process with computers, the effects of magnetic field, temperature field, airflow distribution, and other factors on film orientation and properties can be detailed analyzed. This approach facilitates the industrial preparation of cost-effective thin films of Mg-based materials and enhances their properties.

MgNi/Pd multilayers were fabricated on Ni substrates using dual-target DC magnetron sputtering, with the working pressure maintained at 1.4×10^{-1} Pa in the chamber, the deposition power was 200 W, and the deposition rate was kept at 2 nm/s [87]. The thickness of MgNi layer is ca. 40 nm, while the thickness of each Pd layer is ca. 16 nm. The total thickness of the MgNi/Pd multilayer film was ca. 1.7 μm. The microstructure analysis shows that MgNi layer consists of a mixture of amorphous and nanocrystals, while the Pd layer exhibits a fine crystalline without selective orientation. After three cycles of hydrogen absorption and desorption process at room temperature, the hydrogen storage capacity reached 4.6 wt.% H_2, and the hydrogen desorption was 3.4 wt.% H_2. The homogeneous Pd catalyst layer deposited using the MS technique exists as nanoclusters on the surface of the MgNi layer, which promotes the surface reactions, accelerating the dissociation and recombination of H_2 molecules and benefiting the hydrogen absorption and desorption reaction of the MgNi layer, and therefore, the hydriding properties of the MgNi/Pd multi-layers are greatly boosted after 3 cycles of ad-/de-sorption process. It is noted that other uniform metal layer films can serve as catalysts to activate the surface of the Mg substrate and enhance its hydrogenation performance in the multilayer compounds, but the prerequisite is that the composite metal layers and the Mg substrate should have excellent bonding interaction and continue to play a catalytic role even after multiple cycles of hydrogen absorption and desorption.

Chemical vapor deposition (CVD), which is proposed in the 1960s, utilizes reactions of chemical gases or vapors on the surface of a substrate to prepare thin films or nanomaterials. CVD has become an important technique for preparing thin films or nanoparticles due to its merits of simple equipment, easy to control, high purity of the produced films or nanoparticles, narrow particle size distribution, continuous and stable production, and low energy consumption, etc. CVD technology relies on the vapor of volatile metal halides, hydrides, organometallic compounds, etc., where the desired powder is synthesized through chemical gas-phase reactions. This deposition process can involve the thermal decomposition of a single compound or gas-phase reactions between two or more substances [88].

Different from PVD technique, particles in CVD originate from the vapor-phase decomposition of compounds. With increased temperature, the gas mixture interacts with the substrate surface, leading to the decomposition of specific components of the gas mixture. This process forms a solid film or metal/compound coating layers on the substrate. Several conditions are required for the successful of CVD process: First, there must be a sufficiently high vapor pressure at the deposition temperature.

Second, the reaction products must be gaseous except for the targeted deposition materials. Finally, the saturation vapor pressure of the deposition materials must be sufficiently low.

The Mg-based hydrogen storage alloy powders prepared with traditional methods typically require multiple activation steps, which results in the formation of polycrystals due to excessive strain stress during hydrogen absorption and desorption. To address this issue, Saita et al. [89] proposed the Hydriding Chemical Vapor Deposition (HCVD) method, which involves CVD under high-pressure hydrogen to deposit metal hydrides from high-pressure hydrogen and gaseous metal mixtures. The HCVD method possesses several advantages: (1) the high collision frequency between gaseous metal and hydrogen molecules enables rapid synthesis of metal hydrides. (2) The synthesized metal hydrides are rapidly deposited onto a cool substrate, eliminating the solid–gas reaction activation steps, etc. MgH_2 prepared by HCVD forms needle-like single crystals in the [101] direction. All peaks of the resulting product coincide with MgH_2, indicating its high purity (Fig. 3.37). Additionally, MgH_2 prepared by HCVD does not require activation and can reversibly absorb and desorb approximately 7.6 wt.% H_2 at 578 K. Moreover, the hydriding and dehydriding rates are significantly improved versus conventional Mg.

In addition to the application in preparing Mg-based hydrogen storage materials, CVD is also widely used to prepare catalysts, particularly for carbon nanotubes. The

(a) SEM image of MgH_2 prepared by HCVD

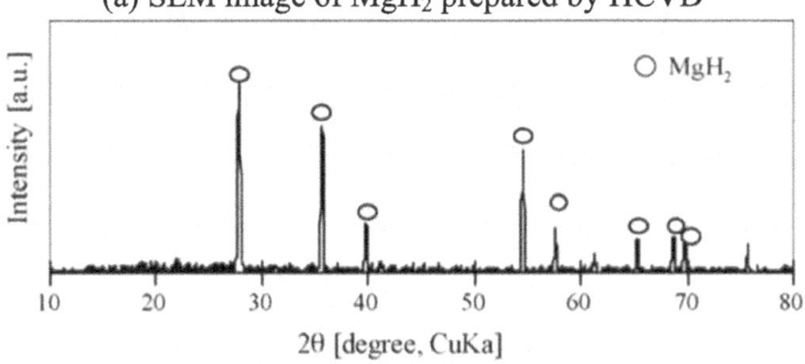

(b) XRD pattern of the product

Fig. 3.37 SEM images and XRD pattern of MgH_2 prepared by HCVD [89]

CVD-synthesized carbon nanotubes can serve as catalysts for Mg-based hydrogen storage materials. These catalysts typically consist of inert material carriers loaded with ultrafine powders of transition metals or their mixtures. Iron, cobalt, and nickel nanoparticles are commonly employed as the active components of the catalyst.

3.2.3 Liquid-Phase Reductions

Liquid phase reduction (LPR) is a method used to prepare metal nanoparticles by directly reducing raw materials into monomers following specific chemical reactions either in the liquid or semi-liquid phases. This method is widely employed in practical production due to the advantages of being simple to operate, easy to control, and high economic efficiency, however, abundant by-products and pollutants are generated. The process for preparing Mg nanomaterials for hydrogen storage via liquid-phase reduction is depicted in Fig. 3.38. Different Mg nanoparticles are obtained by reducing Mg dichalcogenide (Cp_2Mg) in tetrahydrofuran or glycol dimethyl ether liquid with various reducing agents, the results show that the Mg nanoparticles produced via the liquid phase reduction method exhibit exceptional activity, with excellent hydrogen absorption and desorption kinetics [89, 90]. The LPR is a promising method for preparing Mg nanoparticles due to its simple preparation process and the ability to control particle size, morphology, and nucleation process, leading to the narrow particle size distribution. The Rieke method is one of the most successful approaches in LPR [91]. This method involves the reduction of anhydrous $MgCl_2$ powder with lithium naphthalene in tetrahydrofuran or glycol dimethyl ether. However, since anhydrous $MgCl_2$ is powder, the liquid–solid chemical reaction proceeds slowly, resulting in low nucleation rates and large-sized Mg particles. On the other hand, liquid–liquid reactions, such as the reaction between Mg dichalcogenide and lithium naphthalene in a tetrahydrofuran solution, proceed rapidly with high nucleation rates, leading to almost instantaneous and explosive reactions and producing fine Mg nanoparticles. Although anhydrous $MgCl_2$ is nearly insoluble in tetrahydrofuran at room temperature, it can be dissolved in tetrahydrofuran at temperatures ranging from 333 to 343 K to form highly concentrated, homogeneous, and transparent solutions [92].

Rieke's method was improved by utilizing the LPR technique to prepare Mg nanoparticles. Anhydrous Mg chloride was employed instead of Mg dichalcogenide to reduce costs [93]. Lithium naphthalene was used as the reducing agent, lithium metal acts as the principal reducing agent, and naphthalene is set as an ion carrier. The presence of naphthalene facilitated the dissolution of solid lithium metal in the solution, accelerating the reduction reaction. Fine Mg nanoparticles were prepared with a liquid–liquid phase reaction employing anhydrous Mg chloride tetrahydrofuran solution and lithium naphthalene as reactants. The Mg nanoparticles prepared by the LPR method are illustrated in Fig. 3.39. The Mg nanosheets aggregated to form a needle-like structure with lengths ranging from 50 to 200 nm and thicknesses from 10 to 20 nm. Compared with carbon gel-loaded Mg nanosheets, the Mg nanosheets

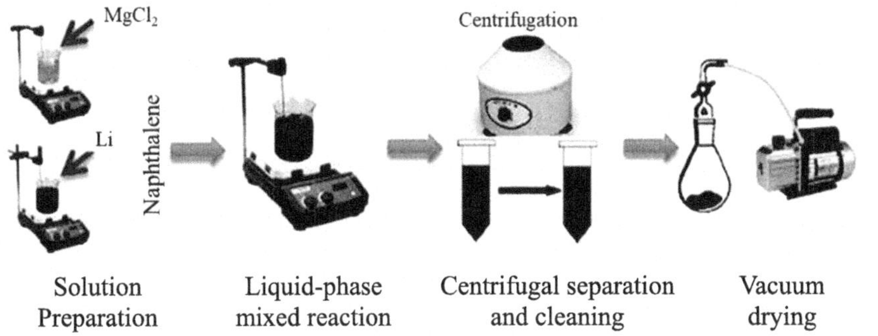

Fig. 3.38 Flow chart for preparing Mg nanomaterials for hydrogen storage by liquid phase reduction [93]

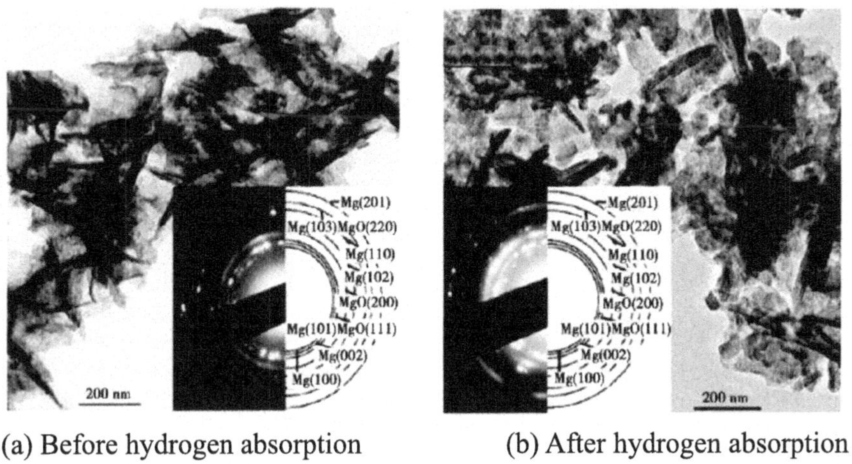

Fig. 3.39 Brightfield TEM image and SAED spectra of liquid-phase reduced Mg nanoparticles before and after hydrogen absorption

prepared with the LPR method maintained a hydrogen storage capacity of 7 wt.% and exhibited improved hydrogen absorption and desorption kinetics due to their large surface area and numerous defects.

3.3 The Preparation of Mg-Based Composites for Hydrogen Storage

3.3.1 Nano Confinement Method

Although theoretical studies indicate that reducing the size to the sub-nanometer (<1 nm) scale is essential for destabilizing the metal hydride phase, the synthesized nanostructures of Mg-based hydrogen storage materials ranging from a few nanometers to several hundred nanometers still yield significant enhancements in both thermodynamic and kinetic properties of hydrogen storage. Nevertheless, several challenges remain in the field of Mg-based hydrogen storage nanomaterials.

① At high temperatures, Mg (MgH_2) nanoparticles agglomerate rapidly, forming larger and more stable particles.
② The strategy to maintain the surface activity and stability of nano-sized Mg (MgH_2) particles is lacking.
③ The strategy to prevent the rapid oxidation of nano Mg (MgH_2) is lacking.

To address the afore-mentioned challenges, the nanoconfinement method aroused significant attention as an appealing strategy.

The nanoconfinement strategy typically combines nanoporous material and Mg-based hydrogen storage nanomaterial, i.e., Mg/MgH_2. The nanoporous materials are used as a scaffold to confine the Mg-based hydrogen storage material. This "bottom-up" approach restricts the particle growth of MgH_2 by the pore size of the scaffold material, which can effectively prevent sintering or agglomeration of the nanoparticles during high-temperature hydrogen absorption and desorption processes. Generally, the nanoporous scaffolds must be chemically inert and maintain good contact with the loaded materials. In some instances, catalytically active scaffold materials may promote the dissociation and formation of hydrogen molecules, enhancing hydrogen storage performance. Additionally, the density of the scaffolds should be as low as possible to preserve the high hydrogen storage capacity during any surface functional modification process [94]. Currently, metal–organic framework materials with two-dimensional layered structures or porous properties, polymeric materials, and mineral porous materials are commonly employed in the nanoconfinement method.

The nanoconfinement method is crucial in preparing Mg-based composite hydrogen storage materials. It combines Mg-based nanoparticles or hydride particles with metal–organic or carbon-porous materials. Commonly employed methods include impregnation and ball milling. In particular, impregnation is preferred over ball milling to preserve nanoporous structures. Impregnation methods typically are subdivided into two categories: melt impregnation and wet impregnation [94]. Figure 3.40 depicts a synthesis schematic illustrating this process.

Melt impregnation fills nanopore channels by melting hydrides at specific temperatures and utilizing capillary properties within the pores of the materials. This method does not require solvents and avoids post-treatment. However, it requires using inert

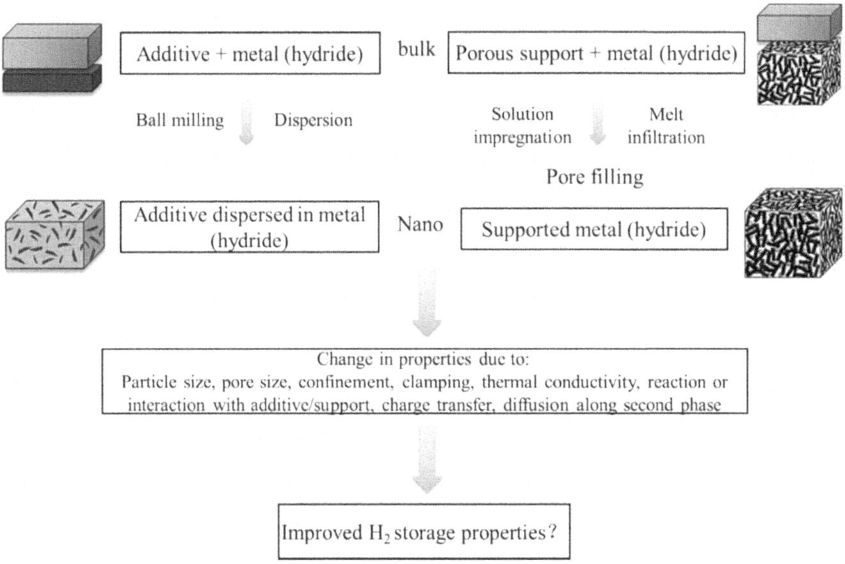

Fig. 3.40 Schematic diagram of nanoconfinement domain synthesis [95]

framework materials due to the reactivity of many metals and their hydrides in the molten state. The melting point of hydrides should be lower than the framework material in this method. The synthesis pressure should be adjusted to prevent the decomposition of hydrides during the heating process. Commonly used inert framework materials include aerogel. When the framework/aerogel mixed with Mg nanoparticles and sealed in a stainless-steel tube, a thermal treatment (ramping rate: 5 K/min, 973 K @ 24 h, cooling rate: 5 K/min) will be carried out to obtain aerogel-Mg nanoparticle composites. Permeate Mg into carbon gel pores, ranging in size from 2 to 30 nm, resulting in a loading amount of only 3.6 wt.% Mg. Modified Ni and Cu on the carbon gel surface can improve Mg content to 9~16 wt.%. This method notably enhances the dehydriding kinetics of the synthesized materials [96].

Wet impregnation is realized by saturating pores with a solution of metal salt, allowing metal ions to permeate and distribute within the pore channels of the material. To facilitate a uniform distribution, wet impregnation with vacuum is preferred, which differs from traditional wet impregnation. The evacuating process removes the excess gases and moistures before mixing with the metal salt solution from the porous materials. Specifically, the precursor $MgBu_2$ (dibutyl Mg) solution was impregnated into the carbon gels. Removing the solvent by centrifugal drying, the wet impregnation process was finished. Subsequently, $MgBu_2$ was hydrogenated under a specific temperature and hydrogen pressure to yield carbon gel-loaded nano-MgH_2. The final carbon gel-loaded nano-Mg (Mg-CA) composites were obtained by desorbing hydrogen under vacuum at higher temperatures [93]. The loading rate of nano-Mg in the composites reached 22.5 wt.%, with a theoretical hydrogen storage capacity of 1.71 wt.%. The maximum hydrogen adsorption of the composite material slightly

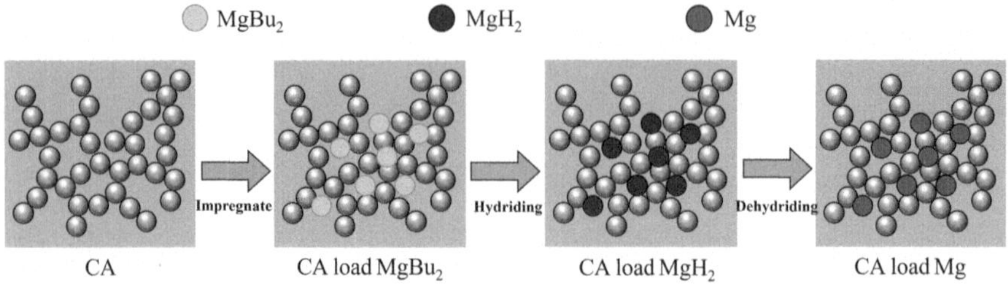

Fig. 3.41 Flow chart for preparing the composite of carbon gel loaded nano-Mg [93]

increased due to the physical absorption from the carbon gel. The hydrogen storage capacity at 623 K, 648 K, 673 K, and 698 K are as high as 2.15 wt.%, 2.17 wt.%, 2.24 wt.%, and 2.35 wt.%, respectively. Enthalpy changes of hydrogen absorption and desorption were decreased by 9.6 kJ/mol H_2 and 5.9 kJ/mol versus pure Mg. The activation energy was reduced by 29.4 kJ/mol H_2 thanks to the nano-size effect of the carbon gel-loaded Mg nanoparticles (Mg-CA) [93] (Fig. 3.41).

For the nanoconfinement method, support materials include metal framework materials, carbon materials, and polymer materials [97]. Metal-organic frameworks (MOFs) are characterized by controllable structure, porosity, and large surface area, which exhibit excellent gas adsorption capacity and show potential for future hydrogen storage applications. MgH_2@Ni-MOF composite nanomaterials were synthesized from Ni-MOF and dibutyl Mg via vacuum impregnation [98] (Fig. 3.42), the grain size of synthesized MgH_2 is ca. 3 nm within the mesoporous structure of Ni-MOF scaffolds. The high chemical stability of Ni-MOF scaffolds makes them suitable for nano-confined domains. Additionally, they serve as size control agents, preventing agglomeration and catalyst media, effectively enhancing the hydrogen storage performance of Mg/MgH_2. However, the mass ratio of inert materials remains high in Mg-MOF composites, leading to a suppressed hydrogen storage capacity. Further investigation is needed to address this issue.

Carbon materials such as carbon nanotubes and graphene have drawn attention in hydrogen storage applications because of their exceptional physicochemical properties like heat resistance and high electrical conductivity. Highly dispersed rGO-Mg (reduced graphene oxide) nanocomposites were synthesized with a simple one-pot reduction method. The rGO layers protect Mg from oxygen and moisture, producing a hydrogen storage capacity of 6.5 wt.%. The thin rGO layer minimizes inactive mass within the composite and improves the hydrogen adsorption kinetics with no significant capacity decrease within 20 cycles. This carbon-coated Mg composite multilayer exhibits promising potential for high hydrogen storage capacity, and can be used as a platform for sensitive nanomaterials [99].

Additionally, polymers can serve as supporting templates for the nano-confining method. Researchers synthesized a Mg/PMMA (poly(methyl methacrylate)) nanocomposite in a homogeneous solution of tetrahydrofuran (THF) at room temperature, achieving high-capacity and rapid hydrogen absorption (~6 wt.% of Mg or ~4 wt.% of overall Mg/PMMA) at 473 K for 30 min without adding heavy metal

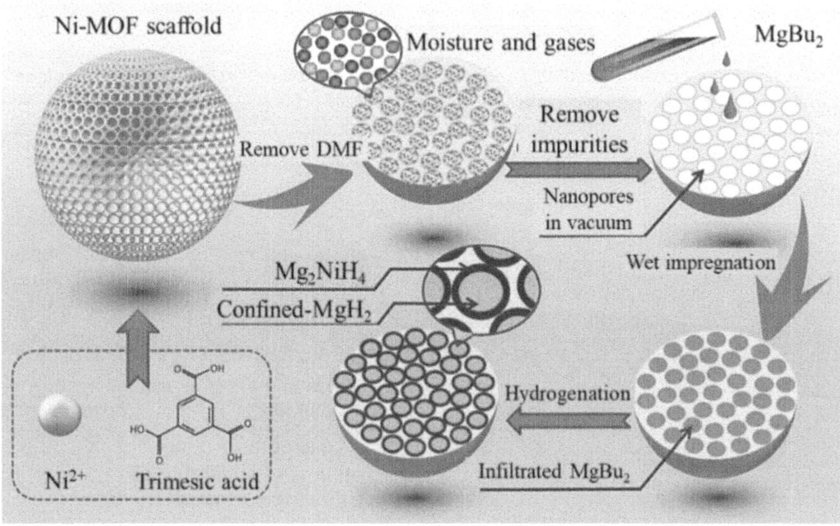

Fig. 3.42 Schematic diagram of the preparation process of MgH_2@ Ni-MOF composite nanomaterials [98]

catalysts [90]. Benefit from the selective filtration ability towards hydrogen and air of PMMA polymer, the hydrogen storage material exhibits excellent air stability.

Table 3.3 shows the properties of various Mg-based hydrogen storage materials prepared using the nano-confined method. These materials host nano-sized Mg/MgH_2 particles and their thermodynamics and kinetics properties are significantly enhanced by the nano-size effects. However, despite the advantages, the nano-confined method also suffers from relatively low Mg content, low reversible capacity, and high hydrogen desorption temperature in some systems. In the near future, the following problems should be solved:

① Insufficient overall loading factor leads to a high proportion of inert support, reducing the overall hydrogen storage capacity.
② Increased hydrogen desorption temperature of some systems does not meet the requirements of practical application.

Comprehensively, the following issues should be addressed:

① The proportion of Mg should be increased, which can increase the overall hydrogen storage capacity of the composites.
② The addition of catalysts should be helpful to further improve hydrogen storage properties in addition to the nano-size effect. Additionally, the cycling stability of the composites should be enhanced by suppressing the side reaction that produces by-products

Table 3.3 Hydrogen storage properties of Mg-based hydrogen storage materials prepared by the nanoconfinement method

Mg-based hydrogen storage materials	Load factor/ wt%	Enthalpy change of hydrogen absorption and desorption/ (kJ/mol H_2)	Hydrogen absorption and desorption activation energy/(kJ/ mol H_2)	Hydrogen absorption kinetics	Hydrogen desorption kinetic properties	References
Mg-CA	22. 5	−65.1/68.8	29.4/153.4	1.3 wt.%/ 4000 s/573 K/ 3 MPa	–	[93]
MgH_2@Ni-MOF	–	−65.7/69.7	41.5/144.7	1.25 wt.%/ 1000 s/573 K/ 3.2 MPa	1.5 wt.%/ 150 s/623 K	[98]
rGO-Mg	–	–	60.8/92.9	5.0 wt.%/30 min/473 K/ 1.5 MPa	6.0 wt.%/ 40 min/ 573 K	[99]
Mg-NCs/PMMA	–	–	25/79	5.3 wt.%/20 min/473 K/ 3.5 MPa	—	[90]
MgH_2@CMK-3	37. 5	−55.4/–	–/125.3	4.0 wt.%/7 min/573 K/ 2 MPa	5.3 wt.%/8 h/573K	[100]

3.3.2 *Catalytic Synthesis*

The hydrogen absorption reaction often occurs at the surface of materials. Therefore, the surface active sites and microstructure are crucial factors affecting the hydrogenation process. Passivation layers like $MgO/Mg(OH)_2$ tend to form on the surface of Mg and its alloys, suppressing their hydrogen adsorption and dissociation capabilities. Introducing catalysts can accelerate hydrogen adsorption, dissociation, and diffusion while maintaining the fine microstructure with internal stability. Therefore, adding catalysts is applied to various Mg-based hydrogen storage alloys or material systems. To ensure the catalytic effect of the catalysts on the hydrogen absorption and desorption kinetics, it is necessary to maintain the original structure or properties of the catalysts during the introducing process. Usually, the two most commonly used introducing methods are mechanical ball milling and chemical reduction methods.

The catalyst is typically introduced via mechanical ball milling, in which the catalyst and the Mg-based hydrogen storage material are put in a ball mill tank at a specific ratio and ball-milling for a short while, typically around 0.5 to 1 h of discontinuous ball milling. During this process, the catalyst is dispersed onto the Mg/MgH_2 surface through the impact of the balls. Currently, various types of catalysts have been investigated. The primary research target includes transition metals and their compounds, such as Ni, Co, Ti, Nb, V, and Mo. MgH_2-TM (TM = Ti, V, Mn, Fe,

Ni) nanocomposites have been successfully prepared using mechanical ball milling [101]. Ti and V were found to show higher catalytic activity than Ni. In MgH_2-TM composites, the presence of Ti or V enabled rapid hydrogen desorption kinetics above 523 K and hydrogen absorption kinetics at 302 K. However, Ti and V exhibit strong oxygen sensitivity, leading to surface oxidation during the conventional preparation of Mg-Ti and Mg-V composites. Once oxidized, the catalytic effect diminishes, resulting in Ni being more effective than Ti and V in catalyzing conventional Mg-based alloys.

Alternatively, Mg-based composite hydrogen storage materials can be synthesized by incorporating a catalyst into the reduction reaction of Mg powder with $TMCl_x$ in tetrahydrofuran solution. This method involves encapsulating Mg powders with transition metals such as Ti, Nb, V, Co, Mo, or Ni at the nanoscale, forming a core–shell structure. Mg-TM (TM = Ti, Nb, V, Co, Ni, and Mo) composites were produced by reacting Mg powders with $TMCl_x$ in a tetrahydrofuran solution. The Mg-TM exhibited significantly improved hydrogen absorption and desorption performances compared with corresponding ball-milled Mg-TM samples. This enhancement was attributed to the ability of TM-encapsulated Mg/MgH_2 composites to offer a more significant number of hydrogen desorption sites compared with ball-milled samples [102]. Additionally, the electronegativity properties of the transition elements were found to be correlated with the activation energy of dehydrogenation. Except for the Mg-Ni system, the dehydrogenation activation energy increases with the electronegativity, leading to a decreased desorption rate, but the Mg-Ni system differs from other Mg-TM systems due to the formation of Mg_2Ni (Fig. 3.43).

Introducing transition metal elements as catalysts to form Mg-based composite hydrogen storage materials can significantly enhance hydrogen storage performance. High-valent metal oxides, metal compounds, and carbon materials have been found to produce superior catalytic effects. Nanoscale FeCo alloys were synthesized using

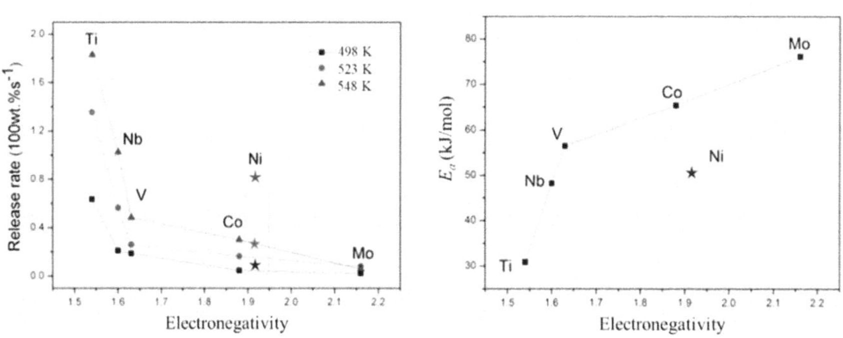

(a) Relationship between the electronegativity of TM and the average desorption rate in Mg-TM systems

(b) Relationship between electronegativity of TM and desorption activation energy in Mg-TM systems

Fig. 3.43 Effect of transition metal elements on hydrogen release rates and activation energies [102]

the wet chemical ball milling technique, and the resulting MgH_2 and FeCo nano-alloys were subsequently ball-milled to produce nano FeCo-doped MgH_2 composite hydrogen storage materials. Compared to ball-milled MgH_2, the hydrogen storage properties of the composite of MgH_2 and FeCo nanocomposites are greatly boosted [103]. The MgH_2 + nano FeCo composites can absorb 1.8 wt.% of hydrogen at 373 K and desorb 6.0 wt.% of hydrogen at 573 K for 9.5 min. Comparatively, pure MgH_2 did not exhibit any hydrogen absorption or desorption behavior under identical conditions. Moreover, FeCo nanodoped MgH_2 composites lowered the hydrogen desorption temperature from 648.3 to 523.9 K with reduced activation energy. When the catalytic effects of FeCo nanoparticles are compared with Fe, Co monometallic nanoparticles, and the physical mixtures of Fe and Co nanocomposites, FeCo nanoparticles exhibit superior catalytic effects, even after 10 cycles, MgH_2 + nano FeCo composites retain a hydrogen capacity of 6.4 wt.%, which is ca. 91.4% of the original hydrogen storage capacity, demonstrating the outstanding stability performance (Fig. 3.44).

MgH_2-VTi-CNTs composites were prepared via the ball milling method [73], and their hydrogen absorption and desorption properties were compared with those of MgH_2-VTi, MgH_2-FeTi, MgH_2-CNTs, and MgH_2-FeTi-CNTs composites, as depicted in Fig. 3.45. It was observed that transition metals significantly enhance hydrogen adsorption kinetics, while CNTs increase hydrogen adsorption capacity. VTi catalyst shows an enhanced effect on the absorption kinetics than FeTi, and the

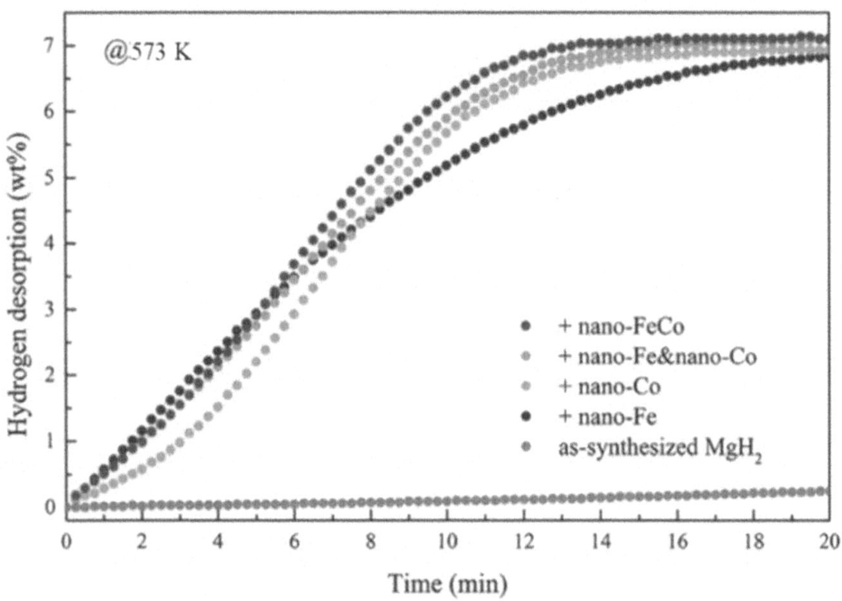

Fig. 3.44 Isothermal hydrogen desorption curves of MgH_2, MgH_2 + nano Fe, MgH_2 + nano Co, MgH_2 + nano Fe and Co and MgH_2 + nano FeCo samples at 573 K [103]

combination of two catalysts was more effective than the single one. Additionally, MgH_2-VTi-CNTs composites exhibit faster hydrogen desorption kinetics at 573 K, releasing 6.0 wt.% H_2 within 350 s, surpassing MgH_2-VTi and MgH_2-FeTi-CNTs composites. The comparison of various catalyst-doped MgH_2 systems with DOE and IEA targets was shown in Fig. 3.46 [73], we can see that the hydrogen storage properties of MgH_2-VTi-CNTs composites achieved the target of IEA, it can absorb 5.1 wt.% H_2 and the hydriding rate is 1.0 wt% H_2/min at 423 K.

Transition-metal-oxide-doped MgH_2 also exhibits improved kinetic properties. TiO_2 nanosheets catalyzed MgH_2 composites were prepared via the ball milling method [104]. TiO_2 nanosheets with {001} planes exhibit superior catalytic effects. The MgH_2-5 wt.% TiO_2 absorbs 6.1 wt.% H_2 at 423 K for 10 s, and desorbs 6 wt.% H_2 at 553 K for 3.2 min and even desorbs 5.8 wt.% H_2 at 513 K within 6 min. Furthermore, MgH_2-TiO_2 nanosheets exhibit lower hydrogen desorption temperatures and faster hydrogen release rates compared to MgH_2 and MgH_2-TiO_2 nanoparticles. The doping of TiO_2 nanosheets reduces the hydrogen desorption temperature of MgH_2

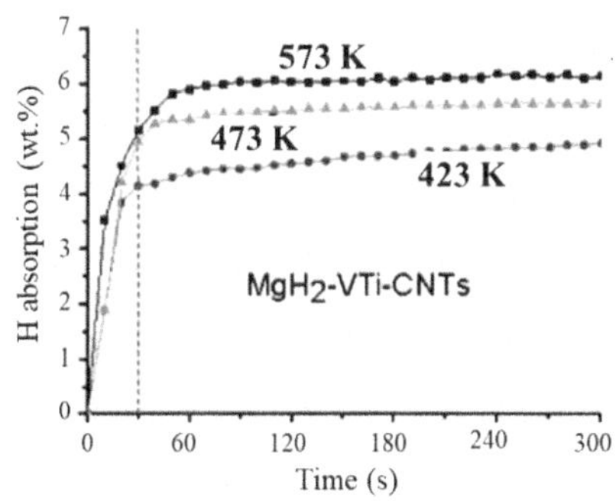

(a) Hydrogenation kinetics curves of MgH_2-VTi-CNTs at different temperatures

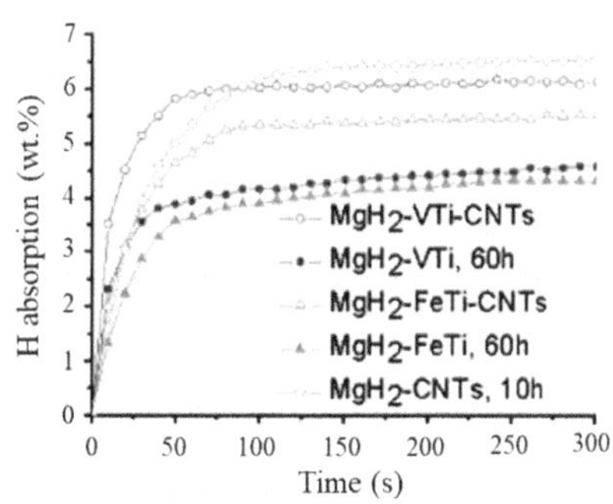

(b) Hydrogenation kinetics curves of MgH_2 systems with different doping at 573K

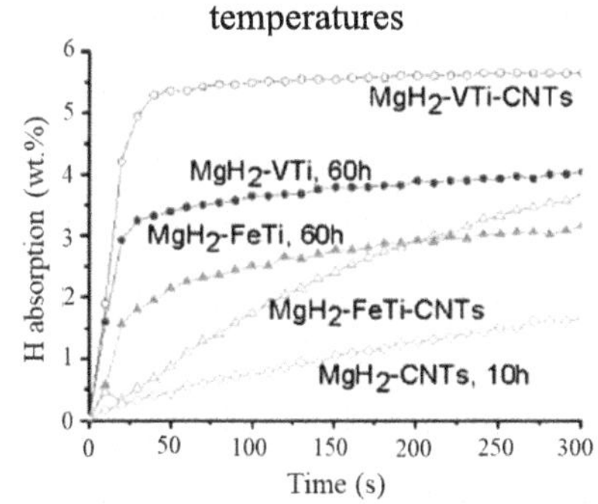

(c) Hydrogenation kinetics curves of MgH_2 systems with different doping at 473 K

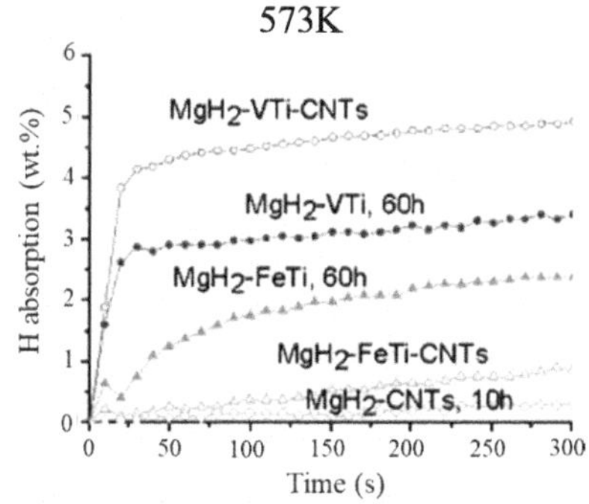

(d) Hydrogenation kinetics curves of MgH_2 systems with different doping at 423 K

Fig. 3.45 Hydrogen uptake kinetics of catalyst-doped MgH_2 samples under 2 MPa hydrogen pressure [73]

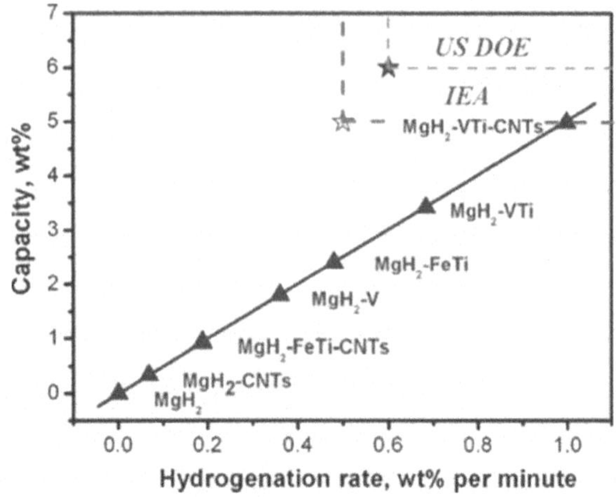

Fig. 3.46 Comparison of maximum capacity and 5 min average hydrogen uptake kinetics of Mg-based nanocomposites at 423 K [73]

from 628.4 to 493.4 K and decreases the activation energy (E_a) of hydrogen release from 186.78 kJ/mol H_2 to 67.64 kJ/mol H_2 (for MgH_2-TiO_2 nanosheets).

3.4 Methods of Powdering and Hydrogenation of Mg-Based Hydrogen Storage Materials

3.4.1 Powdering Methods

The various preparation methods for Mg-based hydrogen storage materials are discussed in the section above; each method has its advantages and drawbacks. Except for the ball milling process, others include ingot casting, forging into blocks, thin strips, and films, the resulting products cannot be directly applied and must be Pulverized to the powders. For instance, the particle size of battery anode materials should be smaller than 200 mesh. Consequently, the industry has adopted different powdering methods, including ball milling, hydrogenation powdering, and gas atomization techniques.

(1) Ball milling

Ball milling of Mg-based hydrogen storage materials is generally categorized into dry ball-milling and wet ball-milling techniques. Dry ball-milling pulverizing materials into powder by the impact and ground of the balls with a specific ball-to-material ratio under a protective atmosphere. The final particle size of the obtained powder is typically controlled by factors of the ball-to-material ratio, rotational speed, milling time, etc. The diameter ratio of the balls or rods also plays an important role. Generally, the optimal parameters of the process can be obtained by experimental methods. Before the ball-milling process, large-size alloys (typically with particle sizes of ca.

30–40 mm) are crushed to 1~3 mm or 3~6 mm using a jaw crusher. Subsequently, the crushed material is further smashed down to ca. 1 mm using a roller machine before undergoing ball milling. During the ball milling process, avoiding excessive ball milling time is essential, which can lead to catalyst adhesion to the wall and difficulty in collecting material. The correct approach is to perform ball milling for 10–20 min followed by one-time sieving. After sieving the fine powder, supplement the coarse particles and mill again. This operation is relatively troublesome and is also prone to contamination.

In modern industry, the function of grinding and sieving is commonly integrated into a single unit. This type of ball mill machine features a dual-layer barrel wall consisting of an inner and outer layer. The inner layer is fitted with a porous plate where balls and materials are loaded, while the outer layer is equipped with a mesh to screen particle size. During the grinding process, when the material reaches the mesh screen size, it automatically passes through the filter under the rotating force and is collected. The rest is still in the inner barrel and continues being milled, enabling continuous powder production. This method of milling and filtering is user-friendly, allowing for constant sample production. It also minimizes contamination and can boast high production efficiency.

The difference between wet and dry ball milling lies in the composition inside of the milling tank. In the wet ball milling process, the tank is filled with liquid media such as water, petrol, or alcohol, rather than inert gas. Typically, the ball milling operates with vertical agitation by the agitator propelled by a paddle. The rotation facilitates collision between the balls and materials, resulting in grinding and crushing. The intensity of ball milling is determined by the factors of mixing speed, ball-to-material ratio, ball size ratio, and milling time. These parameters are typically determined through preliminary experiments. The operational procedures for wet ball milling are similar to dry ball milling. Alloy blocks are crushed to ca. 1 mm and then introduced into the tank. After milling for a specific time, the alloy is collected and filtered, followed by either direct usage or vacuum drying. The wet milling method is simple and excludes wall sticking and contamination. Additionally, it also helps remove ultra-fine powder and some surface oxide skin from the ingot. However, the cost may be increased because the prepared samples should be dried before the practical application [2].

(2) Hydrogenated powder

The alloy hydrogenation pulverizing method has a relatively long history, this process utilizes the volume dilatation and contraction during the hydrogen absorption and desorption process, which induces cracks and new surfaces within the alloy ingot, facilitating hydrogen absorption. The cycles depend on the desired particle size, typically, only 1–2 cycles, the 30~40 mm bulk can be pulverized to powders lower than 200 mesh.

The specific process for the hydrogenation pulverizing method is as follows: (1) the samples are placed in the aluminum crucible and then sealed in an autoclave. The autoclave is evacuated to a pressure of 1~5 Pa. High-purity hydrogen (0.1 MPa) is introduced 2~3 times to purge the atmosphere within the autoclave. (2) High-purity (99.99%) hydrogen with a pressure of 1~2 MPa is introduced, launching a hydrogenation reaction. (3) The temperature is raised to 423 K, and vacuuming the autoclave for 15 min to remove excess hydrogen from the alloy. (4) The above-mentioned processes are repeated in 1–2 cycles. Finally, the alloy is collected from the furnace after cooling. The following points should be paid more attention during the processes:

① The alloy must be loaded into the aluminum box to prevent excessive concentration of exothermic heat during hydrogen absorption. The autoclave should be water-cooled to facilitate heat dissipation.
② The last time hydrogen desorption from the alloy should be as clean as possible and cooling under Ar atmosphere to prevent spontaneous combustion during subsequent operations, including sieving, subpackage, and applications.
③ Immediately cooling should be taken once the alloy powders heat up, preventing spontaneous combustion.
④ The further step operation on the obtained samples should be conducted within an argon-filled glove box to ensure safety.

The advantages of this method include ease of operation, higher hydrogen storage capacity (10~20 mA·h/g) than ball-milled powder, and faster activation. However, it also suffers from the requirement for high-pressure equipment and heat generation when hydrogen is not well-desorbed, which is unfavorable for large-scale applications [105].

(3) Gas atomization method

Gas atomization technology was proposed in the 1920s by Hall, who realized that copper alloy powders could be prepared using this method. Early atomization technology utilized non-restricted nozzles, also known as free-falling-type atomization nozzles. These nozzles are located at a considerable distance from the metal liquid, and the design of free-falling-type nozzles is simple, but their atomization capacity is limited, making them only suitable for producing alloy powders with large particle sizes. For the chemically active metals and alloys, the free-falling-type nozzle is still the best choice, because, during the atomization process, the liquid metal falls freely, avoiding direct contact with intermediate packages and nozzles. Consequently, the alloy composition of the molten metal remains unaffected [106].

Subsequently, the restricted atomization nozzle was developed based on the non-restricted nozzle. The key feature of the restricted atomization nozzle is the compact structure, which significantly shortens the distance between the gas and the melted samples, reducing the kinetic energy loss of the gases going through the nozzles, resulting in finer powder particles and significantly improved atomization efficiency. However, this design is more complicated and suffers from backflow and blockage of the guide pipe.

Before the 1950s, air was commonly used as the atomizing medium for alloy powders, resulting in relatively high oxygen content in the obtained powders. Subsequently, an inert gas atomization process was developed. Inert gases, such as nitrogen and argon, were employed to reduce their oxygen content. To solve the problem that exists in gas atomization technique and enhance efficiency, a series of gas atomization technologies have been then developed, including ultrasonic gas atomization, high-pressure gas atomization, laminar flow gas atomization, hot gas atomization, centrifugal pressure atomization, and electrode induction gas atomization [107]. Powders produced by gas atomization exhibit small particle size, high sphericity, low oxygen content, and good fluidity. This technique can also be applied to large-scale industrial production. Through continuous development, gas atomization technology has emerged as the primary method for producing high-performance spherical metal and alloy powders.

The principle of gas atomization utilizes high-speed gas flow to crush a liquid metal stream into tiny droplets that rapidly condense into metal powders. This process is also known as the high-pressure gas atomization method. Its mechanism includes three stages: (i) the formation of thin fluid layers, (ii) the breaking of thin layers into metal liquid metal wires, and (iii) the contraction of liquid metal wires to form micro-droplets. High-pressure gas atomization is employed to obtain metal powders with diameters ranging from 50 to 100 μm. The resulting powder is smooth and spherical, with a cooling rate typically ranging from 10^2 to 10^3 K/s [108]. In the gas atomization process for preparing Mg-based hydrogen storage materials, the Mg-based alloy ingot is melted under inert gas and then rapidly condensed into spherical powder. During powder preparation, high-pressure atomizing media and rapidly cooling water spray prevent chemical reactions between Mg and oxygen. This combined approach, known as water and gas atomization, significantly reduces the reaction time of molten Mg alloy with oxygen. The Mg alloy powder prepared by water–gas atomization exhibits greater homogeneity and has a porous powder shape, which enhances the kinetic properties of hydrogen absorption and desorption properties. Additionally, gas atomization for preparing Mg-based hydrogen storage material powder reduces the need for grinding, improving production efficiency and lowering production costs [109].

In gas atomization technology, the crucial parameters determining particle size, distribution, oxygen content, and yield include the atomizer structure, atomizing medium, and metal-liquid characteristics (Fig. 3.47) [110]. Optimizing the nozzle and liquid nozzle structure and positioning is critical for gas atomization technology, which can optimize gas flow characteristics and ensure maximum kinetic energy transfer into powder surface energy. Additionally, optimizing the atomization medium and metal-liquid for some specific nozzles is crucial for powder production maintenance.

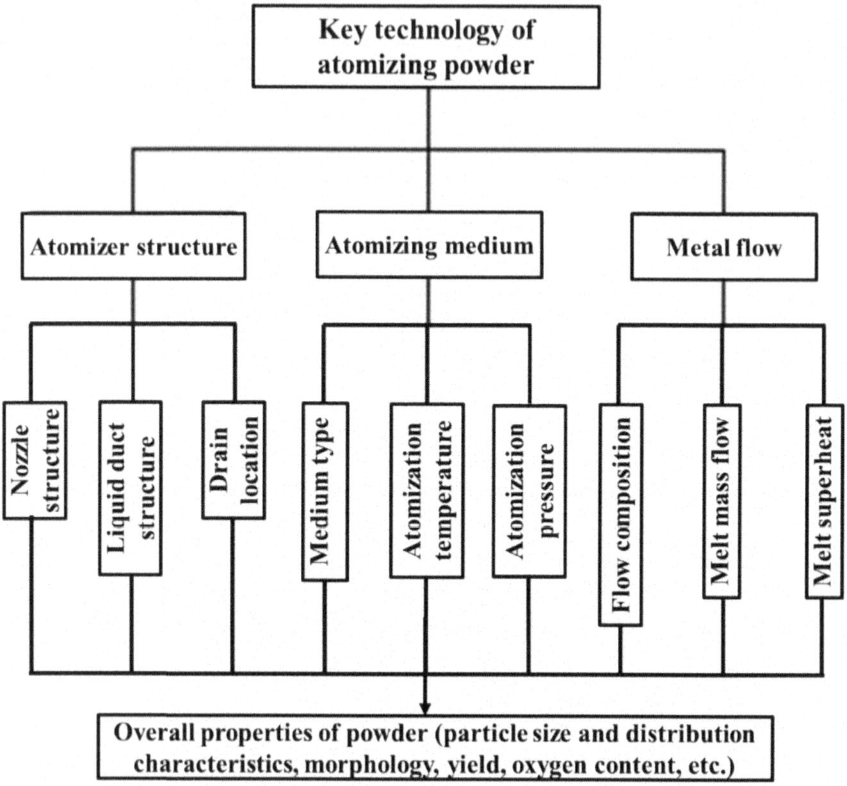

Fig. 3.47 Schematic diagram of critical technologies for aerosolized powder production [110]

3.4.2 Hydrogenation Methods

(1) Hydrogenation reaction method

The hydrogenation reaction method involves the direct reaction of metals with hydrogen at a specific temperature. Most metal hydrides are typically prepared using this method. The general formula for the reaction can be written as follows.

$$M(s) + \frac{n}{2}H_2(g) \xrightarrow{\Delta} MH_n(s) + \Delta X_1, \ldots, X_nH \quad (3.9)$$

Similarly, MgH_2 can be prepared by hydrogenation reaction, as shown below.

$$Mg(s) + H_2(g) \xrightarrow{>573\text{K}} MgH_2(s) + 75kJ/mol \quad (3.10)$$

At ambient temperature and pressure, metals generally do not react with hydrogen, however, when the temperature is raised, hydrogen is absorbed by the metals. The hydrogen absorption temperature varies even for the same kind of metal, depending on their hysteresis effect, usually, the thicker the oxide and pro-oxide layer (passivation layer) on the metal surface, the higher the temperature required for hydrogen absorption. The passivation layer on its surface should be removed to maintain a clean and highly active state and facilitate hydrogenation reactions. Once hydrogen absorption is initiated, the hydrogenation reaction proceeds rapidly. Activation is the process of removing the passivation layer from a metal surface through physical or chemical means before hydrogen absorption by treating materials with high temperature incubation or repeated hydrogen absorption and desorption [105].

Zou et al. [111] developed a method to prepare Mg hydride by hydrogenation reaction whose schematic diagram of the device is shown in Fig. 3.48. The Mg block was put into an open crucible and heated under vacuum at 923~1023 K, leading the Mg block to be evaporated. Part of the Mg vapor is condensed to Mg powder above the reactor and hydrogenated under 5~8 MPa H_2. Using the second heating coil, Mg can be fully hydride by reacting at 623~693 K for 10~12 h. After the reaction, the resulting hydrides can be stabilized by introducing air, forming a passivation layer on the material surface.

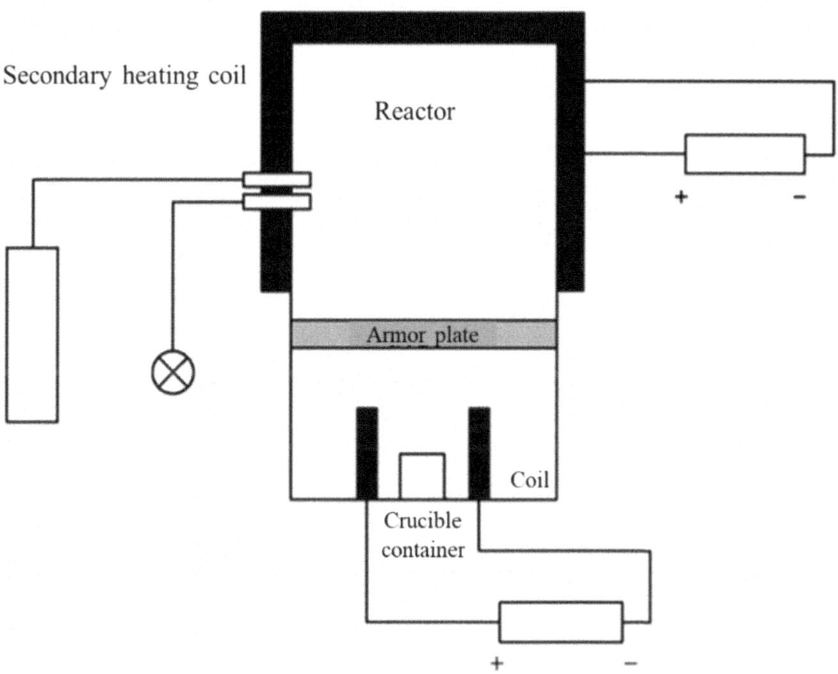

Fig. 3.48 Schematic illustration of the plant for the preparation of Mg hydride by hydrogenation [111]

(2) Thermal decomposition method

Thermal decomposition is one of the common methods for the preparation of MgH_2 nanocrystals. In the recent studies, di-n-butyl Mg was mainly used as the raw material for preparing MgH_2 nanocrystals. It was found that di-n-butyl Mg could produce structure stable nanocrystalline MgH_2, which can absorb 6.8 wt.% H_2 with good kinetic properties. Hydrogen could be fully desorbed in less than 2 h at 523 K, and only 15 min is needed to complete hydrogen desorption process at 573 K [112].

$$(C_4H_9)_2Mg \rightarrow 2C_4H_8 + MgH_2 \tag{3.11}$$

MgH_2 was first produced by hydrogenolysis using Grignard reagent by Becker and Ashby in 1964 with the following reaction [113].

$$2RMgX \rightarrow 2RH + MgX_2 + MgH_2 \tag{3.12}$$

where X is a halide, however, the reaction conditions are harsh. The reaction temperature depends on the Grignard reagent used, which is generally between 348 and 503 K and the hydrogen pressure must be higher than 10 MPa. Subsequent study found that this method can be utilized at 473 K under an inert atmosphere. Using $R_1CH_2CH_2MgR_2$ as the starting material, the $RCH = CH_2$ and $HMgR_2$ mixture can be obtained using a Grignard reagent. However, the reaction mechanism remains unclear. To avoid the tedious separation process of MgH_2 from the products, di-n-butyl Mg has attracted attention in recent years.

The decomposition conditions of di-n-butyl Mg influence the product morphology. Nanorods with lengths of 150~200 nm and widths of 10~50 nm were prepared when the reaction was carried out under argon atmosphere at 473 K. Meanwhile, spherical Mg hydride particles with sizes in the range of 25~170 nm were prepared under hydrogen atmosphere at the same pressure. A rectangular shape of Mg particles around 100 nm was obtained when polystyrene was added to cyclohexane solvent treated at 3 MPa hydrogen pressure and 453 K for 24 h [114]. In addition to the reaction atmosphere, the effect of temperature on the morphology of di-n-butyl Mg hydrolysis products was investigated at a certain hydrogen pressure. The MgH_2 with an average size of ca. 100 nm was obtained at 473 K, while nanorods with diameters of 50~80 nm and 30~50 nm were obtained at higher temperatures of 493 K and 523 K, respectively (Fig. 3.49) [115].

The pyrolysis of di-n-butyl Mg is also commonly used in the nano confinement of Mg. Typically, a solution of di-n-butyl Mg is impregnated into a porous scaffold material and decomposed at elevated temperature to ca. 573 K to achieve nano-confinement. Mg/MgH_2 particles with sizes ranging from 1.3 to 20 nm have been successfully loaded/confined in porous carbon by this method. Another method to achieve Mg nano-refinement is the melt penetration method, where Mg/MgH_2 is first mixed with a porous carbon matrix, and then the mixture is heated up to the melting point of Mg, i.e., higher than 922 K, under hydrogen pressure, therefore,

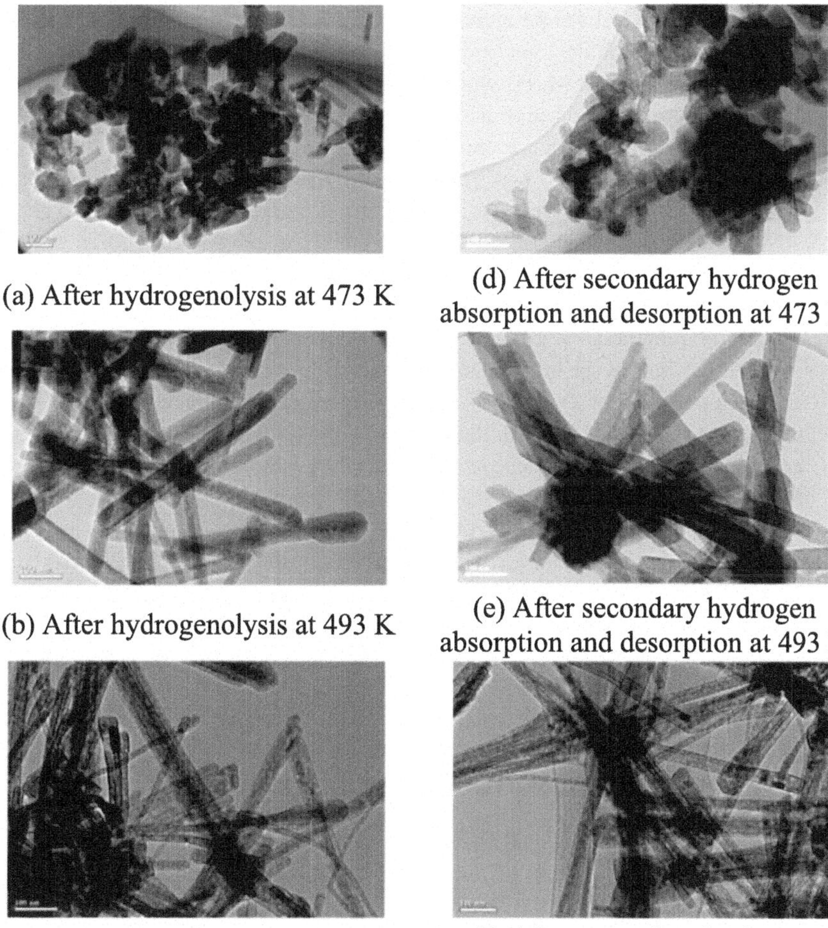

(a) After hydrogenolysis at 473 K

(d) After secondary hydrogen absorption and desorption at 473 K

(b) After hydrogenolysis at 493 K

(e) After secondary hydrogen absorption and desorption at 493 K

(c) After hydrogenolysis at 523 K

(f) After secondary hydrogen absorption and desorption at 523 K

Fig. 3.49 MgH_2 morphology obtained by hydrogenolysis at different temperatures for 4 h under 2 MPa hydrogen pressure and after secondary hydrogen absorption and desorption [115]

the Mg is permeated into the pore structure of the porous materials, and the size of nano-confinement of the Mg particles is determined by the pore size of the matrix [116].

(3) Catalytic synthesis

The catalytic synthesis method is mainly based on the organic Mg complexes, such as MgA-$(THF)_3$, with the help from the catalyst and under certain conditions to produce MgH_2, the A in MgA-$(THF)_3$ is anthracene ($C_{14}H_{10}$). MgA-$(THF)_3$ can be prepared from Mg and anthracene in tetrahydrofuran (THF), and the rate of formation depends on the surface area of Mg and the concentration of anthracene. The whole reaction is reversible and temperature-controlled. The formation of MgA-$(THF)_3$ is favored at the temperature lower than 333 K, while the reaction proceeds in the opposite direction to obtain MgH_2 when the temperature higher than 333 K. Based on the above-mentioned different reaction equilibria at various temperatures, MgH_2 particles with a specific surface area of 100 ~ 130 m^2/g and reversible hydrogen storage capacity of ca. 6.6 wt.% were successfully prepared using catalysts, e.g., $CrCl_3$, and absorbed and desorbed hydrogen at 503 K and 623 K, respectively [117]. MgA-$(THF)_3$ compounds can also decompose to Mg at temperatures higher than 523 K under vacuum. However, the low solubility of MgA-$(THF)_3$ in THF (<0.01 mol/L) limits its application as a feedstock for the preparation of Mg metal [118].

Using a similar method to hydrogenate Mg powder in THF solution at atmospheric pressure with titanium tetrachloride and anthracene as catalysts [119], pure MgH_2 powders were prepared at 293~333 K. The reaction temperature was 573~603 K, and the reaction rate increased with the increased temperature. The activated MgH_2 powders prepared by this method have a fast dehydrogenation performance between 573 and 603 K and the reaction rate increases with the increased temperature, e.g., 21 min is needed to achieve 50% hydrogen desorption at 573 K but only needs 5 min at 603 K, the activation energy of the hydrogen desorption reaction of this Mg hydride was reduced to 125 kJ/mol H_2 (Fig. 3.50). MgH_2 was prepared by thermal decomposition of di-n-butyl Mg under different reaction environments [120], and the decomposition of di-n-butyl Mg was carried out under argon or hydrogen atmosphere at 473 K after removing the solvent from di-n-butyl Mg in vacuum, resulting in MgH_2 nanorod or rounded particles with size of ca. 32 nm. Thermal decomposition of a heptane solution of di-n-butyl Mg by adding ether or cyclohexane under a hydrogen atmosphere yielded scale-like MgH_2 with a size of ca. 40 nm and nanoparticles with a size of ca. 17 nm, respectively, of which the latter had a high hydrogen storage capacity of 7.1 wt.% and could be completely released within 10 min at 573 K.

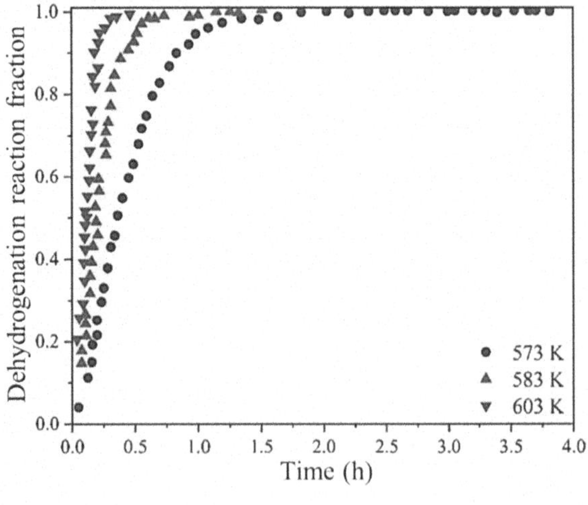

Fig. 3.50 Hydrogen desorption curves of activated Mg hydride at different temperatures under 0.101 MPa H_2 [119]

(4) Reactive ball milling method

The reactive ball milling method is based on mechanical alloying, in which the ball milling atmosphere or reactant is modified, generating new compounds. In the case of Mg-based hydrogen storage materials, hydrogen is typically introduced into the ball milling tank. During the milling process, reduces the oxide components will be reduced by hydrogen to form interstitial solid solutions or hydrides with the metal. Therefore, the mechanical alloying process involving the hydrogenation process is also denoted as hydrogen alloying process [121].

Conventional Mg hydrogenation typically occurs at the temperature higher than 573 K. For instance, typical hydrogenation at 613~623 K and under 3 MPa require 6~336 h of reaction time. However, this process often yields incomplete hydrogenation, with the conversion rate of Mg to MgH_2 lower than 0.9. In contrast, hydrogenation via reactive ball milling occurs at lower temperatures with shorter reaction time, and resulting in the complete conversion of Mg to MgH_2. This high efficiency may be attributed to several factors, including plastic deformation induced by milling, the creation of fresh metal surfaces, and the homogenization of local cold-welded structures due to milling ball impacts. The reactive ball milling method combines the advantages of mechanical alloying and chemical synthesis, offering the potential to produce hydrogen storage materials with superior properties.

Mg_2FeH_6 ternary hydride hosts high bulk hydrogen storage density (150 kg H_2/m^3) and a remarkable mass hydrogen storage density of 5.6 wt.%. However, achieving high-purity Mg_2FeH_6 presents a significant challenge due to the near absence of solid solution of Fe in Mg (up to 0.00041 at.%) and the absence of the thermodynamically stable intermetallic compound Mg_2Fe [122]. Previous attempts were carried out at preparing Mg_2FeH_6 via high-temperature pyrolysis of Mg and Fe powders at 773 K with H_2 pressure of 2~12 MPa, but the conversion rate is only ca. 50% [123]. In contrast, the preparation of Mg_2FeH_6 through reactive ball milling optimizes the preparation conditions, enhances yield, and improves the kinetics of the Mg_2FeH_6 phase. For instance, the synthesis yield of Mg_2FeH_6 reached 83.7% via direct ball milling of Mg and Fe powders in hydrogen with a molar ratio of 3:1. The Mg_2FeH_6 prepared via reactive ball milling of Mg and Fe powders at a chemical stoichiometric ratio (2:1) exhibited the lowest onset of hydrogen release temperature, reaches 477.4 K [124].

The reactive ball milling method is a promising alternative to traditional high-temperature and high-pressure preparation techniques, simplifying the preparation process of Mg-based hydrogen storage materials that typically need activation. This simplification significantly enhances the preparation process for hydrogen storage materials. For instance, the La-Mg-Ni ternary hydrogen storage alloy was synthesized via reactive ball milling, using a ball-to-material ratio of 10:1 with different milling times of 20 h, 40 h, and 60 h under high-purity hydrogen at 3 MPa [125], this approach effectively combines material preparation and hydrogenation processes. XRD spectra of $LaMg_{17}Ni$ with different milling times reveal a significant broadening of diffraction peaks as milling time increases, indicating that amorphization happens during the ball-milling process. The continual high-energy collisions result in fine

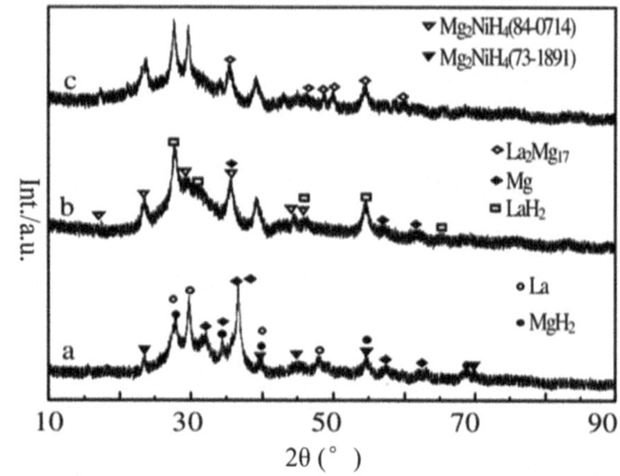

Fig. 3.51 XRD patterns of $LaMg_{17}Ni$ hydride synthesized by reactive ball milling [124]

particle crushing and amorphization, increasing the number of phase boundaries that significantly improve kinetic properties. Integrating preparation and hydrogenation processes enables direct testing of obtained materials for hydrogen absorption and desorption properties without activation. The prepared material achieves a hydrogen storage capacity of 5.23 wt.%, and >90% of hydrogen absorption is finished within 1 min between 423 and 573 K with hydrogen pressure of 3.0 MPa. Similar to the materials prepared by mechanical alloying method, the material can absorb and desorb hydrogen directly without requiring activation. However, prolonged ball milling processes can increase production costs and the risk of contamination. Moreover, the uneven particle size and microstructure of prepared materials may pose challenges for industrial mass production, limiting the potential applications (Fig. 3.51).

This chapter systematically summarized the preparation methods for Mg-based hydrogen storage materials and hydrides. Based on the latest progress in recent years, various techniques are introduced for rapid hydrogen absorption, desorption rates, and long cycle life. For traditional Mg-based hydrogen storage alloys, melting, sintering, diffusion, and ball milling are commonly employed. The melting method is suitable for large-scale industrial production, but it suffers from the challenge of too-stable hydrides, difficult to activation, and rapid saturation of hydrogen absorption. Additionally, the low melting point and volatile nature of Mg lead to significant material loss during the melting process. The materials prepared by replacement-diffusion method and the ball milling-diffusion method possess advantages of reduced activation energy for hydrogen diffusion, improved kinetic and thermodynamic properties of the alloys, and enhanced cycle properties. Notably, the ball milling-diffusion method introduces numerous defects during ball milling, increasing the number of active sites of the alloys. Compared with the melting method, the powder sintering method is characterized by lower reaction temperatures that can address high Mg vapor pressure. In recent years, advancements in powder sintering technology led to the development of more advanced sintering techniques for preparing Mg-based hydrogen storage materials. Amongst the traditional preparation methods, the ball milling method is the most widely used, particularly in synthesizing amorphous/nanocrystalline Mg-based hydrogen storage alloys and catalyst-doped Mg-based hydrogen storage composites. Furthermore, the ball milling method has been combined with other advanced techniques to create new preparation methods, such as discharge plasma ball milling, hydrogen combustion synthesis combined with mechanical ball milling, and the ball milling-diffusion method, etc.

Nanocrystalline and amorphous materials provide promising avenues for advancing Mg-based hydrogen storage materials due to the merits of their unique microstructures, high diffusion coefficients and activity. The hydrogen storage properties of these materials can be enhanced via catalyst doping, the nanostructure of Mg crystals, and incorporation into porous scaffolds, etc. Recently, the preparation strategies for nanostructured or amorphous materials have been developed, remarkably, deposition methods are most widely used for nanowires, nanofibers, and nanofilm materials. Equal-channel angular pressing (ECAP), based on large plastic deformations, has proven to be an effective industrial technique for producing bulk ultrafine crystalline or nanostructured Mg alloys. ECAP offers numerous control factors for designing the final microstructure. The future research will focus on maximizing the positive impact of microstructural coordination on the hydrogen storage behavior of Mg-based materials, aiming to realize greater hydrogen capacity, improved kinetics, and enhanced cycling stability.

Despite the significant progress in hydrogen storage performance of Mg-based materials, high temperatures are still needed for hydrogen absorption and desorption, which is the greatest hurdle for practical applications. The preparation of sub-nanometer Mg crystals (<1 nm) is considered a potential effective way to weaken the strength of Mg-H bonds and achieve low-temperature hydrogen absorption and desorption, but the conceive needs to be verified experimentally. Additionally, nano-confinement techniques have emerged as essential strategies for improving the kinetics, reversibility, and thermodynamics of Mg-based hydrogen storage materials. However, the underlying mechanisms need further investigation. Future research in this field should focus on developing Mg scaffold nanocomposites and improving scaffold filling efficiency. It is also important to control Mg/MgH_2 nanoparticle size at the monodispersed level and functionalizing scaffold surfaces. Furthermore, the combination of several synthetic strategies to develop Mg-rich composites with fine structure and highly active surfaces should be the endeavor directions in the future.

References

1. YUNHAO L. Induction heating in the foundry industry [M]. Metalworking: Thermal working, 2020.
2. ZILONG H. Hydrogen storage materials [M]. Beijing: Chemical Industry Press, 2002.
3. NIE C, YANG H, MOU X. The Development and Trends of Vacuum Induction Melting Technology [J]. Vacuum, 2015, 52(5): 52–57.
4. REILLY J J, WISWALL R H. Reaction of hydrogen with alloys of magnesium and copper [J]. Inorganic Chemistry, 1967, 6(12): 2220–2223.

5. REILLY J J, WISWALL R H. Reaction of hydrogen with alloys of magnesium and nickel and the formation of Mg_2NiH_4 [J]. Inorganic Chemistry, 1968, 7(11): 2254–2256.
6. CHEN Y, ZHOU S, DING P. Research progress on preparation methods of magnesium based hydrogen storage alloys [J]. Material Guide, 2003, 10: 20–23.
7. YUZHEN J, YUN Z, GAOGAO B. Development of Vacuum Arc Furnace Equipment and Melting Technology [J]. Casting technology, 2008, 6: 827–829.
8. CHEN C, WANG J, WANG H, et al. Improved kinetics of nanoparticle-decorated Mg-Ti-Zr nanocomposite for hydrogen storage at moderate temperatures [J]. Materials Chemistry and Physics, 2018, 206: 21–28.
9. MAHMOUDI N, KAFLOU A, SIMCHI A. Synthesis of a nanostructured MgH_2-Ti alloy composite for hydrogen storage via combined vacuum arc remelting and mechanical alloying [J]. Materials Letters, 2011, 65(7): 1120–1122.
10. AGARWAL S, AURORA A, JAIN A, et al. Catalytic effect of ZrCrNi alloy on hydriding properties of MgH_2 [J]. International Journal of Hydrogen Energy, 2009, 34(22): 9157–9162.
11. WANG P, WANG A M, ZHANG H F, et al. Hydriding properties of a mechanically milled Mg-50 wt% $ZrFe_{1.4}Cr_{0.6}$ composite [J]. Journal of Alloys and Compounds, 2000, 297(1–2): 240–245.
12. GIZA K, ADAMCZYK L, HACKEMER A, et al. Preparation and electrochemical properties of $La_2MgNi_8Co_{1-x}M_x$ (M=Al or In; x=0 or 0.2) hydrogen storage alloys [J]. Journal of Alloys and Compounds, 2015, 645: S490–S495.
13. LEE S-L, HSU C-W, HSU F-K, et al. Effects of Ni addition on hydrogen storage properties of $Mg_{17}Al_{12}$ alloy [J]. Materials Chemistry and Physics, 2011, 126(1–2): 319–324.
14. HOU X, SHI H, YANG L, et al. H_2 generation kinetics/ thermodynamics and hydrolysis mechanism of high-performance La-doped Mg-Ni alloys in NaCl solution—A large-scale and quick strategy to get hydrogen [J]. Journal of Magnesium and Alloys, 2021, 9(3): 1068–1083.
15. BUDHANI R C, GOEL T C, CHOPRA K L. Melt-spinning technique for preparation of metallic glasses [J]. Bulletin of Materials Science, 1982, 4(5): 549–561.
16. PAVUNA D. Production of metallic glass ribbons by the chill-block melt-spinning technique in stabilized laboratory conditions [J]. Journal of Materials Science, 1981, 16(9): 2419–2433.
17. HOU X, HU R, ZHANG T, et al. Synergetic catalytic effect of MWCNTs and TiF_3 on hydrogenation properties of nanocrystalline Mg-10wt%Ni alloys [J]. International Journal of Hydrogen Energy, 2013, 38(29): 12904–12911.
18. ZHANG Y, LI X, CAI Y, et al. Improved hydrogen storage performances of Mg-Y-Ni-Cu alloys by melt spinning [J]. Renewable Energy, 2019, 138: 263–271.
19. WU Y, HAN W, ZHOU S X, et al. Microstructure and hydrogenation behavior of ball-milled and melt-spun Mg-10Ni-2Mm alloys [J]. Journal of Alloys and Compounds, 2008, 466(1–2): 176–181.
20. SI T Z, LIU Y F, ZHANG Q A. Hydrogen storage properties of the supersaturated $Mg_{12}YNi$ solid solution [J]. Journal of Alloys and Compounds, 2010, 507(2): 489–493.
21. KALINICHENKA S, ROENTZSCH L, KIEBACK B. Structural and hydrogen storage properties of melt-spun Mg-Ni-Y alloys [J]. International Journal of Hydrogen Energy, 2009, 34(18): 7749–7755.
22. LIU J, SONG X, PEI P, et al. Hydrogen storage properties of Mg–50vol.%$V_{7.4}Zr_{7.4}Ti_{7.4}Ni$ composite prepared by spark plasma sintering [J]. International Journal of Hydrogen Energy, 2009, 34(10): 4365–4370.
23. SONG X, ZHANG P, PEI P, et al. The role of spark plasma sintering on the improvement of hydrogen storage properties of Mg-based composites [J]. International Journal of Hydrogen Energy, 2010, 35(15): 8080–8087.
24. LIU J, SONG X P, PEI P, et al. Hydrogen storage performance of Mg-based composites prepared by spark plasma sintering [J]. Journal of Alloys and Compounds, 2009, 486(1-2): 338–342.
25. LIU D M, SI T Z, WANG C C, et al. Phase component, microstructure and hydrogen storage properties of the laser sintered Mg-20 wt% $LaNi_5$ composite [J]. Scripta Materialia, 2007, 57(5): 389–392.

26. SI T Z, LIU D M, ZHANG Q A. Microstructure and hydrogen storage properties of the laser sintered Mg_2Ni alloy [J]. International Journal of Hydrogen Energy, 2007, 32(18): 4912–4916.
27. SI T, LI Y, LIU D, et al. Phase component and microstructure of laser-sintered Mg-Ni alloys [J]. Rare Metals, 2008, 27(4): 400–404.
28. CHONG W. Preparation and modification of magnesium aluminum based hydrogen storage alloys [D]. Chongqing: Chongqing University, 2014.
29. LI Q, PAN Y, LENG H, et al. Structures and properties of Mg-La-Ni ternary hydrogen storage alloys by microwave-assisted activation synthesis [J]. International Journal of Hydrogen Energy, 2014, 39(26): 14247–14254.
30. MENG J, WANG X L, CHOU K C, et al. Hydrogen storage properties of graphite-modified Mg-Ni-Ce composites prepared by mechanical milling followed by microwave sintering [J]. Metallurgical and Materials Transactions a-Physical Metallurgy and Materials Science, 2013, 44A(1): 58–67.
31. LI Q, LIU J, CHOU K C, et al. Synthesis and dehydrogenation behavior of Mg–Fe–H system prepared under an external magnetic field [J]. Journal of Alloys and Compounds, 2008, 466(1): 146–152.
32. LIU J, LI Q, CHOU K C. Preparation of $La_{0.67}Mg_{0.33}Ni_3$ hydrogen storage alloy by magnetic field assisted sintering method [J]. Rare Metal Materials and Engineering, 2013, 42(02): 392–395.
33. JING L. Magnetic field assisted synthesis of magnesium based hydrogen storage alloys and their hydrogen absorption and desorption kinetics mechanism [D]: 2009.
34. LI PENG, WEIHAO X. The principle and application of selective laser sintering [J]. Material Guide, 2002, (06): 55–58.
35. ZHANG Q, YUAN X, SI T, et al. Method for preparing magnesium based hydrogen storage alloys and their composites by laser sintering, CN101029358 [P/OL].
36. CHENG C W, CHEN J K. Femtosecond laser sintering of copper nanoparticles [J]. Applied Physics A, 2016, 122(4): 289.
37. SI T Z, ZHANG Q A. Phase structures and electrochemical properties of the laser sintered $LaNi_5$-xwt% Mg_2Ni composites [J]. Journal of Alloys and Compounds, 2006, 414(1–2): 317–321.
38. ALEM S A A, LATIFI R, ANGIZI S, et al. Microwave sintering of ceramic reinforced metal matrix composites and their properties: a review [J]. Materials and Manufacturing Processes, 2020, 35(3): 303–327.
39. OGHBAEI M, MIRZAEE O. Microwave versus conventional sintering: A review of fundamentals, advantages and applications [J]. Journal of Alloys and Compounds, 2010, 494(1–2): 175–189.
40. LI Q, YE L, ZHANG X, et al. A method for synthesizing La-Mg hydrogen storage alloy by microwave-assisted heating, CN101886202A [P/OL].
41. SHI H, ZHOU P, LI J, et al. Functional Gradient Metallic Biomaterials: Techniques, Current Scenery, and Future Prospects in the Biomedical Field [J]. Frontiers in Bioengineering and Biotechnology, 2021, 8.
42. FENG H, ZHOU Y, JIA D. The principle and application of discharge plasma sintering technology [J]. Materials Science and Technology, 2003, (03): 327–331.
43. JUNG Y G, HA C G, SHIN J H, et al. Abrication of functionally graded ZrO_2/NiCrAlY composites by plasma activated sintering using tape casting and it's thermal barrier property [J]. Materials Science and Engineering a-Structural Materials Properties Microstructure and Processing, 2002, 323(1–2): 110–118.
44. AKIYAMA T, ISOGAI H, J Y. Combustion synthesis of Mg nickel [J]. International Journal of Self-Propagating High-Temperature Synthesis, 1995, (4): 69.
45. SAITA I, LI L Q, SAITO K, et al. Pressure-Composition-Temperature properties of hydriding combustion-synthesized Mg_2NiH_4 [J]. Materials Transactions, 2002, 43(5): 1100–1104.
46. LI L Q, AKIYAMA T, YAGI J. Production of hydrogen storage alloy of Mg_2NiH_4 by hydriding combustion synthesis in laboratory scale [J]. Journal of Materials Synthesis and Processing, 2000, 8(1): 7–14.

47. LI S, ZHU Y, LIU Y, et al. Nano-inducement of Ni for low-temperature dominant dehydrogenation of Mg-Al alloy prepared by HCS plus MM [J]. Journal of Alloys and Compounds, 2020, 819.
48. LIU X, ZHU Y, LI L. Hydriding and dehydriding properties of nanostructured Mg_2Ni alloy prepared by the process of hydriding combustion synthesis and subsequent mechanical grinding [J]. Journal of Alloys and Compounds, 2006, 425(1–2): 235–238.
49. SHEN P, ZHANG Y, YUAN H, et al. Research on a new synthesis method of hydrogen storage materials: a Mg_2Cu synthesis by displacement-diffusion method [J]. Chem J Chinese Universities, 1982, (04): 580–582.
50. SHEN P, ZHANG Y, YUAN H, et al. Research on a new synthesis method of hydrogen storage materials (II.) - displacement-diffusion synthesis Mg_2Ni [J]. Journal of Inorganic Chemistry, 1985, (03): 197–200.
51. SHEN P, ZHANG Y, YUAN H, et al. Research on New Synthesis Methods of Hydrogen Storage Materials: III. Substitution-Diffusion Synthesis $Mg_2Ni_{0.75}Cu_{0.25}$ [J]. Journal of Inorganic Chemistry, 1986, (02): 1–7.
52. YUAN H, YANG H, ZHOU Z, et al. Pressure-composition isotherms of the $Mg_2Ni_{0.75}Fe_{0.25}$-Mg system synthesized by replacement-diffusion method [J]. Journal of Alloys and Compounds, 1997, 260(1): 256–259.
53. YUNSHI Z, JIANHUA J, HUATANG Y, et al. Synthesis of ternary alloys $Mg_2Ni_{0.75}Pd_{0.25}$ and studies on its surface properties [J]. Journal of Materials Research, 1990, 5(7): 1431–1434.
54. HOUSE S D, VAJO J J, REN C, et al. Effect of ball-milling duration and dehydrogenation on the morphology, microstructure and catalyst dispersion in Ni-catalyzed MgH_2 hydrogen storage materials [J]. Acta Materialia, 2015, 86: 55–8.
55. DOPPIU S, SOLSONA P, SPASSOV T, et al. Thermodynamic properties and absorption-desorption kinetics of $Mg_{87}Ni_{10}Al_3$ alloy synthesised by reactive ball milling under H_2 atmosphere [J]. Journal of Alloys and Compounds, 2005, 404: 27–30.
56. AU M. Hydrogen storage properties of magnesium based nanostructured composite materials [J]. Materials Science and Engineering: B, 2005, 117(1): 37–44.
57. SHAO H, XIN G, ZHENG J, et al. Nanotechnology in Mg-based materials for hydrogen storage [J]. Nano Energy, 2012, 1(4): 590–601.
58. ZHANG Y, ZHANG W, BU W, et al. Improved hydrogen storage dynamics of amorphous and nanocrystalline Ce-Mg-Ni-based $CeMg_{12}$-type alloys synthesized by ball milling [J]. Renewable Energy, 2019, 132: 167–175.
59. HOU X, HU R, ZHANG T, et al. Hydrogenation thermodynamics of melt-spun Mg rich Mg-Ni nanocrystalline alloys with the addition of multiwalled carbon nanotubes and TiF_3 [J]. Journal of Power Sources, 2016, 306: 437–447.
60. CHEN D, XIAO J, ZHAN J. Preparation and characterization of ultrasonic-assisted ball milling nano $Mn_xMg_{(1-x)}Fe_2O_4$ [J]. Journal of Hunan University (Natural Science Edition), 2018, 45(06): 51–55.
61. DAI L, ZENG M, TONG Y, et al. Field-assisted mechanical alloying studies [J]. Functional materials, 2005, (08): 22–25.
62. MIN Z. Introduction to Advanced Hydrogen Storage Materials [M]. Beijing: Science Press, 2015.
63. OUYANG L Z, CAO Z J, WANG H, et al. Enhanced dehydriding thermodynamics and kinetics in Mg(In)-MgF_2 composite directly synthesized by plasma milling [J]. Journal of Alloys and Compounds, 2014, 586: 113–117.
64. SURYANARAYANA C. Mechanical Alloying: A Novel Technique to Synthesize Advanced Materials [J]. Research, 2019, 2019.
65. ZHU X, LIN Q. Research and progress on mechanical alloying [M]. 1999.
66. IVANOV E, KONSTANCHUK I, STEPANOV A, et al. Magnesium mechanical alloys for hydrogen storage [J]. Journal of the Less Common Metals, 1987, 131(1): 25–29.
67. OUYANG L, LIU F, WANG H, et al. Magnesium-based hydrogen storage compounds: A review [J]. Journal of Alloys and Compounds, 2020, 832: 154865.

68. KOU H, HOU X, ZHANG T, et al. On the amorphization behavior and hydrogenation performance of high-energy ball-milled Mg_2Ni alloys [J]. Materials Characterization, 2013, 80: 21–27.
69. HOU X. Microstructure, hydrogen absorption and release behavior and catalytic modification of Mg-Ni-based alloys [D]: Xi'an: Northwestern Polytechnical University, 2016.
70. ZHONG H C, WANG H, OUYANG L Z. Improving the hydrogen storage properties of MgH_2 by reversibly forming Mg-Al solid solution alloys [J]. International Journal of Hydrogen Energy, 2014, 39(7): 3320–3326.
71. SONG M Y, KWON S N, PARK H R, et al. Improvement of hydriding and dehydriding rates of Mg via addition of transition elements Ni, Fe, and Ti [J]. International Journal of Hydrogen Energy, 2011, 36(20): 12932–12938.
72. YONG H, WEI X, ZHANG K, et al. Characterization of microstructure, hydrogen storage kinetics and thermodynamics of ball-milled $Mg_{90}Y_{1.5}Ce_{1.5}Ni_7$ alloy [J]. International Journal of Hydrogen Energy, 2021, 46(34): 17802–17813.
73. YAO X, WU C, DU A, et al. Metallic and carbon nanotube-catalyzed coupling of hydrogenation in magnesium [J]. Journal of the American Chemical Society, 2007, 129(50): 15650–5654.
74. ZHANG Y, ZHANG W, WEI X, et al. Catalytic effects of TiO_2 on hydrogen storage thermodynamics and kinetics of the as-milled Mg-based alloy [J]. Materials Characterization, 2021, 176.
75. XIA T, LI X, DONG H, et al. Preparation of La-Mg-Ni hydrogen storage alloy films by electrodeposition and their properties [J]. Rare Metal Materials & Engineering, 2010, 39(03): 545–548.
76. SHAN X, LIU W. Effect of electrodeposition process on the electrochemical properties of Mg-Ni hydrogen storage alloys [J]. Chemical Research, 2005, (01): 55–58.
77. LI X, LI S, WANG L, et al. Electrodeposition of amorphous Ni-Mg alloy and its hydrogen storage performance [J]. Plating and finishing, 2010, 32(06): 6–10.
78. TATIPARTI S S V, EBRAHIMI F. Electrodeposition of Al–Mg Alloy Powders [J]. Journal of The Electrochemical Society, 2008, 155(5): D363.
79. XU J L, ZHANG X, SHI Z N, et al. Electrodeposition of Mg-Ni-Al Alloy in low temperature molten salts, Cancun, MEXICO, F, 2014 [C]. 2014.
80. ZHANG Q. Preparation of magnesium-based alloys by electrodeposition in low-temperature dissolved salts [D]: Shenyang: Northeastern University, 2014.
81. CHEN X. Preparation and characterization of multivariate nano-magnesium-transition metal hydrogen storage materials [D]: Shanghai: Shanghai Jiao Tong University, 2017.
82. XIE X, CHEN M, HU M, et al. Recoverable Ni_2A_{13} nanoparticles and their catalytic effects on Mg-based nanocomposite during hydrogen absorption and desorption cycling [J]. International Journal of Hydrogen Energy, 2018, 43(48): 21856–21863.
83. LUO Q, LI J, LI B, et al. Kinetics in Mg-based hydrogen storage materials: Enhancement and mechanism [J]. Journal of Magnesium and Alloys, 2019, 7(1): 58–71.
84. LI F, ZHU Y, LI L, et al. Magnetron sputtering technology and its development [J]. Vacuum electronics, 2011, (3): 49–54.
85. YANG W, LIU Y, XU L, et al. The development of sputter deposition technology and its current status [J]. Journal of Vacuum Science and Technology, 2005, (03): 204–210.
86. YE S, OUYANG L, ZHU M. Hydrogen storage properties of preferentially orientated Mg-Ni multilayer film prepared by magnetron sputtering [J]. Rare Metals, 2006, 25: 295–299.
87. OUYANG L Z, WANG H, CHUNG C Y, et al. MgNi/Pd multilayer hydrogen storage thin films prepared by dc magnetron sputtering [J]. Journal of Alloys and Compounds, 2006, 422(1): 58–61.
88. LIU Z, ZHANG S, LIU Z, et al. Principle and research progress of chemical vapor deposition preparation of powder materials [J]. Powder Metallurgy Materials Science and Engineering, 2009, 14(06): 359–364.
89. SAITA I, TOSHIMA T, TANDA S, et al. Hydriding chemical vapor deposition of metal hydride nano-fibers [J]. Materials Transactions, 2006, 47(3): 931–934.

90. JEON K J, MOON H R, RUMINSKI A M, et al. Air-stable magnesium nanocomposites provide rapid and high-capacity hydrogen storage without using heavy-metal catalysts [J]. Nature Materials, 2011, 10(4): 286–290.
91. NORBERG N S, ARTHUR T S, FREDRICK S J, et al. Size-Dependent Hydrogen Storage Properties of Mg Nanocrystals Prepared from Solution [J]. Journal of the American Chemical Society, 2011, 133(28): 10679–10681.
92. RIEKE R D. Preparation of highly reactive metal powders and their use in organic and organometallic synthesis [J]. Accounts of Chemical Research, 1977, 10(8): 301–306.
93. LIU Y. Preparation and hydrogen storage performance of magnesium based nanocomposite hydrogen storage materials [D]: Shanghai: Shanghai Jiao Tong University, 2015.
94. JIA Y, SUN C, SHEN S, et al. Combination of nanosizing and interfacial effect: Future perspective for designing Mg-based nanomaterials for hydrogen storage [J]. Renewable & Sustainable Energy Reviews, 2015, 44: 289–303.
95. ADELHELM P, DE JONGH P E. The impact of carbon materials on the hydrogen storage properties of light metal hydrides [J]. Journal of Materials Chemistry, 2011, 21(8): 2417–2427.
96. GROSS A F, AHN C C, VAN ATTA S L, et al. Fabrication and hydrogen sorption behaviour of nanoparticulate MgH_2 incorporated in a porous carbon host [J]. Nanotechnology, 2009, 20(20): 204005.
97. ZHANG Q, ZOU J, REN L, et al. Research progress on core-shell nano-magnesium-based composite hydrogen storage materials [J]. Materials Science and Technology, 2020, 28(03): 58–67.
98. MA Z, ZHANG Q, PANDA S, et al. In situcatalyzed and nanoconfined magnesium hydride nanocrystals in a Ni-MOF scaffold for hydrogen storage [M]. 2020.
99. CHO E S, RUMINSKI A M, ALONI S, et al. Graphene oxide/metal nanocrystal multilaminates as the atomic limit for safe and selective hydrogen storage [J]. Nature Communications, 2016, 7(1): 10804.
100. JIA Y, SUN C, CHENG L, et al. Destabilization of Mg–H bonding through nano-interfacial confinement by unsaturated carbon for hydrogen desorption from MgH_2 [J]. Physical Chemistry Chemical Physics, 2013, 15(16): 5814–5820.
101. LIANG G, HUOT J, BOILY S, et al. Catalytic effect of transition metals on hydrogen sorption in nanocrystalline ball milled MgH_2–Tm (Tm=Ti, V, Mn, Fe and Ni) systems [J]. Journal of Alloys and Compounds, 1999, 292(1): 247–252.
102. CUI J, LIU J, WANG H, et al. Mg–TM (TM: Ti, Nb, V, Co, Mo or Ni) core–shell like nanostructures: synthesis, hydrogen storage performance and catalytic mechanism [J]. Journal of Materials Chemistry A, 2014, 2(25): 9645–9655.
103. YANG X, JI L, YAN N, et al. Superior catalytic effects of FeCo nanosheets on MgH_2 for hydrogen storage [J]. Dalton Transactions, 2019, 48(33): 12699–12706.
104. ZHANG M, XIAO X, WANG X, et al. Excellent catalysis of TiO_2 nanosheets with high-surface-energy {001} facets on the hydrogen storage properties of MgH_2 [J]. Nanoscale, 2019, 11(15): 7465–7473.
105. OSUMI Y. Properties and Applications of Metal Hydrides [M]. Beijing: Chemical Industry Press, 1990.
106. HUANG P. Principles of powder metallurgy [M]. Beijing: Metallurgical Industry Press, 1982: 500.
107. XU L, ZHOU X, LI J. New progress in the research of metal powder gas atomization technology [J]. Thermal spray technology, 2018, 10(02): 1–7.
108. ZHANG Y, DONG B. Research on the method of preparing 3D printing metal powder by gas atomization method [J]. Mechanical Research and Applications, 2016, 29(02): 203–205.
109. ZHAO Y, LIU S, SUN Y. Development of atomized aluminum-magnesium alloy powder [J]. Hunan non-ferrous metals, 1994, (03): 168–172.
110. YE S, ZHANG P, QIU K, et al. Key technologies and development trends of gas atomization for the preparation of metal spherical powder for 3D printing [J]. Sichuan non-ferrous metals, 2017, (02): 51–54.

111. ZOU J, GUO H, ZENG X, et al. A method and device for preparing magnesium hydride, CN102583244A [P/OL].
112. SETIJADI E J, BOYER C, AGUEY-ZINSOU K-F. Remarkable hydrogen storage properties for nanocrystalline MgH_2 synthesised by the hydrogenolysis of Grignard reagents [J]. Physical Chemistry Chemical Physics, 2012, 14(32): 11386–11397.
113. BECKER W E, C A E. Hydrogenolysis of the grignard reagent [J]. Journal of Organic Chemistry, 1964, 29(4): 954–955.
114. SETIJADI E J, BOYER C, AGUEY-ZINSOU K-F. Switching the thermodynamics of MgH_2 nanoparticles through polystyrene stabilisation and oxidation [J]. Rsc Advances, 2014, 4(75): 39934–39940.
115. KONAROVA M, TANKSALE A, BELTRAMINI J N, et al. Porous MgH_2/C composite with fast hydrogen storage kinetics [J]. International Journal of Hydrogen Energy, 2012, 37(10): 8370–8378.
116. DE JONGH P E, WAGEMANS R W P, EGGENHUISEN T M, et al. The preparation of carbon-supported magnesium nanoparticles using melt infiltration [J]. Chemistry of Materials, 2007, 19(24): 6052–6057.
117. BOGDANOVIĆ B, RITTER A, SPLIETHOFF B. Active MgH_2 Mg-systems for hydrogen storage [J]. Angewandte Chemie International Edition in English, 1990, 29(3): 223–234.
118. BARTMANN E, BOGDANOVIČ B, JANKE N, et al. Active magnesium from catalytically prepared magnesium hydride or from magnesium anthracene and its uses in the synthesis [J]. Chemische Berichte, 1990, 123(7): 1517–1528.
119. LIN G, ZHOU G, LI Q, et al. Catalytic synthesis of magnesium hydride under atmospheric pressure [J]. Rare Metal Materials & Engineering, 2006, 35(5): 802–805.
120. SETIJADI E J, BOYER C, AGUEY-ZINSOU K-F. MgH_2 with different morphologies synthesized by thermal hydrogenolysis method for enhanced hydrogen sorption [J]. International Journal of Hydrogen Energy, 2013, 38(14): 5746–5757.
121. WANG H, NIAN H E, REN X, et al. Preparation of Mg Ni M (M=Ni, Nb, Y, Ti) hydrogen storage materials by reactive ball milling method and their hydrogen release performance [J]. Powder metallurgy technology, 2011, 29(4): 259–262.
122. KHAN D, PANDA S, MA Z, et al. Formation and hydrogen storage behavior of nanostructured Mg_2FeH_6 in a compressed $2MgH_2$-Fe composite [J]. International Journal of Hydrogen Energy, 2020, 45(41): 21676–21686.
123. DIDISHEIM J, ZOLLIKER P, YVON K, et al. DiMg iron (II)hydride, Mg_2FeH_6, containing octahedral Feanions [J]. Inorganic Chemistry, 1984, 23(13): 1953–1957.
124. LI S, LIU Y, CUI J, et al. Reactive mechanical alloying synthesis of Mg_2FeH_6 hydrogen storage materials and their hydrogen release properties [J]. Journal of Central South University(Natural Science Edition), 2008, 39(1): 1–6.
125. LI Q, JIANG L, LIN Q, et al. Mechanically alloyed La-Mg-Ni ternary hydrogen storage alloys [J]. Rare Metal Materials & Engineering, 2004, 33(9): 941-944.

Chapter 4
Performance Testing Technique and Characterization of Mg-Based Hydrogen Storage Materials

With the expanding scope of research and advancements in the preparation technologies on Mg-based hydrogen storage materials, the synthesized materials possess much more complex microstructures. These structures have a more complex influence on the properties of hydrogen storage materials. Therefore, advanced analysis and testing technologies are essential to uncover the relationship between the microstructures and properties, which can reveal the reaction mechanisms and promotes the practical applications of Mg-based hydrogen storage materials. The progress in science and technology has yielded various physical and chemical characterization methods, many of which are now integral to evaluating hydrogen storage properties. These techniques greatly enhance the preparation strategies and structural analysis of hydrogen storage materials.

In this chapter, we focus on introducing the performance testing techniques and methods for magnesium-based hydrogen storage materials, as well as crystal and microstructure characterizations. Some techniques with distinctive features will be introduced in detail.

4.1 Performance Indicators of Mg-Based Hydrogen Storage Materials

Magnesium-based hydrogen storage materials should meet the following requirements:

① Easy to activate.
② High reversible gravimetric and volumetric hydrogen storage capacity.
③ Fast hydrogen absorption/desorption kinetics, fast hydrogen diffusion rate, and good reversibility.

Q. Li and F. Pan, *Magnesium-based Hydrogen Storage Materials*,
https://doi.org/10.1007/978-981-96-5531-1_4

④ A relatively broad equilibrium pressure plateau, with moderate equilibrium decomposition pressure. The decomposition pressure near room temperature should be between 0.2 and 0.3 MPa.
⑤ Minimal hysteresis between hydrogen absorption and desorption equilibrium pressures.
⑥ Low enthalpy of hydride formation.
⑦ Long cycling lifetime and low pulverization to ensure minimal capacity loss after hydrogen absorption/desorption cycles.
⑧ Stable in air, good safety, and tolerance to impurities of N_2, O_2, H_2O, and H_2S, etc.
⑨ Cost-effective, environmentally benign, and easy to manufacture.

These benchmarks serve as criteria for evaluating materials in hydrogen storage applications, which can be summarized as activation performance, reversible hydrogen capacity, Pressure-Composition-Temperature (P–C–T) characteristics, equilibrium plateau pressure, hysteresis, hydride formation and decomposition enthalpy, the rates and activation energy of hydrogen absorption and desorption, cycle life, and impurity tolerance.

(1) Activation performance

Generally, Mg-based hydrogen storage materials cannot absorb hydrogen due to the existence of a passivation layer on the surface. This passivation layer can be removed by exposing the material to hydrogen at high temperature and high pressure, followed by vacuuming. The hydrogen storage ability can be gradually improved through multi-cycles of hydrogen absorption/desorption process. These cycling processes are known as activation treatment, which is a necessary step before the practical application of hydrogen storage materials. The hydrogen temperature/pressure, number of absorption/desorption cycles, and the time required for fully activation represent the difficulty level of the activation process. Additionally, there is no absorption or desorption reaction in the initial stage of the activation process; this period is known as the incubation period. After the incubation period, the absorption and desorption reaction of the alloy occurs rapidly. The incubation time is related to the material properties, particularly the surface state of the materials. Furthermore, the incubation time is also influenced by the activation conditions of the alloys, such as hydrogen pressure and temperature. Under the same activation conditions, a longer activation time indicates a higher degree of difficulty of the materials [1].

(2) Reversible hydrogen storage capacity

The reversible hydrogen storage capacity of a material is the hydrogen capacity that can be absorbed and desorbed between the highest and lowest operating pressures. The shape of the isothermal pressure-hydrogen composition curve is an important indicator, as it determines the main pressure range of the reversible hydrogen absorption/desorption. Furthermore, the hydrogen absorption/desorption kinetics also affects the hydrogen storage capacity since the poor absorption kinetics prevent the full hydrogenation of the materials, and some of the adsorbed hydrogen may not be desorbed in a limited time due to slow kinetics. For Mg-based hydrogen storage material hydrides, the hydrogen storage capacity is clearly defined as the ratio between the mass of hydrogen absorbed and the total mass of the material after hydrogen absorption, which is technically more important than the total amount of hydrogen absorbed in the material or the maximum amount of hydrogen absorbed. The difference between the two indicators of hydrogen storage capacity and the maximum hydrogen storage capacity is determined by the hydrogen absorption behavior of the material under different conditions.

(3) Pressure-composition-temperature characteristics (P–C–T Curve)

The P–C–T curve contains important information for evaluating the thermodynamic properties of hydrogen storage materials. It provides the maximum hydrogen content that can be absorbed in the metal hydride and the decomposition pressure at a given temperature. The platform pressure, width, slope, and hysteresis effect in the P–C–T curve are the main criteria for evaluating conventional hydrogen storage materials and developing advanced hydrogen storage materials. The hysteresis effect refers to the phenomenon that the equilibrium pressure (P_a) during the metal hydriding process is generally higher than the hydride dissociation pressure (P_d). This can be expressed as $H_f = \ln (P_a/P_d)$. A smaller H_f represents a smaller pressure difference and a larger amount of hydrogen absorbed or desorbed. Therefore, a broader platform region, smaller platform slope and negligible hysteresis effects are desirable for hydrogen storage materials. In the characterization of hydrides, the formation and decomposition enthalpy of hydrides (ΔH) are the two most important aspects in terms of thermodynamics. The platform pressure of the hydrides is the function of the enthalpy change (ΔH) and entropy change (ΔS) of hydrogenation or dehydrogenation reaction. Therefore, the calculation of reaction enthalpy and entropy in hydrogen storage materials is important for both the theoretical study and practical applications of hydrogen storage materials. The formation entropy of a hydride indicates the tendency of hydrogenation reaction; a larger formation entropy corresponds to a lower the equilibrium decomposition pressure and a more stable hydride. The formation enthalpy is the heat generated during the hydride formation process, and a larger negative value indicates a more stable hydride. The formation and decomposition enthalpy of hydrides can be obtained from the natural logarithm of the plateau pressure against the reciprocal of the temperature on the pressure-composition isotherm curve.

(4) Hydrogen absorption and desorption kinetics

The hydrogen absorption/desorption reactions of materials involve several steps: surface-catalyzed H_2 dissociation, hydrogen diffusion from the material surface to bulk, hydride generation, and heat transfer. The kinetic performance of hydrogen storage materials determines the hydrogen absorption/desorption rate during the storage process. Therefore, the hydrogen absorption/desorption rate is commonly used to evaluate the kinetic performance. The P–t curve in hydrogen absorption/desorption records the amount of hydrogen absorbed and desorbed over time, and the hydrogen absorption/desorption rate can be calculated using the P–t curve. In addition to the hydrogen absorption/desorption rate, the activation energy (energy barrier of the reaction) is another pivotal parameter that can describe the kinetics of hydrogen absorption/desorption. Many classical kinetic models have been developed to calculate the activation energy, e.g. the Kissinger method is one of the most widely used in previous researches.

(5) Cycle life of hydrogen storage materials

Cycle life is a critical indicator for hydrogen storage materials, as it describes the cyclic stability of the materials. During the cyclic processes, hydrogen absorption and desorption reactions are repeated multiple times (n cycles). The cyclic stability is commonly evaluated using S_n/S_{max}, where S_n is the hydrogen storage capacity of the nth cycle and S_{max} is the maximum hydrogen storage capacity. The number of cycles with a capacity greater than 80% of S_{max} is used to denote the cycling stability of the material.

The anti-poisoning effect of Mg-based hydrogen storage materials against impurity gases is also crucial when assessing the hydrogen storage materials for reversible hydrogen storage, as susceptibility to poisoning is a significant drawback of Mg-based materials. Therefore, the third critical parameter of hydrogen storage materials is their anti-poisoning properties against gaseous impurities such as carbon monoxide (CO), carbon dioxide (CO_2), humidity, hydrogen sulfide (H_2S), ammonia (NH_3), oxygen (O_2), hydrocarbons, formaldehyde, formic acid, and halides.

4.2 Performance Testing Techniques for Mg-Based Hydrogen Storage Materials

In laboratory settings, hydrogen storage property tests are usually conducted by measuring the pressure change in a constant-volume reactor, the sample mass change in a constant-pressure reactor, or the heat change caused by the hydrogen absorption and desorption reactions. There are generally two methods to measure the hydrogen storage properties: volumetric method and gravimetric method. The volumetric method is the most commonly used technique for hydrogen absorption and desorption measurements. It assesses the relationship between hydrogen pressure, hydrogen absorption capacity, temperature, and time in a constant volume, using the relationship between hydrogen pressure and the amount of hydrogen in the ideal gas law. The volumetric method is more convenient in practical applications compared to the gravimetric method due to the availability of commercial gas control systems, making it easier to set up the operation systems. The gravimetric, on the other hand, quantifies the amount of hydrogen absorbed or desorbed by measuring the change in sample mass during a variable or constant temperature process.

4.2.1 Volumetric Method

The volumetric method is easily implemented in a gas control system with a known volume where temperature and pressure can be accurately measured. The hydrogen absorption capacity of the sample is calculated using the ideal gas equation:

$$PV = nZRT \tag{4.1}$$

where P is the pressure, V is the volume, n is the amount of substance, Z is the gas compression coefficient, R is the gas constant, and T is the temperature.

Volumetric testing can be conducted using several methods, including the Sieverts method (also known as pressure measurement method), the flow volume method, the differential volume method, and the dynamic volumetric determination, etc.

The principle of the pressure measurement method is as follows: At a specific temperature, the container with a known volume is filled with hydrogen at a known pressure. The hydrogen storage material reactor and the container are then connected to facilitate the hydrogen absorption/desorption reaction. Once the system pressure stabilizes, the amount of hydrogen absorbed in the hydride can be calculated based on the pressure difference before and after the reaction. A schematic illustration of the pressure measurements system is shown in Fig. 4.1. The system comprises the reaction chamber, hydrogen source, gas storage tank, vacuum system, pressure detection and control system, temperature detection and control system, and data collector. Thanks to advancement in computer science and AI technology, modern Sievert systems now feature automated control and data collection. In addition, a manual mode is available to accommodate specific conditions.

As shown in Fig. 4.1, for a single-step adsorption experiment, valves B and C are opened to evacuate chambers V_1 and V_2. After fully evacuating, valves B and C are closed, and valve A is opened. Once V_1 is filled with hydrogen at a specific pressure P_i, valve A is closed, and valve C is opened to charge V_2 with hydrogen. Any additional pressure decreases beyond the volume change between V_1 and $V_1 + V_2$ is due to the hydrogen absorption by the samples in chamber V_2. Therefore, assuming that the experiment is conducted at a constant temperature T and the final pressure is P_f, the molar fraction of hydrogen absorbed can be expressed as:

$$\Delta n = \frac{P_i V_1}{Z_{i,T} RT} - \frac{P_f (V_1 + V_2)}{Z_{f,T} RT} \tag{4.2}$$

where $Z_{i,T}$ and $Z_{f,T}$ are the compression factors of hydrogen at the test temperature T and pressures P_i and P_f, respectively.

Taking Δn and P_f as starting values, and repeating the procedures mentioned above, an isothermal hydrogen absorption curve can be obtained. After m steps of hydrogen absorption, the total hydrogen absorption of the sample can be expressed as:

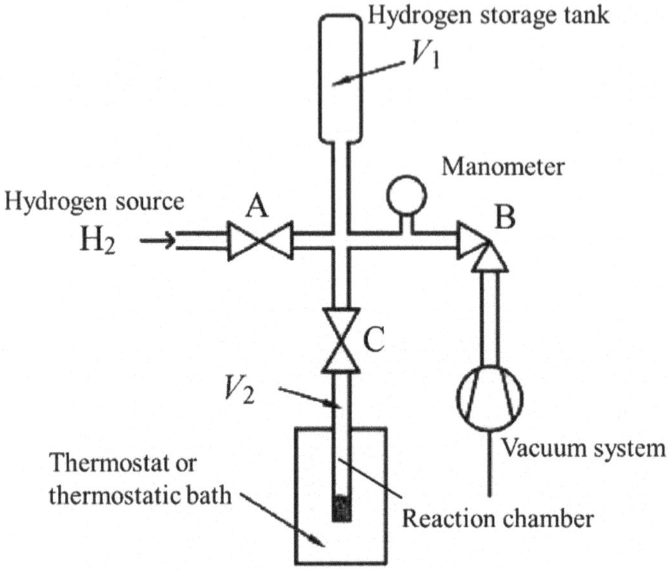

Fig. 4.1 Sievert schematic diagram of the device

$$n_m = \sum_{j=1}^{m}\left[\frac{P_{j,j-1,T}V_2}{Z_{j,j-1,T}RT} + \frac{P_{i,j,T}V_1}{Z_{i,j,T}RT} - \frac{P_{f,j,T}(V_1 + V_2)}{Z_{i,j,T}RT}\right] \tag{4.3}$$

where $P_{i,j,T}$ and $P_{f,j,T}$ are the starting and final pressures at point j on the adsorption isotherm, and $Z_{i,j,T}$, $Z_{f,j,T}$ are the compression rates of hydrogen at $P_{i,j,T}$ and $P_{f,j,T}$, respectively. T is the measured temperature.

In Eqs. (4.2) and (4.3), V_2 is the volume of the sample reaction chamber, which is the space not occupied by the sample or other contaminants. Accurately measuring the volume of the sample reaction chamber is critical because the mass and volume of the loaded sample are not identical. V_2 needs to be re-measured before each test, which can be determined using either the "direct" or "indirect" volumetric methods proposed by Rouquerol et al. [2]. Direct measurements are performed by using a non-interaction gas, such as helium, determining its volume, V_2. For the direct measurement method, V_2 equals to the dead volume of the sample reactor in Eq. (4.3), which can be determined via a helium calibration test with an activated sample in the reactor. The volume calibration process of the reaction chamber in the "Direct method" is as follows: ① Open valves B and C before calibration to evacuate the system until the pressure is less than 5.0×10^{-3} Pa; ② After the temperature stabilizes in the sample chamber, close valve B, and open valves A and C to fill the sample chamber with N_2 until the pressure (P_1) is between 0.5 and 1.0 MPa. ③ Close valves A and C, and open valve B to evacuate the system again until the pressure is less than 5.0×10^{-3} Pa; ④ Close valve B, open valve C, and record the stabilized pressure P_2. The sample volume V_{sample} can be calculated using the following equation:

$$P_1\left(\frac{V_{sam}}{T_{sam}}+\frac{V_{con}}{T_{con}}\right) = P_2\left(\frac{V_{sys}}{T_{sys}}+\frac{V_{sam}}{T_{sam}}+\frac{V_{con}}{T_{con}}\right) \tag{4.4}$$

where V_{sys}, V_{sam} and V_{con} represent the system volume, sample chamber volume, and pipe volume, respectively. Correspondingly, T_{sys}, T_{sam}, T_{con} represent the temperatures of the system, sample chamber, and tube, respectively.

The indirect measurement method is typically performed by subtracting the estimated volume of the sample from the measured volume of the empty reactor. The expression for the indirect measurement method is:

$$n_m = \sum_{j=1}^{m}\left[\frac{P_{f,j+1,T}\left(V_{cell}-\frac{m_s}{\rho_s}\right)}{Z_{f,j-1,T}RT}+\frac{P_{i,j,T}V_1}{Z_{i,j,T}RT}-\frac{P_{f,j,T}\left(V_1+\left(V_{cell}-\frac{m_s}{\rho_s}\right)\right)}{Z_{f,j,T}RT}\right] \tag{4.5}$$

where V_{cell} is the volume of the empty reaction chamber, and m_s and ρ_s are the mass and density of the sample, respectively.

Equation (4.5) is applicable only when the temperature of the sample reaction chamber and the entire test system is the same. In practice, however, a temperature gradient may exist between the reaction chamber, gas delivery piping, and other parts of the systems. Therefore, a coefficient ($f < 1$) is typically used in calculations, assuming a clear temperature boundary exists in the system. Further details are beyond the scope of this chapter.

The pressure method can be used to measure the activation performance, hydrogen absorption/desorption capacity, thermodynamic (P–C–T) properties of hydrogen absorption/desorption, and hydrogen absorption/desorption kinetics of Mg-based hydrogen storage materials.

(1) Activation properties

Activation treatment is an essential procedure for measuring the thermodynamics and kinetics of hydrogenation and dehydrogenation in Mg-based hydrogen storage materials. It directly affects the accuracy of the experimental results, specifically the hydrogen storage properties. The activation procedures and mechanism involve placing Mg-based powder samples in a high-temperature and high-pressure hydrogen environment. The powders are exposed fresh metal surfaces in this environment, allowing hydrogen molecules to dissociate and promoting hydrogen absorption. When the hydrogen absorption amounts no longer changes over time, the samples are depressurized and vacuumed to fully desorb hydrogen. These hydrogenation and dehydrogenation processes constitute one cycle of the activation process. More cycles gradually enhance the hydrogen absorption and desorption capacity of the material until reaches the maximum.

According to the schematic illustration of the Sievert device shown in Fig. 4.1, the specific steps for the activation process are as follows:

① Evacuate the sample chamber at a specific temperature (T_{sam}).
② Close valves B and C, and keep the sample chamber temperature constant.
③ Open valve A, and purge high-pressure hydrogen gas until the pressure reaches P_{sys}.
④ Open valve C to allow hydrogen into the sample chamber until the pressure stabilizes; record this pressure as P_{eq}.
⑤ Use Eq. (4.2) to calculate the hydrogen absorption capacity of the sample for this activation cycle.
⑥ Repeat steps ①–⑤ for 3–5 times. Generally, the sample is considered fully activated when the hydrogen absorption reaches 90% of the theoretical value or when the absorption and desorption curves nearly overlap.

The essence of the activation performance test is to evaluate the hydrogen absorption/desorption behavior of Mg-based hydrogen storage materials under high temperatures and high hydrogen pressures. The resulting curves show the relationship between the hydrogen content absorbed and time. The activation performance of $Mg_{10}Ni$ and catalyst-doped $Mg_{10}Ni$ are shown in Fig. 4.2. The activation of the $Mg_{10}Ni$ alloy was completed after three cycles of hydrogen absorption/desorption. The hydrogen capacities of $Mg_{10}Ni$ for the first two cycles were 1.51 wt.% and 5.48 wt.%, respectively, and reached 5.75 wt.% in the third cycle, which is close to the maximum value [3].

(2) Hydrogen absorption/desorption capacity tests and thermodynamic analysis

The thermodynamic properties of hydrogen absorption and desorption in hydrogen storage materials can be expressed via the pressure-composition-temperature (P–C-T) curves, also known as the binary phase diagram of metal and hydrogen. It represents the relationship between hydrogen pressure and alloy composition at a given temperature. The abscissa of the P–C–T curve represents the atomic ratio of hydrogen to metal or the hydrogen capacity in the solid phase, while the ordinate represents the hydrogen pressure. Based on the P–C–T curves, the reversible hydrogen storage capacity, hydrogen absorption/desorption plateau pressure, plateau width and slope, and hysteresis effect of Mg-based hydrogen storage materials can be determined. Additionally, the formation and decomposition enthalpy/entropy of hydrides can be obtained by plotting the natural logarithm of the plateau pressure against the reciprocal of the temperature, a graph known as the van't Hoff diagram.

The automatic P–C–T testing system, based on Sievert's principle, can automatically conduct data acquisition (Fig. 4.1) and calculate the hydrogen absorption and desorption amounts at any temperature and hydrogen pressure. The principle of the P–C–T test typically involves measuring the hydrogen absorption and desorption amount (W_1) under isothermal and constant volume conditions by varying the step hydrogen pressure (ΔP_1) from a vacuum or an initial hydrogen pressure of approximately 100 Pa to setting pressure. The specific operation for testing the hydrogen absorption P–C–T curve is as follows:

First, open valves B and C and vacuum chambers V_1 and V_2 chamber. After fully vacuuming, close valves B and C and open valve A to purge hydrogen into

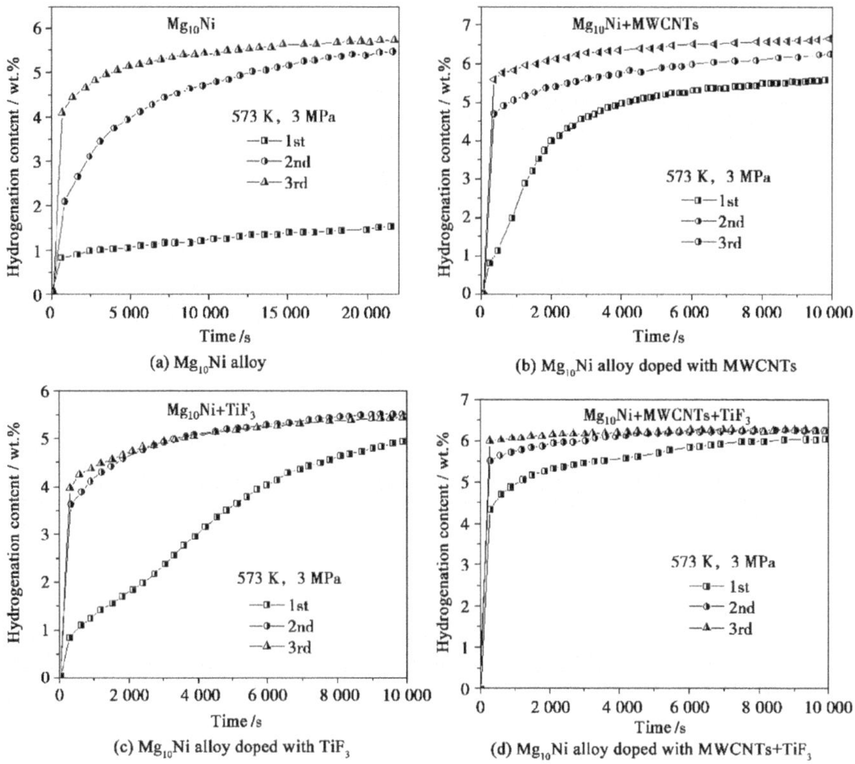

Fig. 4.2 Activation curves of $Mg_{10}Ni$ alloy and the alloys doped with MWCNTs, TiF_3 and MWCNTs + TiF_3, respectively [3]

V_1 until the pressure reaches ΔP_1. Then, close valve A and open valve C again to feed hydrogen into V_2. The system automatically corrects the volume change of V_2 caused by opening and closing valve C. If the final measured equilibrium pressure is ΔP_{f_1}, the mole fraction of hydrogen absorbed is:

$$\Delta n_1 = \frac{\Delta P_1 \mathrm{V}_1}{\mathrm{Z}_{\mathrm{i,T}}\mathrm{RT}} - \frac{\Delta P_{\mathrm{f}_1}(\mathrm{V}_1 + \mathrm{V}_2)}{\mathrm{Z}_{\mathrm{f,T}}\mathrm{RT}} \tag{4.6}$$

and the hydrogen content per unit mass of metal (ΔW_1, wt.%) can be obtained:

$$\Delta W_1 = \frac{2\Delta n_1}{m} \times 100\% \tag{4.7}$$

where m is the mass of the hydrogen storage material.

Following the principle mentioned above, the pressure step size is ΔP_1. After n steps of hydrogen absorption, the new equilibrium pressure is ΔP_{f_n} and the amount of hydrogen involved in the reaction is Δn. By combining the hydrogen absorption and desorption amounts of ΔW_{n} of the alloy, the hydrogen absorption P–C–T curves can be obtained by plotting ΔW_{n} against equilibrium pressure ΔPf_{n} on a coordinate system with hydrogen content (abscissa) and pressure (ordinate).

The principle and procedures of the P–C–T tests for hydrogen desorption are similar to those of thermodynamic absorption but start with a high initial pressure and proceed to a low final pressure. If the thermodynamic desorption experiments are carried out directly after the thermodynamic absorption measurements, the hydrogen remaining in the sample chamber results in high hydrogen pressure. By releasing hydrogen with a step pressure of ΔP, the new equilibrium pressure will be obtained. The P–C–T desorption curve can then be obtained by repeating the process of decreasing hydrogen pressure of ΔP.

When testing the thermodynamics of hydrogen absorption/desorption of the sample, a smaller ΔP should be selected if the plateau pressure is not known, which provides a smooth curve to provide more accurate plateau pressure data.

Figure 4.3 provides the P–C–T curves of $Mg_{90}Y_{1.5}Ce_{1.5}Ni_7$ alloy (Fig. 4.3a). The van't Hoff diagram, obtained from fitting the P–C–T curves of different Mg alloys, as shown in Fig. 4.3b. The abscissa of the P–C–T curve is the hydrogen storage capacity, representing the reversible hydrogen storage capacity of the material, while the ordinate is the hydrogen pressure, providing information on the absorption and desorption plateau pressures at different temperatures and the hysteresis effect. In Fig. 4.3a, it is evident that $Mg_{90}Y_{1.5}Ce_{1.5}Ni_7$ alloy has two plateau pressures: the lower plateau pressure corresponds to the Mg/MgH_2 equilibrium, and the higher plateau pressure corresponds to the Mg_2Ni/Mg_2NiH_4 equilibrium. Moreover, the $Mg_{90}Y_{1.5}Ce_{1.5}Ni_7$ alloy has a wide and flat pressure plateau, with an almost reversible hydrogen absorption and desorption plateau pressure curve, indicating minimal hysteresis. Figure 4.3b is the van't Hoff plot based on the natural logarithm of the plateau pressures against the reciprocal of the temperature at three different temperatures. The reaction enthalpy can be calculated from the slope of the fitting line, and the reaction entropy is the intercept of the fitting line with the ordinate. The hydrogen absorption and desorption plateaus and the enthalpy/entropy changes of the $Mg_{90}Y_{1.5}Ce_{1.5}Ni_7$ alloy at different temperatures are provided in Table 4.1 [4].

(3) Kinetic properties

The kinetic properties of hydrogen storage materials are crucial parameters that can be determined by measuring the hydrogen absorption/desorption rates. Time (t) is a critical factor in the kinetic studies as it involves examining the relationship between the amount of hydrogen absorption/desorption or the reaction fraction over time under specified temperature and pressure conditions.

The amount of hydrogen absorption/desorption is measured by using the Sievert method mentioned above. By introducing the time parameter (t), the kinetic performance of hydrogen storage materials can be obtained by calculating the relationship between the hydrogen content absorbed or desorbed and time. Note that for the first isothermal hydrogen desorption kinetics test, hydrogen at a certain pressure should be filled into the sample chamber to ensure that the pressure remains above the hydrogen

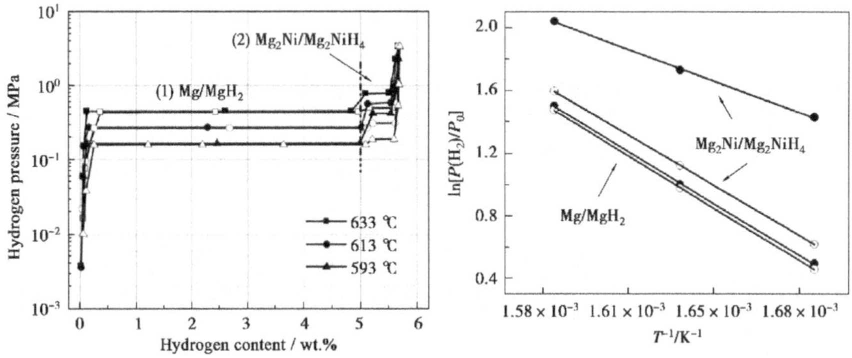

(a) Hydrogen absorption/desorption P-C-T curves of $Mg_{90}Y_{1.5}Ce_{1.5}Ni_7$ sample at different temperatures

(b) Van't Hoff curves of different phases

Fig. 4.3 Hydrogen absorption/desorption P–C–T curves and van't Hoff curves of $Mg_{90}Y_{1.5}Ce_{1.5}Ni_7$ samples

Table 4.1 Enthalpy and entropy changes of hydrogen absorption/desorption and plateau pressure of $Mg_{90}Y_{1.5}Ce_{1.5}Ni_7$ alloy at different temperatures

$Mg_{90}Y_{1.5}Ce_{1.5}Ni_7$ alloy		Platform pressure/MPa			ΔH ($kJ \cdot mol^{-1}$ H_2)	ΔS ($J \cdot K^{-1} \cdot mol^{-1}$ H_2)
		633 K	613 K	593 K		
Mg/MgH_2	Absorption	0.4535	0.2741	0.1650	−78.9	−137.0
	Desorption	0.4407	0.2678	0.1593	79.4	137.6
Mg_2Ni/Mg_2NiH_4	Absorption	0.7780	0.5692	0.4207	−47.9	−92.7
	Desorption	0.4978	0.3091	0.1874	76.2	133.7

desorption equilibrium pressure at the set temperature during the heating process, preventing partial or complete hydrogen desorption before the measurement.

The hydrogen desorption process of Mg-based hydrogen storage materials is generally governed by the nucleation and growth mechanism. Given the low diffusion energy barrier of hydrogen atoms in the Mg matrix, the activation energy E for hydrogen absorption/desorption reflects the energy required for the entire process. The activation energy for hydrogen absorption and desorption can be calculated using the Johnson–Mehl–Avrami (JMA) model based on the hydrogen absorption and desorption kinetic curves. The specific equation is as follows:

$$\ln[-\ln(1-\alpha)] = \eta \ln k + \eta \ln t \tag{4.8}$$

where α is the reaction ratio, t is the reaction time, k is the rate constant, and η is the Avrami index.

From the intercept and slope of the fitted curves of $\ln[-\ln(1-\alpha)]$ versus $\ln t$ at different temperature ranges, $\ln k$ can be obtained. According to the Arrhenius equation, the activation energy for the hydrogen absorption and desorption can then be calculated:

$$\ln k = -\frac{\Delta E_{abs/des}}{RT} + \ln A \tag{4.9}$$

where k is the rate constant, $\Delta E_{abs/des}$ is the activation energy, R is the gas constant (8.314 J/(mol·K)), T is the temperature and A is the frequency factor.

Figure 4.4 illustrates the hydrogen absorption/desorption kinetics of Ni-catalyzed MgH_2 synthesized via nano-confinement method. The results show that MHGH-75 can absorb 5.7 wt.% hydrogen within 10 min and desorb 5.4 wt.% within 2 h at 473 K. However, under identical conditions, MgH_2 does not exhibit any hydrogen absorption or desorption behavior. At 323 K, Ni-MHGH-75 can absorb 2.3 wt.% hydrogen within 60 min, and the calculated activation energies for hydrogen absorption/desorption are 22.7 kJ/mol H_2 and 64.7 kJ/mol H_2, respectively, which are significantly lower than those of MgH_2 (99.0 kJ/mol H_2 for absorption and 158.5 kJ/mol H_2 for desorption). Figure 4.5 presents the relationship between $\ln[-\ln(1-\alpha)]$ and $\ln t$, as well as the relationship between $\ln k$ and $1000/T$ for Mg-5Ni-15La and Mg-15Ni-5La alloys at different temperatures. By linearly fitting the data (symbol) in Fig. 4.5c, the hydrogen desorption activation energies for Mg-5Ni-15La and Mg-15Ni-5La alloys are determined to be 118.6 and 94.0 kJ/mol H_2, respectively, based on the slope of the fitting line.

(4) Cycle life

In the practical application of hydrogen storage materials, cyclic stability is a critical property alongside appropriate equilibrium pressures, excellent kinetics, and proper temperatures of hydrogen absorption and desorption. The testing method for cyclic performance is generally similar to that used for kinetic measurements. The tested materials undergo repeated hydrogen absorption and desorption processes in a constant-volume reactor. By observing the changes of hydrogen pressure in the reactor during these processes, the amount of hydrogen adsorbed or desorbed can be determined using the gas equation under these conditions. Consequently, the reaction rates and hydrogen absorption/desorption capacities of the hydrogen storage materials under different cycling conditions can be evaluated. Based on these reaction rates and capacities after different cycles, the cycling performance and lifespan can be assessed. Currently, there is no standard method for evaluating cycling stability, but weight capacity loss or the percentage of maximum reversible capacity after a certain number of cycles can effectively indicate material degradation. The U.S. Department of Energy (DOE) has set a stability target, stating that hydrogen storage capacity loss should less than 40% after 1500 cycles between full and 1/4 full with 99.99% H_2 [5].

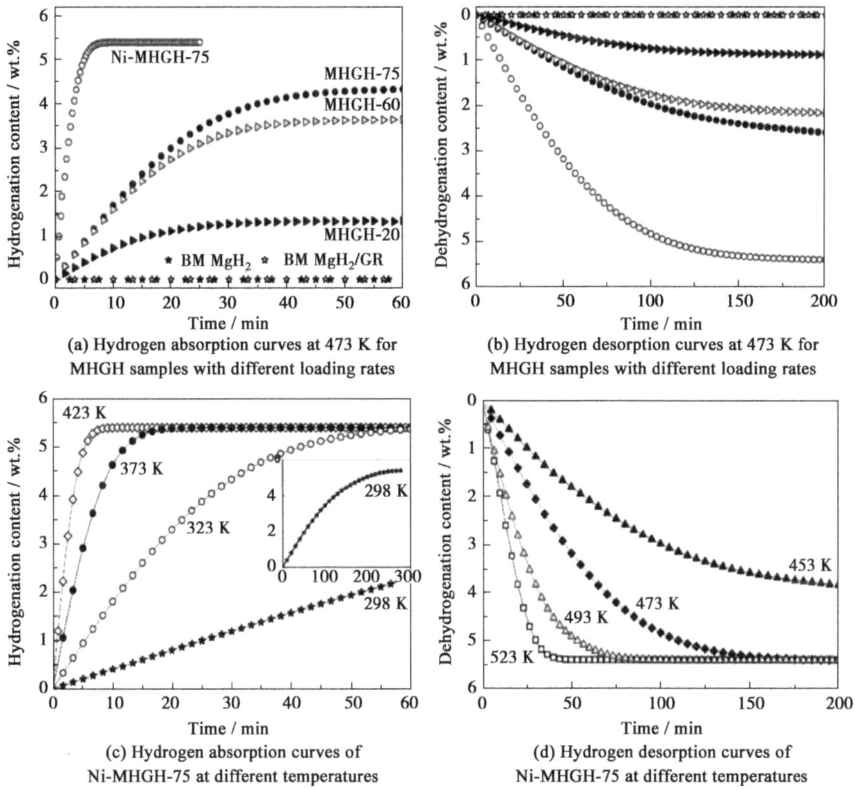

Fig. 4.4 Hydrogen absorption and desorption kinetic curves of MHGH samples

Figure 4.6 presents the relationship between hydrogen storage capacity and kinetics of bulk nanostructured ZK60 alloy at 623 K with the number of cycles and time. The nanostructured ZK60 alloy exhibits excellent cyclic stability, retaining 75% of its maximum hydrogen storage capacity (6.6 wt.% H_2) after 1,000 cycles, with the hydrogen absorption/desorption process completed within 10 min [6]. Figure 4.7 illustrates the varied hydrogen absorption capacity of Mg-Ni-Nd alloys with the number of cycles. It shows that the hydrogen absorption capacity of the $Nd_4Mg_{80}Ni_8$ alloy decreases from 4.773 wt.% to 3.757 wt.% after 819 cycles, while that of the $Nd_{16}Mg_{96}Ni_{12}$ alloy declines from 3.857 wt.% to 3.095 wt.% after 502 cycles [7].

4.2.2 Gravimetric Method

The gravimetric method is conducted by directly detecting the sample's weight changes under different hydrogen pressures. It was primarily developed for gas adsorption tests of solid-state materials. With advancements in various research

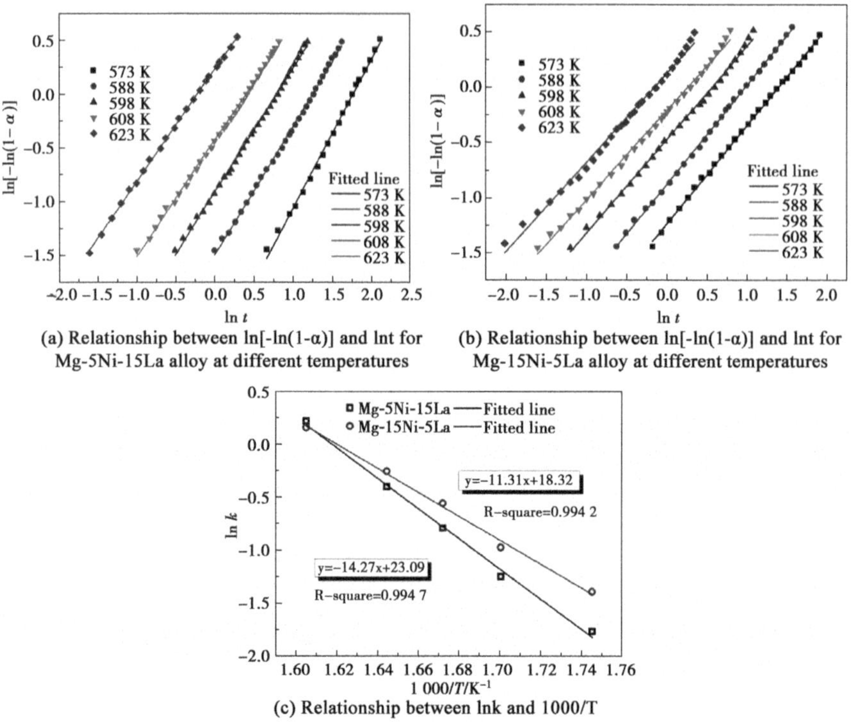

Fig. 4.5 The relationship between reaction fraction-time and rate constant-time for Mg-5Ni-15La and Mg-15Ni-5La alloys

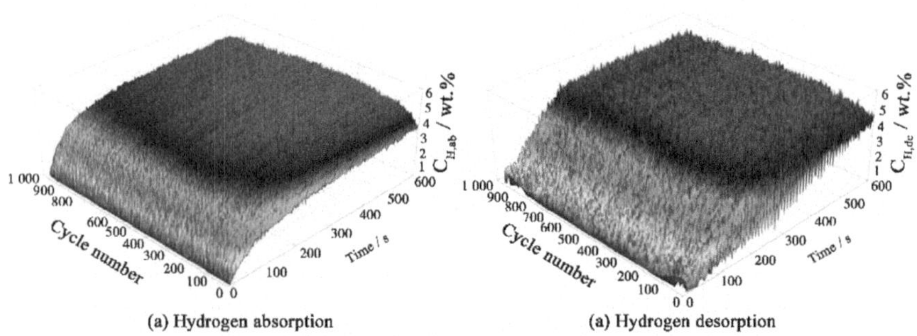

Fig. 4.6 Hydrogen storage capacity and kinetics of bulk nanostructured ZK60 alloy at 623 K with the number of cycles and time

fields, this method has been adapted for various gas absorption tests and is widely accepted for hydrogen adsorption/desorption measurements.

The hydrogen absorption and desorption process of Mg-based hydrogen storage materials involves weight change. Therefore, a high-pressure thermobalance can be

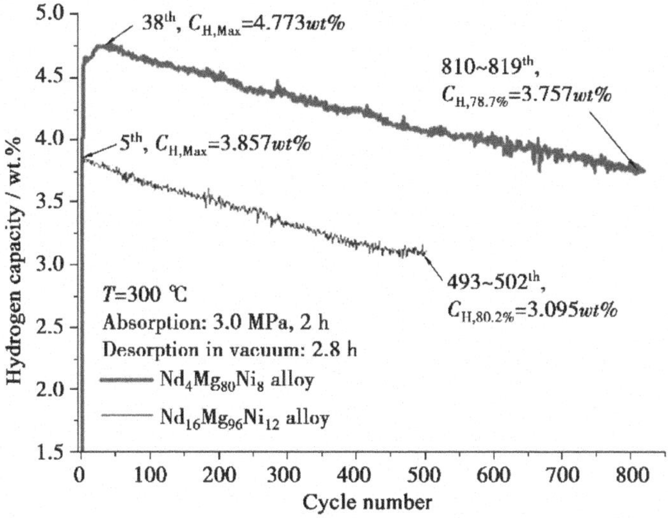

Fig. 4.7 Cycle life of $Nd_4Mg_{80}Ni_8$ and $Nd_{16}Mg_{96}Ni_{12}$ alloys

used to measure the sample's weight change over time at different temperatures and hydrogen pressures. This allows for determining the hydrogen content in the sample, and ultimately, the weight variation is obtained with hydrogen absorption and desorption. For testing hydrogen storage alloys, the experimental pressure is crucial, especially when experiments are conducted under high hydrogen pressure. It is essential to eliminate the side effects of buoyancy and convection in high-pressure hydrogen to accurately measure weight changes during gravimetric tests. High-pressure differential thermobalances with a precision of ± 0.1 μg are typically employed for this purpose. Differential thermal analysis (DTA) and thermogravimetric analysis (TG) can be simultaneously conducted in hydrogen atmospheres up to 50 MPa, which meets the technical the requirements for hydrogen absorption and desorption tests.

As shown in Fig. 4.8, the instrument for gravimetric tests generally comprises a pressure system, a testing system, a temperature control system, and a data acquisition system. The pressure system includes a vacuum system and a gas supply system. The former is used for sample degassing and systematic vacuuming controlled by Valve B, while the latter controls the hydrogen pressure of the chamber with the balance inside via Valve A. The two valves work in coordination to adjust the hydrogen pressure of the system, and the number of required pressure regulators depends on the range of required hydrogen pressure. The testing system consists of a high-pressure differential thermobalance, a temperature control system, and thermostats, among other components. In this system, the sample is placed in the suspended tray of the microbalance. The temperature control system is used to accurately maintain the sample temperature, while the thermostats regulate the system temperature, ensuring stable balance readings through precise temperature control.

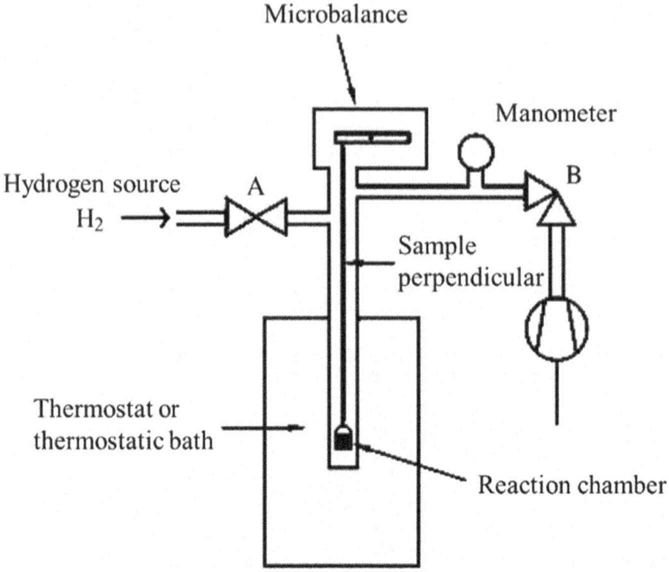

Fig. 4.8 Schematic of the gravimetric hydrogen absorption and desorption testing system

The gravimetric method can be used to determine the P–C–T curve of hydrogen storage materials. The test steps are as follows: the sample is first placed on the microbalance tray. After that, Valve B is opened to vacuum the microbalance chamber, and the tare weight is read on the microbalance. Then, Valve B is closed, and Valve A is opened to purge hydrogen in until the set pressure is reached. Next, the electric furnace is turned on, and the temperature control system maintains a constant temperature. When the sample weight approaches equilibrium, using a dry or blank sample as a reference, the buoyancy effect (Archimedes' principle) is applied for correction. The amount of hydrogen absorbed is obtained by measuring the sample weight, and the P–C–T curve can be derived by measuring the sample weight at various temperatures [8]. Furthermore, the accumulative error can be minimized by calibrating at every test point with the reference samples.

The gravimetric method can also be used to record weight variation over time during the hydrogen absorption and desorption process using a high-precision high-pressure thermobalance, allowing the calculation of the kinetic reaction rate. The most critical parameter in kinetic testing of hydrogen absorption and desorption using the gravimetric method is the hydrogen absorption and desorption content at equilibrium. It can be obtained by measuring the force (F_{abs}) acting on the high-pressure thermobalance when the sample reaches the hydrogen absorption equilibrium state:

$$F_{abs} = m_{abs} g \tag{4.10}$$

where m_{abs} represents the weight of hydrogen absorbed by the sample; g is the gravitational acceleration (9.8 m/s).

The actual force must be corrected based on the buoyancy effect of the sample, and the direction of the buoyancy force ($F_{buoyancy}$) is upward. It can be calculated as follows:

$$F_{buoyancy} = m_0 g\left(\frac{\rho_H}{\rho_{sample}}\right) \tag{4.11}$$

where m_0 is the weight of the sample measured under vacuum; ρ_{sample} is the density of the sample; and ρ_H is the density of hydrogen under the set temperature and pressure.

In practical measurements, it is essential to consider the influence of other gases. When using a beam balance, the suspended counterweight mass and chains have a negative effect on the balance readings, whereas, on the sample side, the chains and the sample holder have positive influences. The weight and density of the sample chain ($m_{sample\ pendant\ chain}$ and $\rho_{sample\ pendant\ chain}$), the counterweight chain ($m_{counterweight\ pendant\ chain}$ and $\rho_{counterweight\ pendant\ chain}$), the counterweight mass ($m_{counterweight}$ and $\rho_{counterweight}$), and the sample disk ($m_{sample\ disk}$ and $\rho_{sample\ disk}$.) must all be considered. The buoyancy calibration can be carried out using the following the equation:

$$\begin{aligned} F_{buoyancy} = & \left[m_{samplependantchain} g\left(\frac{\rho_H}{\rho_{samplependantchain}}\right) \right. \\ & \left. + m_{sampledisk} g\left(\frac{\rho_H}{\rho_{sampledisk}}\right) + m_0 g\left(\frac{\rho_H}{\rho_{sample}}\right) \right] \\ & - \left[m_{counterweightpendantchain} g\left(\frac{\rho_H}{\rho_{counterweightpendantchain}}\right) \right. \\ & \left. + m_{counterweight} g\left(\frac{\rho_H}{\rho_{counterweight}}\right) \right] \end{aligned} \tag{4.12}$$

The total weight of hydrogen absorbed by the sample (F_{total}) can be calculated as:

$$F_{total} = F_{absorb} - F_{buoyancy} \tag{4.13}$$

From the analysis above, it can be seen that the gravimetric method indirectly calculates the amount of absorbed and desorbed hydrogen using a high-pressure microbalance. The measurement must consider sample quantity, buoyancy effect correction, and the interference from balance components [2].

When testing the hydrogen storage properties using the gravimetric method, the sample size and quantity must match the measuring range, graduation value, and long-term weighing stability of high-pressure microbalances. Generally, high-pressure microbalances have a resolution of ± 0.1 μg, a long-term resolution of ± 1 μg, and a measuring range of 200 mg. In practice, the applicable sample amount is determined according to the theoretical hydrogen absorption amount of the sample. The resolution of the balance is related to the accuracy of the readings, so the graduation value of the balance must be selected based on the changes in isothermal points when testing the P–C–T curve. The minimal sample mass for common microbalances is typically in the range of tens of milligrams. If the balance's precision is lower, the sample quantity should be increased accordingly to achieve equivalent measurement accuracy.

Buoyancy effect correction involves accurately expressing the hydrogen pressure-density relationship under different temperatures. As shown in Eq. (4.12), for a sample of known weight, a decrease in sample density increases the buoyancy correction value; an increase in pressure raises the density of hydrogen, resulting in a higher buoyancy correction value. In weight measurements, in-situ measurement of the sample weight at the beginning of the experiment ensures high precision. However, when the correction value exceeds a certain threshold, the gravimetric method may become less accurate than the volumetric method. For instance, in low-pressure hydrogen absorption and desorption tests, the threshold value of the correction depends on specific experimental conditions, operating temperature, and sample density. Under any experimental conditions, it is difficult to establish a consistent relationship between temperature and sample density. Multiple factors influencing buoyancy should be considered when using the gravimetric method.

Gravimetric testing is sensitive to measurement errors caused by the balance disturbances. External influence can be minimized by eliminating vibrations and properly securing the balance. Additionally, the thermal instability of the outer frame of the balance, where the temperature of the instrument's components differs from the rest of the system (known as cold spots or hot spots), can result in pressure error. For example, cold spots cause a pressure drop in the system, which can be mistaken for sample adsorption, and vice versa. Furthermore, two other factors that can disturb the balance are related to the temperature and pressure: the heat dissipation effect at low pressure and the convection of hydrogen at high pressure. These two problems occur when there is a noticeable temperature/pressure gradient between the sample and the balance chamber.

Both volumetric and gravimetric testing methods have their advantages. From a practical perspective, the volumetric method is much easier to implement. On the other hand, the gravimetric method is more precise and accurate under appropriate pressures due to microbalances' precision and the absence of accumulated testing errors. Generally, volumetric testing is faster than gravimetric testing. Given the effects of compression rates and physical interference from the equipment, special care must be taken when operating microbalance testing.

4.3 Crystal Structure and Microstructure Characterization of Mg-Based Hydrogen Storage Materials

The rapid development of science and technology continuously provides new methods for material analysis, leading to innovative analytical techniques. To date, a variety of new characterization methods and analytical techniques have been applied to the field of Mg-based hydrogen storage materials, significantly improving the accuracy and reliability of the analysis. These achievements play an important role in the exploration and development of novel Mg-based hydrogen storage materials, offering a solid foundation for future innovations. This section will focus on the characterization of the crystal structure and microstructure of Mg-based hydrogen storage materials, highlighting the characterization methods such as X-ray diffraction (XRD), neutron diffraction, scanning electron microscopy (SEM), and transmission electron microscopy (TEM).

4.3.1 Crystal Structure Characterization Techniques

(1) X-ray Diffraction

X-ray diffraction (XRD) is a technique used to study the physical phases and crystal structure of materials based on the diffraction effect of X-rays in crystalline materials. By analyzing the diffraction patterns, the composition and the structure or morphology of the materials can be determined.

X-ray diffraction (XRD) analysis is widely used in the structural and compositional analysis of hydrogen storage materials due to its advantages: non-destructive detection, no contamination, quick measurements, and high accuracy. Conventional XRD testing is typically conducted in air at room temperature and under ambient pressure. Therefore, during sample preparation, it is crucial to be aware of the sample's physical and chemical properties, such as flammability, corrosivity, hygroscopicity and toxicity. Figure 4.9 provides the XRD patterns of the $Mg_{80}Ce_{18}Ni_2$ alloy before and after hydrogen absorption and desorption [9]. After fully hydrogenation, the alloy transforms into a composite of $CeH_{2.73}$-MgH_2-Ni. During the hydrogen desorption process, the in-situ formed Ni can promote the transition of the highly stable $CeH_{2.73}$ to CeH_2, and MgH_2 to Mg, with some MgH_2 remaining.

The application of conventional X-ray diffraction (XRD) is limited in Mg-based hydrogen storage materials due to the high temperatures and pressures required for hydrogen absorption and desorption reactions. However, to systematically investigate the dynamic phase or structural changes of these materials during the hydrogen absorption and desorption process, an in situ X-ray diffraction technique is needed. In situ X-ray diffraction is a recent phase identification technique that employs time-resolved X-ray diffraction. When combined with devices for programmed heating/cooling or varying pressure, the XRD testing under different temperature and gas pressure conditions become feasible. Figure 4.10 shows a schematic illustration of

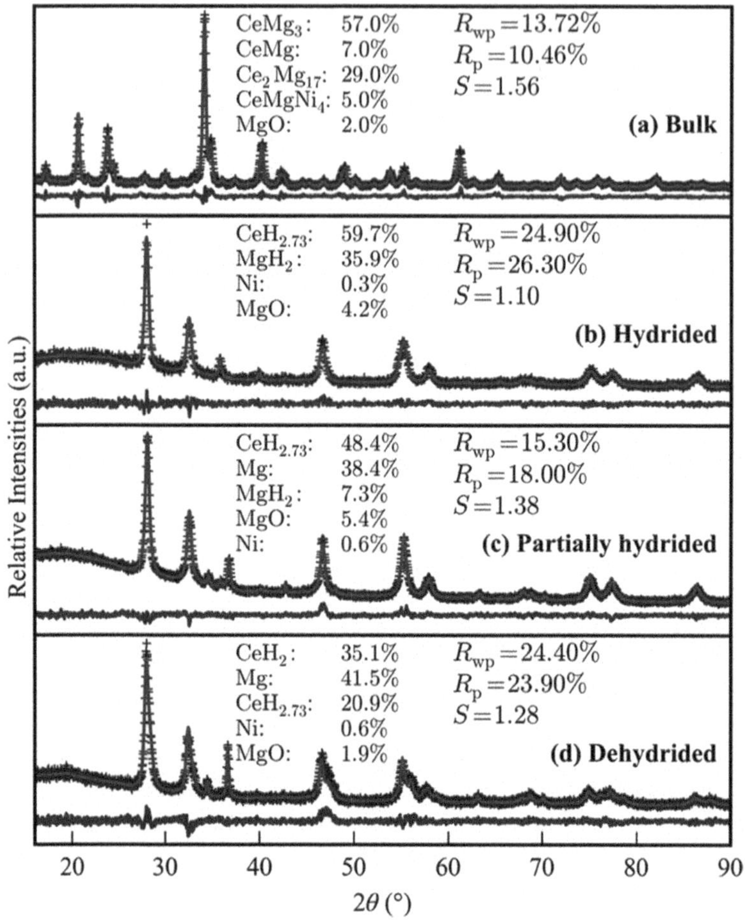

Fig. 4.9 XRD patterns of $Mg_{80}Ce_{18}Ni_2$ alloy before/after hydrogen absorption and desorption [9]

the X-ray diffraction instrument. Given the importance of understanding the different hydrogen absorption and desorption behaviors at various temperatures and pressures for Mg-based hydrogen storage materials, the in-situ X-ray diffraction technique, which can dynamically characterize the physical phases and structural changes over time or with temperature, is widely used in the study of hydrogen storage compounds. Usually, the entire in-situ testing process focuses on the same position of the same material, ensuring that the obtained information (such as cell parameters, peak intensity, or other parameters) is much more accurate than that obtained through ex-situ method.

Figure 4.11 depicts the phase transformation process of as-cast $Mg_{86.1}Ni_{7.2}Y_{6.8}$ alloy under in-situ synchrotron radiation XRD (SR-XRD) characterization at 553 K and 0.05 MPa H_2 [10]. The results indicate that the content of 18R-type LPSO structural phase (denoted as 18R) decreases from approximately 90 wt.% to around

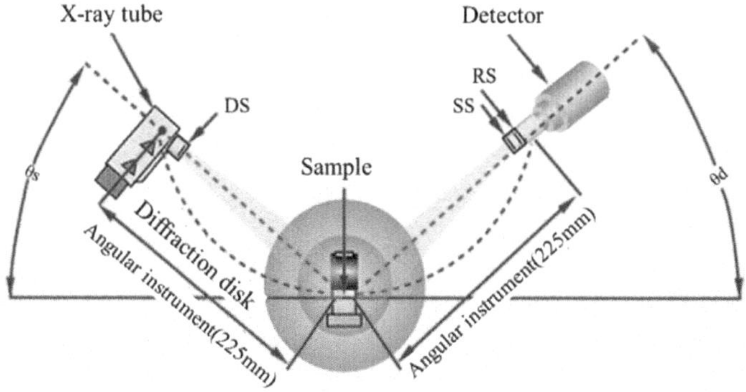

Fig. 4.10 Schematic of the in-situ X-ray diffraction instrument

10 wt.% within 90 min, while the contents of Mg_2Ni and YH_2 increase by about 10 wt.% and 15 wt.%, respectively. The content of the hcp-Mg phase initially rises to around18 wt.% within 10 min, and then falls to approximately 6 wt.% due to the conversion to MgH_2. The content of MgH_2 increases continuously to about 50 wt.% within 90 min, but the content of $MgNi_4Y$ (~4 wt.%) remains constant throughout the process, indicating that it does not decompose. The content of Mg_2NiH_4 reaches 3 wt.% within 90 min. Overall, the phase transformation process of the 18R structure and the generation of hydrogen decomposition products and their content can be monitored in real-time through SR-XRD analysis.

(2) X-ray absorption spectroscopy (XAS)

Synchrotron radiation is generated by high-energy particles moving at the near-light speed ($v \approx c$) in a circular magnetic field accelerator. The radiated photon was initially observed in a synchrotron, hence as the term "synchrotron radiation". The schematic illustration of synchrotron radiation apparatus is shown in Fig. 4.12. Synchrotron radiation offers several advantages, including high radiation intensity, high polarization, and well-collimated beam, making it suitable for studying electronic states, coordination, and other properties of Mg-based hydrogen storage materials.

In-situ synchrotron radiation technique can dynamically collect the data on microstructural changes during the hydrogen absorption and desorption process. Furthermore, it enables the monitoring of rapid changes in the samples during hydrogen absorption and desorption, which is crucial for the development of advanced hydrogen storage materials. Using in-situ XAS, researchers analyzed the evolution of Nb in the high-energy ball-milled MgH_2/Nb_2O_5 system during the hydrogen cycling process [11]. The results showed that Nb_2O_5 was partially reduced during the ball-milling process, and further reduction occurred during the heating and cycling processes, eventually reaching the lowest oxide state. Moreover, the reaction of Nb_2O_5 with Mg/MgH_2 leads to the formation of a new ternary oxide phase, Mg_xNb_yO. The Nb oxidation–reduction reaction was observed repeatedly during the cycling process, indicating that hydrogen can diffuse along the ternary

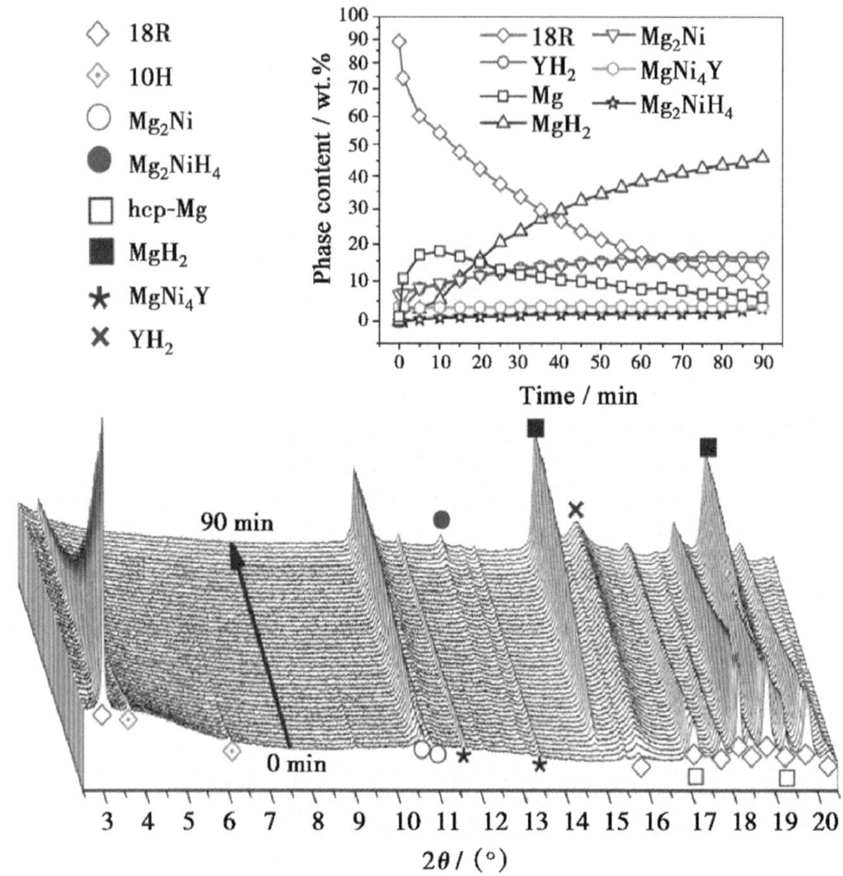

Fig. 4.11 SR-XRD patterns of as-cast $Mg_{86.1}Ni_{7.2}Y_{6.8}$ alloy at 553 K under 0.05 MPa H_2 [10]

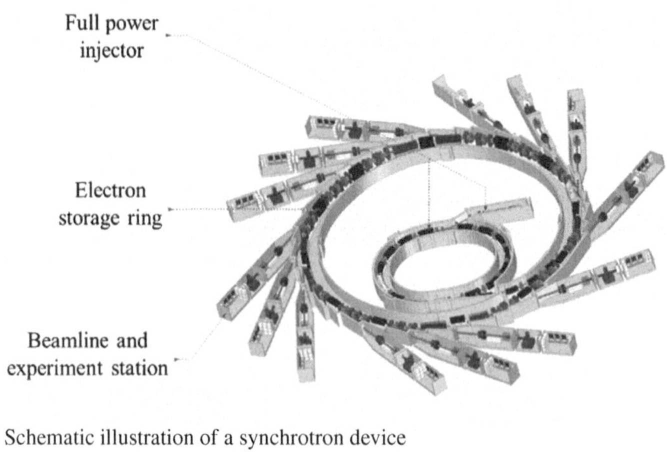

Fig. 4.12 Schematic illustration of a synchrotron device

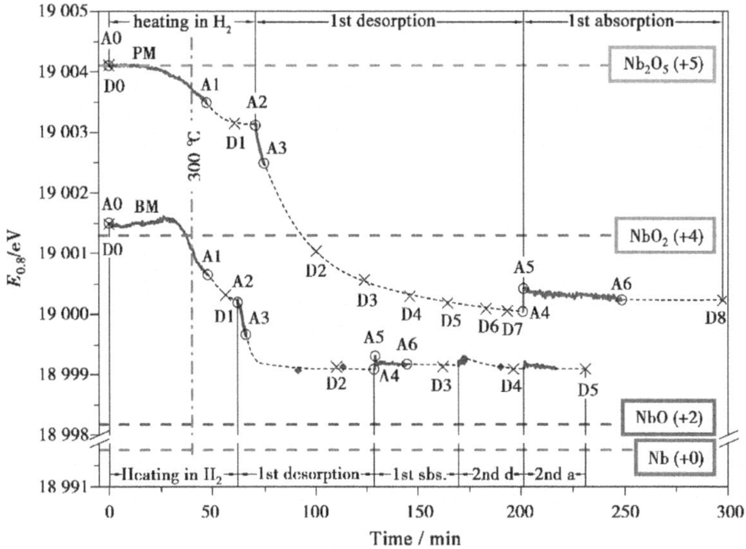

Fig. 4.13 The comparison of normalized K-edge absorption of Nb samples with Nb standards (the number in brackets is the oxidation states of Nb). D is the diffraction points, the selected absorption spectrum points (labeled with A and circles). The dashed line is the predicted energy transfer for the two samples [11]

oxides and form the metastable Nb*x*H*y* phase. Hydrogen diffusion along the Mg–Nb–O ternary oxides improves the kinetic performance of hydrogen absorption and desorption process and may reduce the activation energy of the Mg-MgH_2 transition (Fig. 4.13).

Additionally, due to the powerful energy and detection depth of the synchrotron light source, other techniques like XRD, Raman spectroscopy, and Fourier-Transform Infrared Spectroscopy (FT-IR), can be combined with synchrotron radiation for in-situ or ex-situ studies. This approach provides detailed information on the structural changes during hydrogen absorption and desorption in hydrogen storage materials, paving the way to elucidate the reaction mechanism.

(3) Neutron diffraction technique

Neutron diffraction (Fig. 4.14), also known as elastic neutron scattering, is a common characterization method used in crystallography. Neutrons possess many characteristics that complement those of X-rays, making neutron diffraction particularly suitable for investigating the crystal structures of materials containing abundant hydrogen, such as MgH_2. In the neutron scattering process, the energy of the scattered neutron is nearly identical to or slightly lower than that of the energy of the incident neutrons, similar to X-ray diffraction. The main difference lies in their detection depths. Due to the weak energy source, X-ray analysis is more suitable for surface analysis, while synchrotron-generated X-rays, with their higher energy, are better for the sub-surface of bulk or thin film samples. In contrast, neutron diffraction, due to the electric

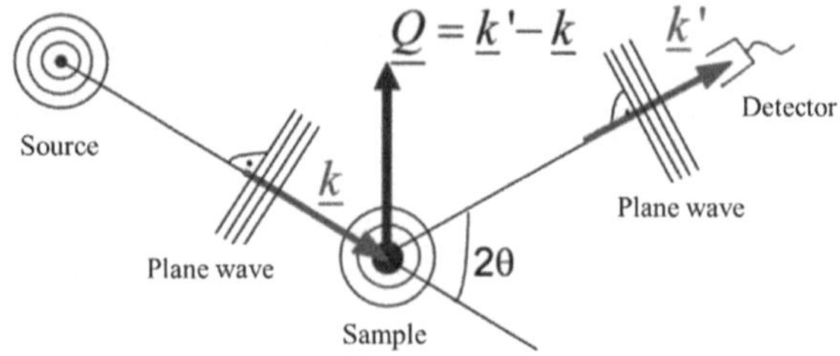

Fig. 4.14 Schematic illustration of the neutron scattering mechanism

neutrality of neutrons, has a much larger detection depth, making it more suitable for analyzing bulk samples.

The neutron diffraction experiments require a neutron source with sufficient flux, meaning an experimental apparatus setup with a nuclear reactor or pulse neutrons source to provide neutrons at a constant wavelength. Although neutron diffraction technique has many advantages over X-ray diffraction, it is not commonly used in laboratory settings, mainly because the neutrons produced by these sources have very high energies, around 1 MeV. In contrast, the laboratory-used neutrons typically have energies in the millielectron volt range. Therefore, the energy of the powerful neutrons generated must be reduced before used in experiments. Compared with other scattering techniques, neutron scattering offers several significant advantages:

① The neutron is electrically neutral, and its interaction with the nuclei of the material is based on the strong nuclear force, making it immune to the sample environment factors such as high pressure, temperature, or magnetic field.
② The magnetic moment of neutrons corresponds to the magnetic variations at the atomic scale, making it suitable for studying magnetic materials.
③ Neutrons do not interface with the detection system and are non-destructive.
④ Neutron scattering originates from interaction of the nuclei of the material, rather than the electron cloud. This means that the atomic scattering power (cross-section) is not affected by the atomic number of the atom. Unlike electrons, whose scattering power increases proportionally with the atomic number, neutron scattering can be used to distinguish light atoms (such as hydrogen) from those with higher atomic numbers.

Since the 1990s, neutron diffraction has been widely used in the investigation of hydrogen storage materials. It is an effective way for studying the relationship between the hydride phase structure and hydrogen storage properties of Mg-based materials, especially in the field of hydride phase structure and hydrogen absorption and desorption properties [12, 13]. The in-situ neutron diffraction technique provides a more accurate reflection of activation processes [14], hydrogen absorption and desorption properties [15], as well as precise prediction of crystal structures [16].

4.3.2 Microstructure Characterization Techniques

The microstructure and morphology of Mg-based hydrogen storage materials directly affect their hydrogen storage performance, such as particle size, surface morphology, catalytic elements, and catalyst distribution. These microstructural features can only be obtained using advanced techniques. To date, the most commonly used characterization techniques for microstructure characterization are scanning electron microscopy (SEM) and transmission electron microscopy (TEM).

(1) Scanning electron microscopy

SEM involves using a finely focused beam of electrons to scan the sample surface, generating secondary or backscattered electrons through interactions between the beam and the sample. By collecting these electrons, SEM provides the information about the surface topography and composition of the samples. The advantage of SEM lies in its ability to directly observe very rough sample surfaces. However, the observation must be carried out in a vacuum environment, which requires special sample preparation, such as ensuring the sample is clean, dry, oil-free, conductive, non-volatile, non-luminous, and weak magnetic. Hydrogen storage materials typically exist in the form of bulk or powders, and the preparation methods for these two forms vary slightly.

For the bulk and conductive samples, no special preparation is required. The sample size should be cut to fit the sample holder, the samples should be mounted on the holder using conductive tape for observation under the SEM. To prevent contamination, the sample should be cleaned with acetone or alcohol before placement and, if necessary, cleaned with sonication. Non-conductive or poorly conductive bulk samples should be coated with conductive membrane or powders to improve image quality and prevent thermal damage caused by the accumulation of electrons.

For the conductive powder samples, conductive tape should be applied to the sample holder first. The powder is then evenly spread on the tape. Unstuck particles are removed with an ear wax removal before observation. Non-conductive or poorly conductive powders also need to be coated with a conductive layer before SEM observation. For efficient testing, several samples can be mounted on one holder. However, when removing the unstuck powders, care should be taken to prevent cross-contamination between samples. Below is a summary for the preparation of powder samples.

① Avoid pressing the sample and keep them in natural states.
② Ensure the powder sample has a uniform thickness, a flat surface, and an appropriate amount (ca. 1 g). Excessive sample amounts may lead to peeling and poor conductive.

The powder morphology and composition of the as-cast $Mg_{24}Y_3$ alloy and ball-milled Mg-Y-C-Ni composites were analyzed by SEM with secondary electron imaging (Fig. 4.15). The morphology of all selected specimens shows irregular shapes, and the particle size of the cast $Mg_{24}Y_3$ powders is approximately 30–50 μm. In

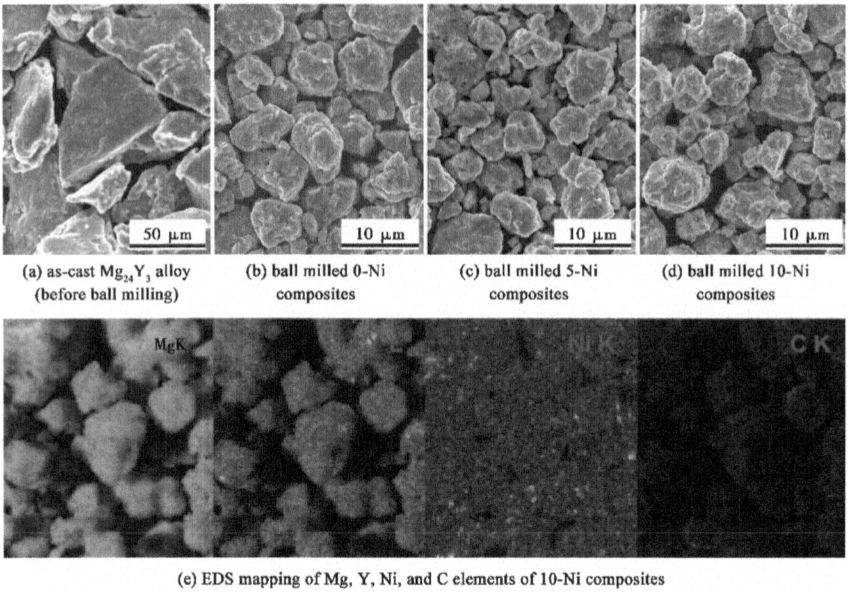

Fig. 4.15 Morphology of $Mg_{24}Y_3$ cast alloy and ball-milled Mg-Y-C-Ni composite powders

contrast, the ball-milled Mg-Y-C-Ni powders have a particle size of about 2–10 μm, exhibiting a significantly different morphology and size distribution compared to the as-cast $Mg_{24}Y_3$. The surface morphology of the Mg-Y-C-Ni composites was relatively rough. EDS mapping shows uniformly distributed elements of Y, Ni, and C in particles of the 10-Ni composites (Fig. 4.15e) [17].

In-situ SEM can characterize the evolution of metal surface morphology and elemental distribution during the hydrogenation and dehydrogenation processes of hydrogen storage materials. Figure 4.16 presents the morphology and structure of MgH_2 before and after heating and hydrogen release with in-situ SEM [18]. The structure and morphology of the particles changed before and after heating. However, like optical microscopy techniques, SEM has a relatively low resolution and cannot obtain detailed information about microstructures, such as crystal structures and defects during the hydrogen storage reaction process. This limitation necessitates the use of another advanced technique, TEM, in hydrogen storage studies.

(2) Transmission electron microscopy

TEM is a technique in which an accelerated electron beam is transmitted through a very thin sample to form an image. The interaction between electrons and atoms in the sample generates scattered electrons at different angles. By collecting these scattered electrons, a light or dark image is obtained and displayed on an imaging device after magnification and focusing. The sample preparation methods for TEM are partially similar to those for SEM, requiring non-toxic non-corrosive properties. Considering that TEM collects electrons transmitted through the sample, and the energy loss

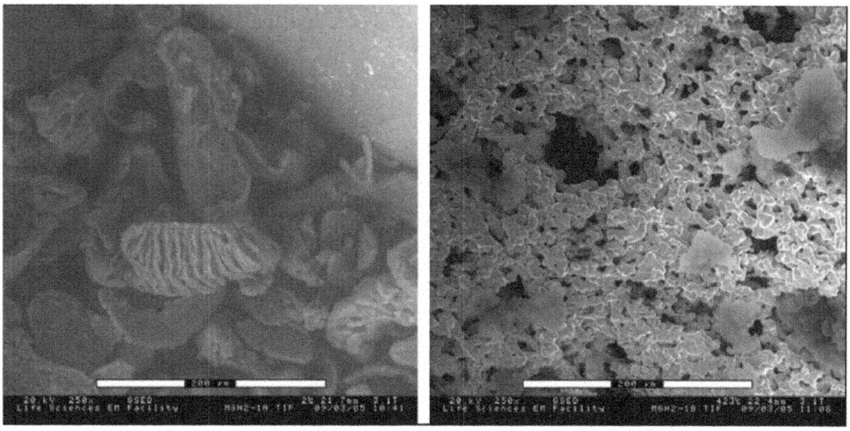

(a) Before heating and hydrogen desorption (b) After heating and hydrogen desorption

Fig. 4.16 SEM images of MgH_2 powder [18]

of electrons on passing through results in a relatively low collected electron beam energy, the sample must be extremely thin, typically around 50–500 nm.

The preparation method for powder materials is relatively simple. A small amount of the powder sample is dispersed into a suspension in a solution (one that does not interact with the powder) using sonication. The suspension is then dropped onto an electron microscope copper mesh covered with a supporting film using a pipette. After the sample is completely try, it is ready for observation. For bulk materials, the samples must be prepared as thin film (mechanical or chemical pre-thinning is required) that allow electron beams to pass through. Two distinct thinning methods are usually used: electrolytic polisher with double spray and ion beam thinning.

Characterizing Mg-based hydrogen storage materials using TEM allows for the observation of the structure and morphology of the samples, as well as determining the elemental and phase composition of the materials. Combined with other characterization tests, the hydrogen storage mechanism of the samples can be further investigated.

The TEM images of metals and Mg-based alloys synthesized by the hydrogen plasma metal reaction (HPMR) are shown in Fig. 4.17. It can be observed that Ni, Cu, Co, Fe and Al particles exhibit a granular structure with an average particle size of 30–50 nm. In contrast, the Mg particles exhibit a hexagonal structure with an average particle size of approximately 300 nm. The TEM images demonstrate that the average particle sizes of 50–200 nm were obtained for the Mg_2Ni, Mg_2Co, and Mg_2Cu, which are smaller than those of the pristine Mg particles [19].

The catalytic effect of Pt nanoparticles on hydrogen storage properties of Mg was investigated collaboratively by Zhou's group and Deng's group [20]. They employed high-angle annular dark field scanning TEM (HAADF-STEM) and energy dispersive X-ray STEM techniques (EDX-STEM) to characterize the morphology and microstructure of the hydrogenated Mg-Pt composites (Fig. 4.18). It can be seen

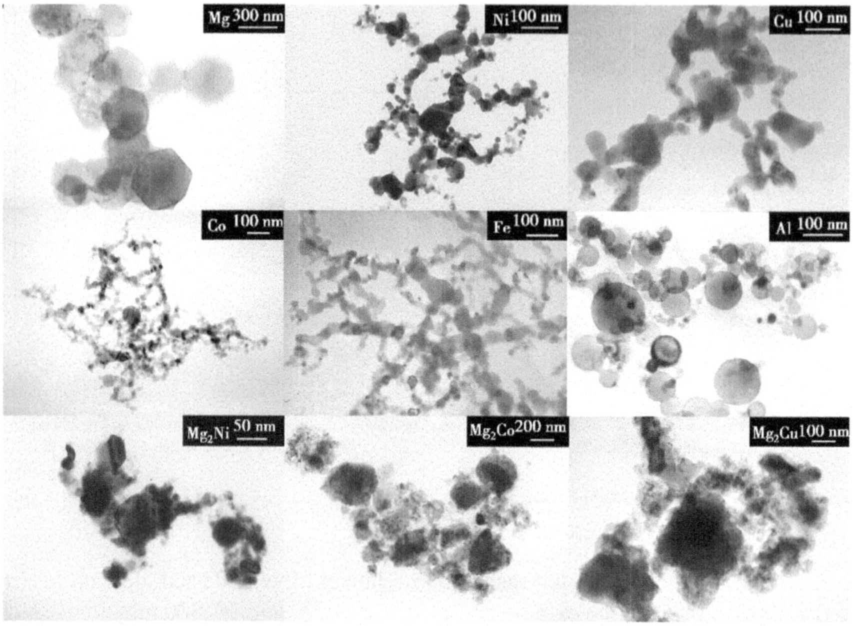

Fig. 4.17 TEM images of Mg, Ni, Cu, Co, Fe, Al metal nanoparticles and Mg_2Ni, Mg_2Co and Mg_2Cu synthesized by hydrogen plasma metal reaction

that the morphology of the Mg particles prepared by the arc plasma method is primarily icosahedral. In addition, the crystal structure of Mg-Pt is observed by HRTEM (Fig. 4.18b and c). Furthermore, the STEM-HAADF image in Fig. 4.18d and the EDX mapping of Pt in Fig. 4.18e demonstrate the good dispersibility of Pt nanoparticles. The selected area electron diffraction (SAED) pattern in Fig. 4.18f confirms that the Mg particles in the Pt-loaded Mg composite material are single crystals.

The phase composition and microstructural evolution of $Nd_4Mg_{80}Ni_8$ were investigated using TEM characterization methods during the hydrogen absorption process, revealing the hydrogen-induced microstructural evolution mechanism [21] (Fig. 4.19). The HR-TEM images show that $Nd_4Mg_{80}Ni_8$ consists of uniform and highly crystallized nanoparticles. When the alloy reacts with hydrogen, $Nd_4Mg_{80}Ni_8$ in-situ generates $NdH_{2.61}$ ultrafine nanoparticles, and $Nd_4Mg_{80}Ni_8$ transforms into $NdH_{2.61}$-Mg-Mg_2Ni nanocomposites. Afterward, Mg reacts with hydrogen to generate MgH_2, and hydrogen further reacts with the generated Mg_2Ni to produce $Mg_2NiH_{0.3}$. The $NdH_{2.61}$-Mg-Mg_2Ni nanocomposites have a high density of $NdH_{2.61}$ nanoparticles, and numerous interfaces and grain boundaries exist between MgH_2 and $Mg_2NiH_{0.3}$.

The traditional TEM can only operate under vacuum and at room temperature. However, in the study of hydrogen storage materials, it is sometimes necessary to analyze the dynamic structural evolution of the hydrogen absorption/desorption

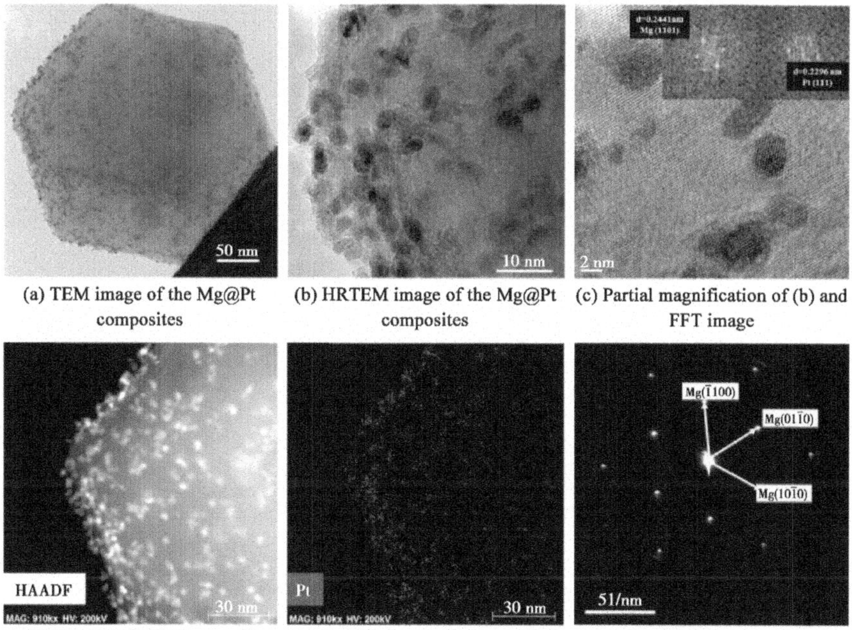

(a) TEM image of the Mg@Pt composites
(b) HRTEM image of the Mg@Pt composites
(c) Partial magnification of (b) and FFT image
(d) HAADF-STEM image of Mg@Pt composites
(e) Pt element distribution corresponding to HADDF-STEM image
(f) SAED image of Mg@Pt composites

Fig. 4.18 TEM images of Mg@Pt composite [20]

process under different temperatures and pressures. To unveil the evolution of morphology, structure, and chemical composition of hydrogen storage materials in real-world hydrogen storage conditions, in-situ TEM has attracted significant interest from researchers.

Based on the structure of TEM, temperature fields can be created by adding a resistive heater near the sample holder, but introducing gases into the TEM operating chamber presents a major challenge. Conventional TEM requires maintaining a high vacuum in the chamber to protect the electron beam gun and minimize scattering from gases, which could result in contamination and low resolution. Currently, there are two main strategies to introduce reactive gases in the TEM chamber: (1) Modification of traditional TEM using a differentially pumped vacuum system of the environmental TEM (Differentially Pumped Environmental TEM, DP-ETEM). As shown in Fig. 4.20, This setup allows the introduction of reaction gases to the sample chamber in two ways: (a) Gases can be introduced through the objective lens chamber directly to the sample chamber, providing uniform gas distribution; (b) The injection method injects gases directly into the sample chamber through the pipeline within the sample holder, ensuring more effective contact between the gases and the samples. The main limitation of commercial ETEM is the low maximum pressure (approximately 10 Pa), which differs significantly from the real hydrogen absorption/

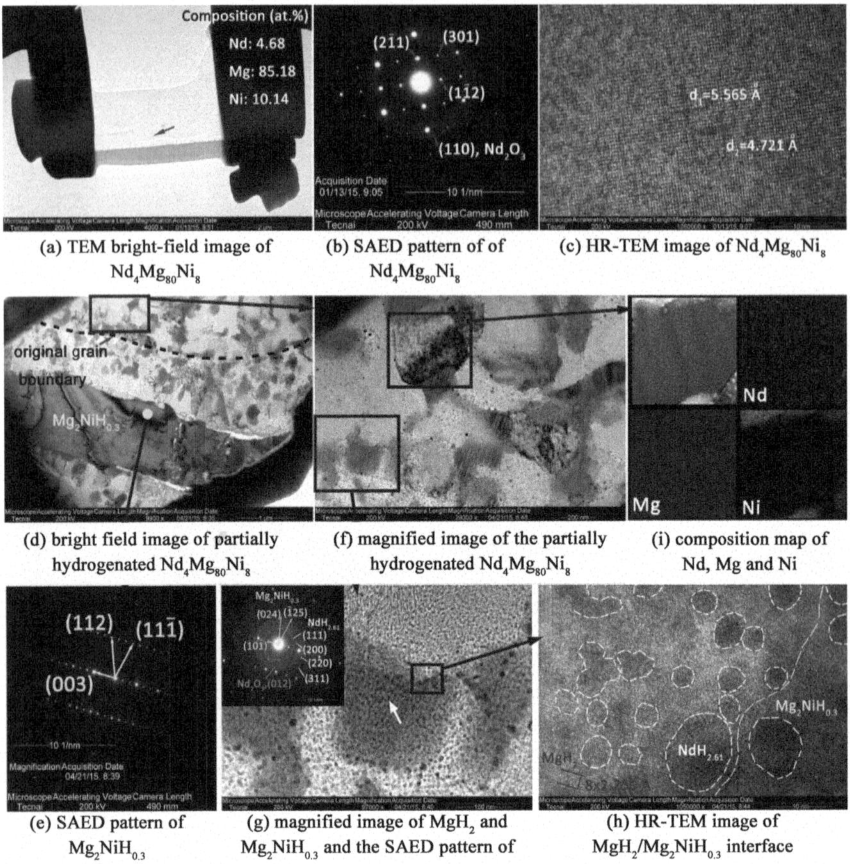

Fig. 4.19 The TEM images of $Nd_4Mg_{80}Ni_8$ before and after hydrogenation

desorption environments, posing a major challenge to studying reaction mechanism under high hydrogen pressure.

The second strategy uses a modified sample holder equipped with two layers of amorphous films (e.g., multi-layer graphene or silicon nitride) with a thickness that is transparent to electron beams. The specimen and reactive gases are encapsulated in the nanoreactor sealed with an O-ring, isolating the reactor from the TEM chamber. Gases are introduced directly into the reactor via a gas pipeline, ensuring that the chamber vacuum remains unaffected (Fig. 4.21).

The MgH_2 sample containing 10 mol% Nb_2O_5 catalyst was analyzed using high voltage transmission electron microscopy (HVEM) [22]. Bright field (BF) images and Fast Fourier Transform (FFT) results of the ball-milled samples at room temperature, 373 K, and 473 K are shown in Fig. 4.22. At room temperature, MgH_2, Mg and NbO were confirmed by the FFT results, but not Nb_2O_5. MgO was not observed at room

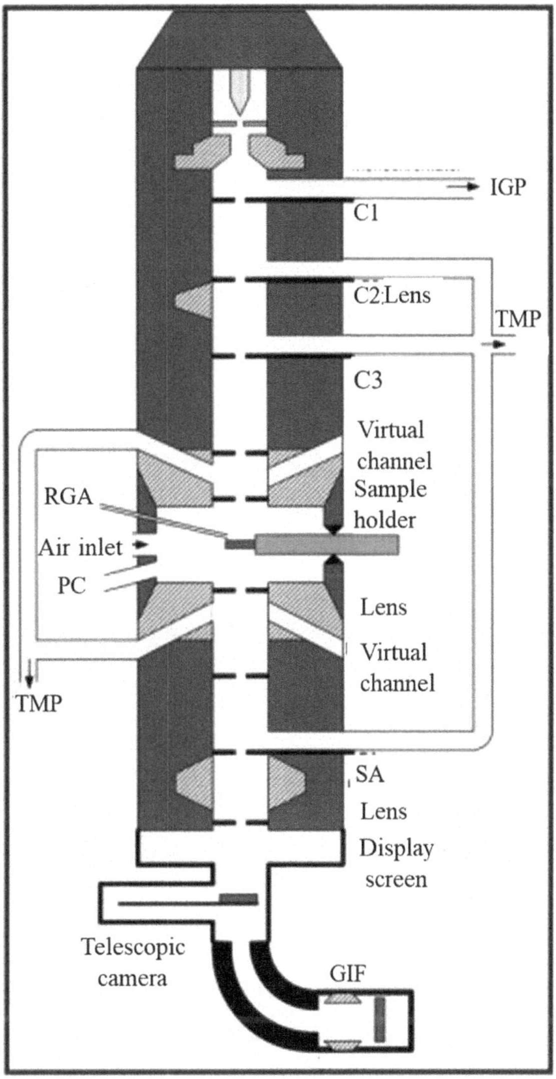

Fig. 4.20 Simplified TEM diagram highlighting the major components (FEG, field emission gun; IGP, ion suction pump; TMP, turbomolecular pump; RGA, residual gas analyzer; PC, plasma cleaning; C1, first condenser aperture; SA, selected area aperture)

temperature but appeared at 373 and 473 K. As the temperature increased, MgH_2 continued to decompose, generating Mg.

To achieve high resolution TEM images, a spherical aberration correction device is added on the concave lens to correct the spherical aberration, resulting in the well-known aberration corrected transmission electron microscope (AC-TEM). AC-TEMs are classified into AC-TEMs (where the spherical aberration corrector is mounted at the objective lens position) and AC-STEMs (where the spherical aberration corrector is mounted at the condenser lens position). Additionally, TEMs with both objective and condenser correctors are called double spherical aberration corrected TEM,

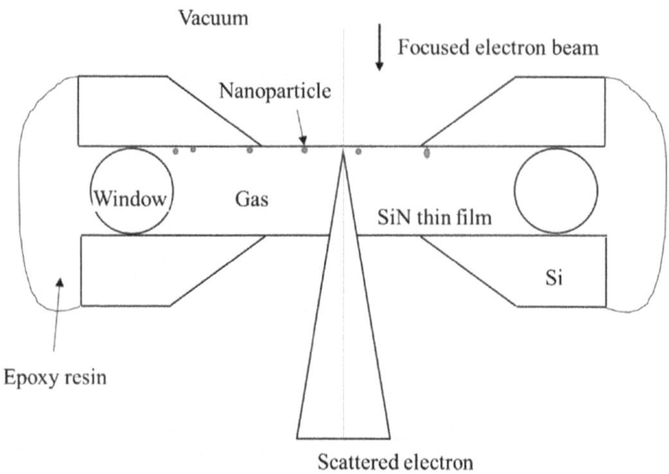

Fig. 4.21 Schematic of the flow system for atmospheric scanning transmission electron microscopy (STEM)

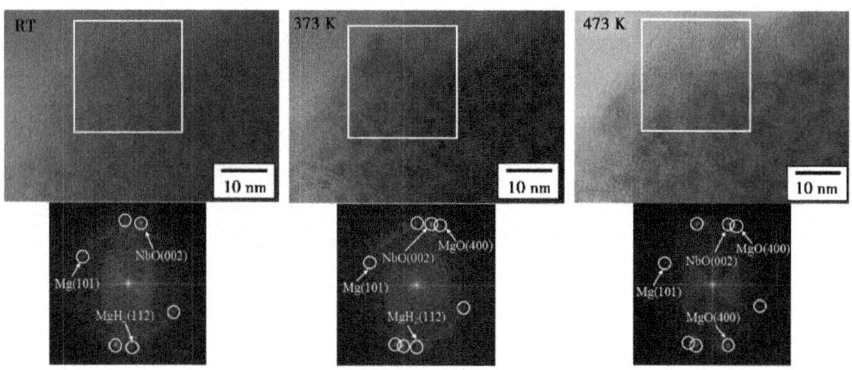

Fig. 4.22 Bright field (BF) images and fast Fourier transform (FFT) results of ball milled samples at room temperature, 373 K and 473 K, respectively [22]

currently the highest resolution TEMs available. The body-centered cubic (BCC) and face-centered cubic (FCC) crystalline surfaces of Mg nanoparticles have been observed using spherical aberration electron microscopy [23], and the arrangement of individual atoms can be clearly seen. Furthermore, sub-nanometer scale and high shear stress Bcc crystal faces, which are not be visible with XRD and traditional TEM, are clearly observed under spherical aberration electron microscopy (Fig. 4.23).

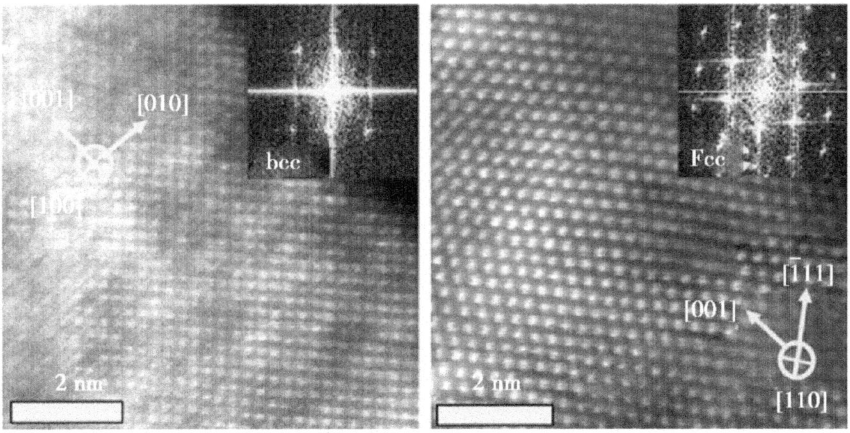

(a) Bcc (b) Fcc

Fig. 4.23 HAADF Cs-corrected TEM images of Mg nanoparticles [23]

References

1. YU X B, WU Z, HUANG T Z, et al. Improvements of activation performance for hydrogen storage alloys [J]. Materials Review, 2004, 18(5): 85–88.
2. DARREN P B. Hydrogen storage materials: the characterisation of their storage properties [M]. Beijing: Machinery Industry Press, 2011.
3. HOU X J, HU R, ZHANG T B, et al. Hydrogenation thermodynamics of melt-spun magnesium rich Mg-Ni nanocrystalline alloys with the addition of multiwalled carbon nanotubes and TiF_3 [J]. Journal of Power Sources, 2016, 306: 437–447.
4. YONG H, WEI X, ZHANG K W, et al. Characterization of microstructure, hydrogen storage kinetics and thermodynamics of ball-milled $Mg_{90}Y_{1.5}Ce_{1.5}Ni_7$ alloy [J]. International Journal of Hydrogen Energy, 2021, 46(34): 17802–17813.
5. KELLER J O, KLEBANOFF L E. Final report for the DOE Metal Hydride Center of Excellence [R]. United States, 2012.
6. KRYSTIAN M, ZEHETBAUER M J, KROPIK H, et al. Hydrogen storage properties of bulk nanostructured ZK60 Mg alloy processed by equal channel angular pressing [J]. Journal of Alloys and Compounds, 2011, 509: S449–S455.
7. LI Q, LUO Q, GU Q F. Insights into the composition exploration of novel hydrogen storage alloys: evaluation of the Mg-Ni-Nd-H phase diagram [J]. Journal of Materials Chemistry A, 2017, 5(8): 3848–3864.
8. JIA Z, WANG Y. Test method for hydrogen absorption of hydrogen storage alloys [J]. Metal functional materials, 2004, (05): 28–31.
9. OUYANG L Z, YANG X S, ZHU M, et al. Enhanced hydrogen storage kinetics and stability by synergistic effects of in situ formed $CeH_{2.73}$ and Ni in $CeH_{2.73}$-MgH_2-Ni nanocomposites [J]. The Journal of Physical Chemistry C, 2014, 118(15): 7808–7820.
10. LI Y, GU Q F, LI Q, et al. In-situ synchrotron X-ray diffraction investigation on hydrogen-induced decomposition of long period stacking ordered structure in Mg-Ni-Y system [J]. Scripta Materialia, 2017, 127: 102–107.
11. FRIEDRICHS O, MARTíNEZ-MARTíNEZ D, GUILERA G, et al. In Situ Energy-Dispersive XAS and XRD Study of the Superior Hydrogen Storage System MgH_2/Nb_2O_5 [J]. The Journal of Physical Chemistry C, 2007, 111(28): 10700–10706.

12. SATO T, MOCHIZUKI T, IKEDA K, et al. Crystal structural investigations for understanding the hydrogen storage properties of $YMgNi_4$-based alloys [J]. Acs Omega, 2020, 5(48): 31192–31198.
13. JARAMILLO D E, JIANG H Z H, EVANS H A, et al. Ambient-temperature hydrogen storage via vanadium (II)-dihydrogen complexation in a metal-organic framework [J]. Journal of the American Chemical Society, 2021, 143(16): 6248–6256.
14. KISI E H, WU E, KEMALI M. In-situ neutron powder diffraction study of annealing activated $LaNi_5$ [J]. Journal of Alloys and Compounds, 2002, 330: 202–207.
15. CUEVAS F, JOUBERT J M, LATROCHE M, et al. In situ neutron-diffraction study of deuterium desorption. from $LaNi_5$+x (x~1) alloy [J]. Applied Physics A, 2002, 74(1): S1175–S1177.
16. KAPELEWSKI M T, RUNCEVSKI T, TARVER J D, et al. Record high hydrogen storage capacity in the metal-organic framework Ni2 (m-dobdc) at near-ambient temperatures [J]. Chemistry of Materials, 2018, 30(22): 8179–8189.
17. YANG T, WANG P, LI Q, et al. Hydrogen absorption and desorption behavior of Ni catalyzed Mg-Y-C-Ni nanocomposites [J]. Energy, 2018, 165: 709–19.
18. BEATTIE S D, SETTHANAN U, MCGRADY G S. Thermal desorption of hydrogen from magnesium hydride (MgH_2): An in situ microscopy study by environmental SEM and TEM [J]. International Journal of Hydrogen Energy, 2011, 36(10): 6014–6021.
19. LUO Q, LI J D, LI B, et al. Kinetics in Mg-based hydrogen storage materials: Enhancement and mechanism [J]. Journal of Magnesium and Alloys, 2019, 7(1): 58–71.
20. LU C, MA Y L, LI F, et al. Visualization of fast "hydrogen pump" in core-shell nanostructured Mg@Pt through hydrogen-stabilized Mg_3Pt [J]. Journal of Materials Chemistry A, 2019, 7(24): 14629–14637.
21. LUO Q, GU Q F, ZHANG J Y, et al. Phase equilibria, crystal structure and hydriding/dehydriding mechanism of $Nd_4Mg_{80}Ni_8$ compound [J]. Scientific Reports, 2015, 5(1): 15385.
22. MORITA E, ONO A, ISOBE S, et al. In-situ TEM observation for dehydrogenation mechanism in MgH_2 with catalyst; proceedings of the 7th Pacific Rim International Conference on Advanced Materials and Processing, Cairns, AUSTRALIA, F Aug 02–06, 2010 [C]. Trans Tech Publications, 2010.
23. EDALATI K, EMAMI H, IKEDA Y, et al. New nanostructured phases with reversible hydrogen storage capability in immiscible magnesium-zirconium system produced by high-pressure torsion [J]. Acta Materialia, 2016, 108: 293–303.

Chapter 5
Thermodynamic and Kinetic Regulations of Mg-Based Hydrogen Storage Materials and the Related Applications

Mg-based alloys are emerging as promising systems for hydrogen storage. MgH_2 offers several advantages, including a high mass hydrogen storage capacity (7.6 wt.% H_2), high energy density (9 MJ/kg), good reversibility, lightweight properties, abundant geological reserves, and low cost. These attributes have attracted significant attention from researchers worldwide. However, the hydrogen absorption/desorption reactions in Mg-based hydrogen storage materials occur only under specific conditions. Typically, magnesium and hydrogen react at the temperatures of 573–673 K and pressure of 2–40 MPa to form MgH_2. MgH_2 remains stable under mild conditions and exhibits low platform pressures, making its decomposition challenging at low temperatures and ambient pressure. High-temperature heating (>573 K) is required for MgH_2 to release hydrogen, and the absorption/desorption rate is relatively slow, significantly limiting its practical applications. The limitations are primarily attributed to the thermodynamic and kinetic properties of the reaction between magnesium and hydrogen. The thermodynamic properties of MgH_2 can be characterized by its enthalpy of formation (ΔH) and entropy (ΔS), while the kinetic barriers to hydrogen absorption/desorption reactions are expressed by the reaction activation energy (E_a). Figure 5.1 illustrates the thermodynamic and kinetic barriers associated with hydrogen absorption/desorption reactions in hydrogen storage alloys. Generally, a higher reaction enthalpy indicates a higher hydrogen desorption temperature under a given hydrogen pressure, while a higher activation energy results in a slower reaction rate and necessitates a higher reaction temperature. Thus, optimizing these two parameters in the hydrogen absorption/desorption reactions of Mg-based hydrogen storage alloys is crucial for enhancing their hydrogen storage performance. Currently, extensive research efforts are focused on optimizing the thermodynamic and kinetic properties of hydrogen absorption and desorption reactions in Mg-based alloys. The following sections will first discuss the research progress on the thermodynamic and kinetic properties of Mg-based hydrogen storage materials, followed by an overview of the primary application fields of magnesium alloy hydrogen storage materials.

Q. Li and F. Pan, *Magnesium-based Hydrogen Storage Materials*,
https://doi.org/10.1007/978-981-96-5531-1_5

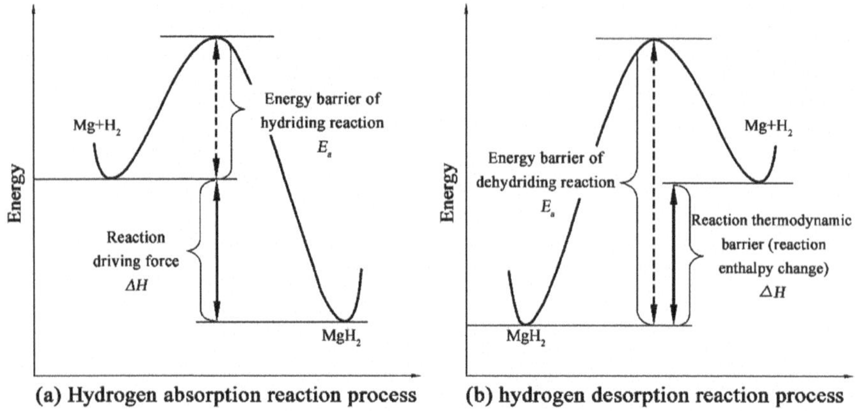

Fig. 5.1 Thermodynamic and kinetic barriers for hydrogen absorption/desorption reactions [1]

5.1 Thermodynamic Regulation of Hydrogenation Reactions in Mg-Based Hydrogen Storage Alloys

5.1.1 Factors Influencing the Thermodynamics of Hydrogen Absorption/Desorption Reactions

The primary thermodynamic challenges in Mg-based hydrogen storage alloys are the significant enthalpy change during the hydrogen absorption/desorption process and the high desorption temperature. As discussed in Chap. 1, based on classical thermodynamics, the key factors influencing the enthalpic changes of reactions include alloying, nanosizing, and external fields. A trace amount of solid solution elements can modify the reaction enthalpy of Mg by altering the activity of magnesium-group elements in the magnesium alloys. Conversely, the introduction of alloying elements that generate new compounds can significantly reduce the enthalpy change in the reactions by directly modifying the type and structure of reactants and products. Additionally, refining the alloy particle size increases their surface energy. When the grain size of magnesium alloys is refined to the nanoscale, the enthalpy changes of the reactions and the hydrogen desorption temperatures decrease. As shown in Eq. (1.10) in Chap. 1, the Gibbs free energy change of reaction for nanomaterials can be expressed as:

$$\Delta G_{nano} = \Delta G^{\circ} + RT \ln \frac{a_{MH_2}}{a_M P_{H_2}} + \Delta G^{surf}_{M \rightarrow MH_2} + \Delta G^{int}_{M \rightarrow MH_2} \tag{5.1}$$

where

$\Delta G^{surf}_{M \rightarrow MH_2}$—Thermodynamic deviation caused by surface free energy.

$\Delta G^{int}_{M \rightarrow MH_2}$—Thermodynamic deviation caused by interfacial free energy.

a_{MH_2}, a_M—Activities of MH_2 and M, respectively.

The surface free energy can be expressed as:

$$\Delta G^{surf}_{M\to MH_2} = \frac{3V_M E_{M\to MH_2}(\gamma, r)}{r} \tag{5.2}$$

where

$E_{M\to MH_2}$ (γ,r)—Surface energy term, a function of surface energy γ (J/m^2) and the radius of metal particles r (m), which can be expressed as:

$$E_{M\to MH_2}(\gamma, r) = \left[\gamma_{MH_2}(r)\left(\frac{V_{MH_2}}{V_M}\right)^{\frac{2}{3}} - \gamma_M(r)\right] + E_{ads} \tag{5.3}$$

Here:

V_i—Molar volume of each phase. The molar volume of MH_2 is typically 10–20% larger than that of metal.

The binding of H_2 on metal or hydride surfaces will reduce their surface energy, in Eq. (5.3), the energy reduction will be amended by the additional energy term of E_{ads}.

The thermodynamic deviation due to interfacial free energy $\left(\Delta G^{int}_{M\to MH_2}\right)$ is given by:

$$\Delta G^{int}_{M\to MH_2} = \frac{\overline{V}_{Mg}}{V}\left(\sum_{int} A_{MgH_2|i} \times \gamma_{MgH_2|i} - \sum_{int} A_{Mgli} \times \gamma_{Mgli}\right) \tag{5.4}$$

where

$\overline{V}_{Mg}$—Molar volume of magnesium extended to all interfaces.

$A_{MgH2} \mid i$, $\gamma_{MgH2} \mid i$—Area and specific free energy of interface i in the hydride phase. When the subscript is Mg, the variables correspond to the area and specific free energy of Mg.

Substituting Eqs. (5.2) and (5.4) into Eq. (5.1), we obtain:

$$\ln P^{eq}_{H_2} = \frac{\Delta H^\circ + \frac{3V_M E_{M\to MH_2}(\gamma,r)}{r} + \frac{\overline{V}_{Mg}}{V}\left(\sum_{int} A_{MgH_2li} \times \gamma_{MgH_2li} - \sum_{int} A_{Mgli} \times \gamma_{Mgli}\right)}{RT} - \frac{\Delta S^\circ}{R} \tag{5.5}$$

This equation shows that surface and interfacial energies depend on the size (r) and surface area ratio of the material. Smaller particle sizes and larger surface areas result in higher introduced surface and interfacial energies, thereby lowering the reaction enthalpy.

From a microscopic perspective, alloying, nanosizing, and external fields aim to reduce the bonding energy of Mg-H. The United States Department of Energy (DOE) suggests an ideal bond energy range of 10–60 kJ/mol for hydrogen storage materials

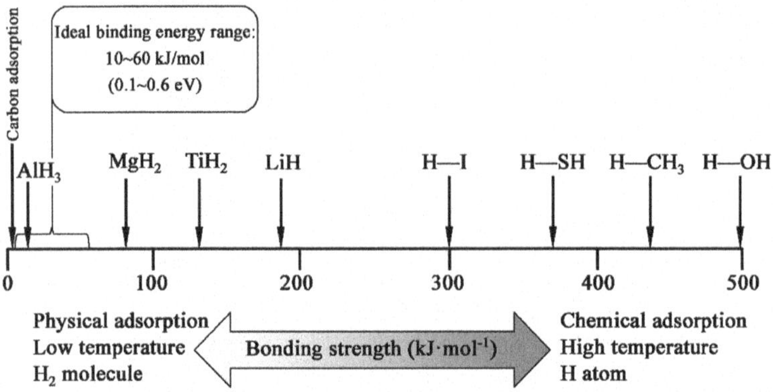

Fig. 5.2 Targeted range of bonding strength for allowing hydrogen desorption at room temperature [2]

[2] (Fig. 5.2). This range facilitates strong hydrogen binding for adsorption without impeding desorption under mild conditions. The enthalpy and entropy changes for hydrogen desorption of pristine MgH_2 are 76 kJ/mol H_2 and 130 J/(mol·K) H_2, respectively, meaning desorption only occurs under ambient pressure at temperatures above 573 K. This is due to the high binding energy of the Mg-H ionic bond. To achieve thermodynamic control over hydrogenation, the Mg-H bond strength must be weakened, reducing the stability of the hydride. By combining classical thermodynamics and micro-thermodynamics, methods such as alloying, nanosizing, and external fields can be applied to adjust the thermodynamic properties of Mg-based hydrogen storage materials.

5.1.2 *Alloying and Compositing*

Alloying is a crucial technique for modifying the thermodynamic properties of Mg-based hydrogen storage alloys. By adding alloying elements that interact with Mg and H to form sub-stable hydrides, the stability of the Mg-H bond is reduced, leading to a decreased enthalpy of hydrogen desorption in MgH_2. These alloying elements combine with Mg to form intermetallic compounds, resulting in new Mg-based hydrides. Consequently, this alters the hydrogen absorption/desorption path in Mg-based hydrogen storage alloys and lowers both the enthalpy changes and desorption temperature of hydrogen. The additives in magnesium alloys can be categorized into solid solution alloying elements, intermetallic compound forming elements, and coordination hydride compounds, which will be introduced in detail as follows.

(1) Alloying and compositing

Some alloying elements exhibit significant solid solubility in Mg, such as indium (In) and rare earth elements (RE). However, RE elements show a strong affinity with hydrogen. Once hydrogenated, they tend to dissociate from Mg, making their influence on the enthalpy change during hydrogen desorption in MgH_2 negligible. Indium (In) possesses several advantageous properties, including a low melting point (429.6 K), an atomic radius (1.66 Å) comparable to that of Mg (1.60 Å), and intermediate electronegativity (1.7 for In, 1.23 for Mg, and 2.20 for H). Due to their similar atomic sizes, the Mg-In system easily forms solid solution alloys. Additionally, the low melting point ensures rapid thermal diffusion of atoms within alloys, enhancing the reversibility of the hydrogenation/dehydrogenation reaction. During hydrogenation, if a Mg-In-H interaction occurs, In atoms may share electrons with H lost by Mg, weakening the Mg-H bond and reducing the stability of MgH_2. Given the potential of Mg(In) solid solution alloys to decrease thermodynamic stability and enhance reaction reversibility, significant research efforts have been devoted to exploring this system.

To understand the phase transition mechanism during the hydrogenation and dehydrogenation process of the Mg-In-H system, researchers analyzed the phase composition of Mg(In) samples. Before fully hydrogenation, the $Mg_{0.95}In_{0.05}$ solid solution consists of MgH_2 and the Mg(In) solid solution. After complete hydrogenation at temperatures exceeding 573 K, intermetallic compounds such as MgIn (β'' phase with L_{10} structure) and Mg_2In are observed [3]. Upon heating to 600 K, the diffraction peaks of the β" phase and Mg_2In disappear, leaving only the Mg-diffraction peak. This indicates that MgH_2 transforms back into Mg(In) solid solution after dehydrogenation, and the phase transition is entirely reversible. A comparison of the hydrogen desorption peaks in the DSC curves of Mg(In) solid solution and pristine Mg shows that the hydrogen desorption temperature of Mg(In) is approximately 35 K lower than that of pristine Mg. The in-situ phase transitions of the hydrogenated Mg(In) solid solution during the desorption process at various temperatures are depicted in Fig. 5.3 [3]. Upon heating from room temperature to 573 K and holding for 10 min, in addition to the initial MgH_2, Mg_2In, and β'' phases, weak Mg diffraction peaks appear. This is attributed to a small amount of MgH_2 decomposition caused by heating under vacuum conditions. After maintaining the temperature at 673 K for 10 min, the Mg_2In and β'' phases disappear, and a disordered β-phase solid solution with a face-centered cubic structure is observed. However, only partial decomposition of MgH_2 occurs at this stage. With a further increase in temperature to 723 K, rapid and reversible decomposition of the remaining MgH_2 takes place, returning to the Mg(In) solid solution state. These findings suggest that the formation of the β-phase promotes the decomposition of MgH_2. Based on XRD analysis, the hydrogen desorption reaction of Mg(In) solid solution at 573 K or higher can be expressed as:

$$MgH_2 + Mg_2In + \beta''(MgIn) \rightarrow MgH_2 + Mg + \beta(MgIn) + H_2 \rightarrow Mg(In) + H_2 \quad (5.6)$$

The enthalpy change of the hydrogen desorption reaction in MgH_2 decreases to 68.1 ± 0.2 kJ/mol H_2, which is about 10 kJ/mol H_2 lower than that of pristine magnesium.

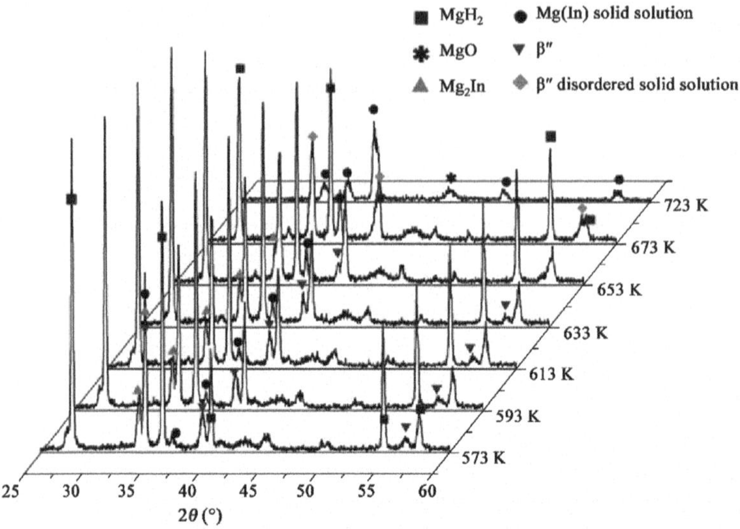

Fig. 5.3 In-situ XRD patterns of $Mg_{0.95}In_{0.05}$ solid solution with hydrogen discharge at different temperatures [3]

Despite the Mg(In) solid solution achieving a reversible hydrogen-absorbing phase transition and reducing the enthalpy change of the hydrogen-desorbing reaction, the high cost of indium limits its large-scale application. Furthermore, the reduction in enthalpy change remains insufficient to significantly enhance its market viability. As a result, researchers have sought more cost-effective alternatives to further optimize the hydrogen storage properties of the solid solution. For example, Mg (In, Al) ternary solid solutions were synthesized by partially substituting In with aluminum (Al). The hydrogenated Mg(In, Al) was observed to decompose into MgH_2 and a β-phase, with Al dissolved in the intermediate of the β-phase. This transformation is also completely reversible. Moreover, the enthalpy of the exothermic reaction in the $Mg_{0.9}In_{0.05}Al_{0.05}$ alloy was reduced to 66.3 kJ/mol H_2 [4].

(2) Addition of intermetallic compound forming elements

When alloying elements exceed their solid solubility limits in Mg, the excess combines with Mg to form intermetallic compounds. Transition metal (TM) elements like Ni, Fe, and Ti form intermetallic compounds with Mg such as Mg_2TM. These intermetallic compounds can form hydrides distinct from MgH_2, thereby reducing the enthalpy change of the reaction. The Mg-Ni alloy is a representative Mg-TM hydrogen storage material. Based on the Mg-Ni phase diagram, the Mg-Ni system has two stable compounds: Mg_2Ni and $MgNi_2$. In 1968, Reilly and Wiswall at Brookhaven National Laboratory first synthesized Mg-Ni alloys under an argon atmosphere using an induction furnace [5]. They discovered that $MgNi_2$, an intermetallic compound, showed minimal hydrogen absorption, making it unsuitable for hydrogen storage. Conversely, Mg_2Ni readily reacts with hydrogen under conditions of 2 MPa and 573 K, forming Mg_2NiH_4 and releasing hydrogen between 520 and 570 K. The hydrogen absorption process involves two steps: initially, Mg_2Ni absorbs hydrogen to form $Mg_2NiH_{0.3}$; subsequently, $Mg_2NiH_{0.3}$ further reacts with hydrogen until saturation, yielding Mg_2NiH_4. Additionally, the hydride exhibits both low-temperature and high-temperature structures. Upon cooling from 518 to 483 K, the high-temperature cubic structure transforms into a low-temperature monoclinic structure. The overall reaction is:

$$Mg_2Ni + 2H_2 \rightarrow Mg_2NiH_4 \tag{5.7}$$

The formation enthalpy of Mg_2NiH_4 is 65.6 kJ/mol H_2, lower than that of MgH_2 (75.6 kJ/mol H_2), indicating that Mg_2NiH_4 is thermodynamically less stable and easier to decompose.

Although the thermodynamic properties of the Mg_2Ni hydrogen absorption and desorption are significantly improved compared to MgH_2, challenges remain for practical applications. To further reduce the hydrogen absorption and desorption temperatures and enhance the performance of Mg-Ni-based hydrogen storage alloys, researchers have developed Mg-Ni-based hydrides with reduced stability by substituting Mg_2Ni with additional elements. In general, hydrogen storage alloys consist of a high-temperature hydride-forming element (A) and a non-hydride-forming element (B), as illustrated in Fig. 5.4. For example, Mg_2Ni comprises Mg on the A side and Ni on the B side. Adding the non-hydride-forming elements can significantly reduce the hydrogenation enthalpy by forming less stable hydrides.

The substitution elements for Mg_2Ni hydrogen storage alloys are shown in Fig. 5.5 [6]. Typically, elements from groups I A to V B, such as Ti, V, Zr, Al, Ca, and rare earth (RE), replace Mg, while heat-absorbent transition metals from groups VIB to VIIIB, such as Mn, Fe, Co, Cr, and Cu, replacing Ni. Substituting Mg often lowers hydride generation enthalpy and hydrogen desorption temperature, albeit at the cost of reduced hydrogen storage capacity. Figure 5.6a illustrates the relationship between hydrogen pressure and temperature of $Mg_{2\text{-}x}Sn_xNi$ ($x = 0, 0.05, 0.1, 0.15, 0.2$) alloys [7]. The partial substitution of Sn for Mg effectively reduces the hydrogenation temperature. While hydrogen absorption temperature slightly increases at low substitution levels ($x = 0.05$), it significantly decreases when $x = 0.15$. The $Mg_{1.85}Sn_{0.15}Ni$ alloy can react with hydrogen at approximately 330 K, 80 K lower than Mg_2Ni. As shown in Fig. 5.6b, the van't Hoff curve slopes vary with substitution content. The optimal substitution amount, around $x = 0.15$, corresponds to the smallest van't Hoff slope, indicating the lowest reaction enthalpy.

Theoretical calculations demonstrate that the substituting A-side elements in Mg_2Ni improve its hydrogenation thermodynamic properties, providing theoretical insights. For example, the effect of partially substituting Mg with Al on the structure and hydrogen storage properties of Mg_2Ni and related hydrides was studied using a plane-wave pseudopotential method based on density-functional theory (DFT) [8]. The results show that the cell volume of Mg_2Ni decreases with increasing Al content, hindering the mass transfer of H atoms into the alloy and reducing the

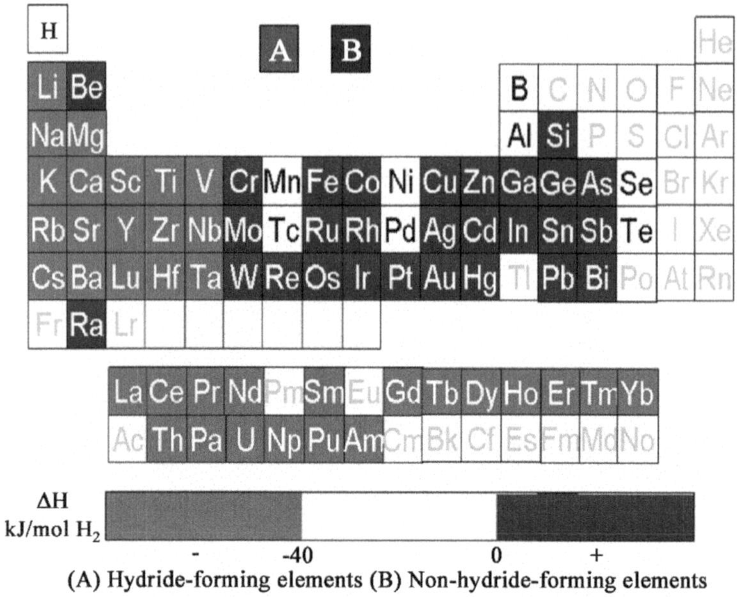

Fig. 5.4 Elements in the periodic table of elements

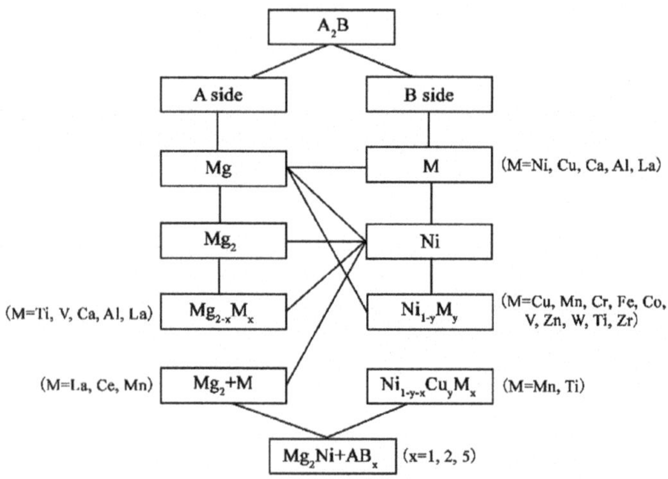

Fig. 5.5 Substitution of A- and B-side elements in Mg_2Ni alloy and the modification by adding other compounds [6]

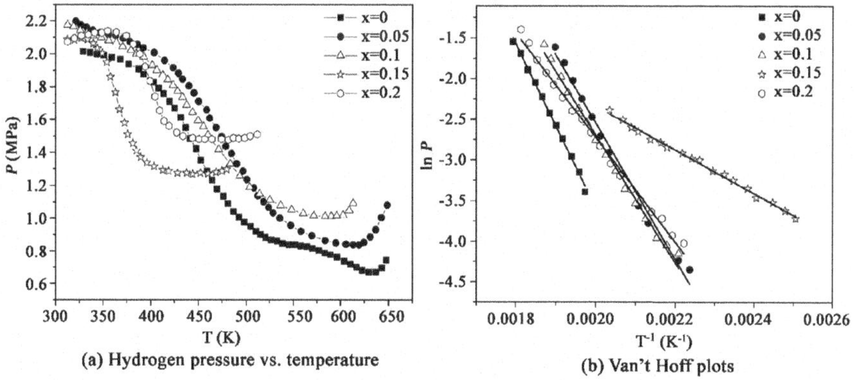

Fig. 5.6 Hydrogen storage thermodynamics of $Mg_{2-x}Sn_xNi$ ($x = 0, 0.05, 0.1, 0.15, 0.2$) alloys [7]

hydrogen storage capacity. Further investigation into Mg_2NiH_4 and related substituted hydrides revealed that Mg–H and Al–H interactions in the hydrides are weaker than Ni–H interactions. Reducing the Ni–H interaction strength is therefore crucial for enhancing the hydrogen desorption performance. A contour map of charge density for $Mg_{2-x}Al_xNiH_4$ ($x = 0, 0.125$ and 0.25) (Fig. 5.7), indicates that the overlap of electronic clouds between Ni and H atoms decreases with higher Al content. This weakens the Ni–H interactions, reducing the thermodynamic stability of hydrides. The formation enthalpies of Mg_2NiH_4, $Mg_{1.875}Al_{0.125}NiH_4$, and $Mg_{1.75}Al_{0.25}NiH_4$ were calculated as −56.1, −45.1, and −32.5 kJ/mol H_2, respectively. These results align well with experimental data for Mg_2NiH_4 (−61 kJ/mol H_2) [9] and $Mg_{1.9}Al_{0.1}NiH_4$ (−53.1 kJ/mol H_2) [10]. This consistency suggests that Al substitution reduces the structural stability of the hydrides, enhancing hydrogen desorption thermodynamics.

The substitution of B-side elements in hydrogen storage alloys also significantly affects hydride formation enthalpy and the reversibility of hydrogen absorption and desorption. Based on molecular orbital theory, the influence of partially substituting Ni with 3*d* transition metals M (M = V, Cr, Fe, Co, Cu, Zn) on the electronic structure of Mg_2Ni was investigated using the DV-Xα method [11]. The findings reveal that

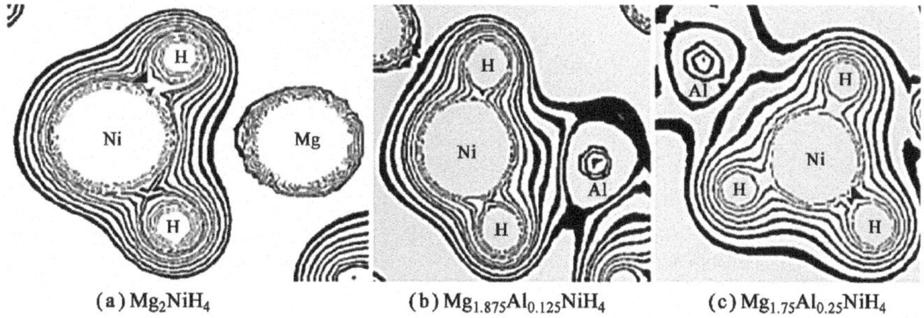

Fig. 5.7 The charge density contour map of $Mg_{2-x}Al_xNiH_4$ ($x = 0, 0.125, 0.25$) [8]

the Ni-Mg bond in Mg_2Ni hydrides is the strongest. Hydrogen atoms preferentially react with nickel atoms rather than Mg, forming strong chemical bonds. Consequently, the Ni–H and Ni-Mg interactions directly impact hydride stability. When different alloying elements M partially substitute Ni, the weaker bonding interactions between M and Mg lead to reduced structural stability of Mg_2Ni hydrides. Experimental studies show that the hydrogen desorption temperature of mechanically alloyed $Mg_{69}M_{26}Ti_5$, $Mg_{64}M_{26}Ti_{10}$ and $Mg_{64}M_{31}Ti_5$ (M = Ni or Fe) powders is approximately 40 K lower than that of Mg_2NiH_4 at the H_2 pressure of 0.1 MPa [12].

Despite advancements in experimental techniques, determining the thermodynamic properties of numerous multi-component alloys experimentally remains labor-intensive. Theoretical calculations based on available experimental data and quantitative analyses using developed models offer a more convenient and efficient approach to investigating the effects of elemental substitutions on thermodynamic properties. By employing stepwise regression and microstructural parameterization methods, a semi-empirical mathematical model has been developed to analyze the enthalpy of hydride formation, entropy change, and hydrogen storage capacity in Mg-based multi-component alloy of $Mg_{2-x}A_xNi_{1-y}B_y$ alloy [13].

$$\begin{aligned}\Delta G = &-2.454 \times 10^6 \times \Delta X^2 + 125.957 \times 10^4 \times \left(\frac{e}{a}\right)^{\frac{2}{3}} \\ &+ 7.100 \times 10^3 \times \Delta X^2 \times \frac{Z}{R} - 2.182 \times 10^4 \times \left(\frac{e}{a}\right)^{\frac{2}{3}} \\ &\times \Delta n^{\frac{2}{3}} - 7.301 \times 10^4 \text{ J/mol}\end{aligned} \tag{5.8}$$

$$\begin{aligned}\Delta H = &-2.454 \times 10^6 \times \Delta X^2 + 3.275 \times 10^4 \times \left(\frac{e}{a}\right)^{\frac{2}{3}} \\ &+ 7.100 \times 10^3 \times \Delta X^2 \times \frac{Z}{R} - 2.182 \times 10^4 \times \left(\frac{e}{a}\right)^{\frac{2}{3}} \\ &\times \Delta n^{\frac{2}{3}} - 7.301 \times 10^4 \text{ J/mol}\end{aligned} \tag{5.9}$$

$$\begin{aligned}C = &1.539 \times 10^2 \times \Delta X^2 - 1.387 \times \left(\frac{e}{a}\right)^{\frac{2}{3}} - 1.476 \\ &\times \frac{Z}{R} + 1.298 \times 10^{-3} \times T + 2.285\text{w[H]}\end{aligned} \tag{5.10}$$

where

ΔX^2—electronegativity difference,

$\left(\frac{e}{a}\right)^{\frac{2}{3}}$—electron concentration,

$\frac{Z}{R}$—ratio of charge and atomic radius,

$\Delta n^{\frac{2}{3}}$—electron-density,

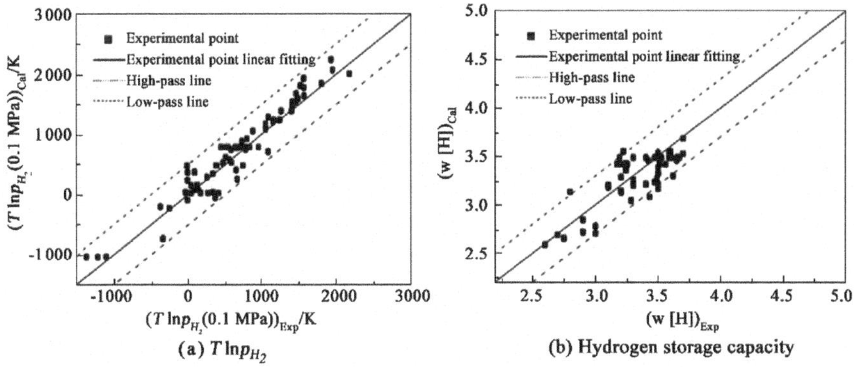

Fig. 5.8 The comparison of predicted and experimental values for $Mg_{2-x}A_xNi_{1-y}B_y$ multicomponent alloys [13]

T—temperature.

The weight of the factors influencing the thermodynamic properties of Mg-based multi-elemental alloys follows the order of $\left(\frac{e}{a}\right)^{\frac{2}{3}} > \left(\frac{e}{a}\right)^{\frac{2}{3}} \times \Delta n^{\frac{2}{3}} > \Delta X^2 > \Delta X^2 \times \frac{Z}{R}$. When the electronegativity difference ΔX^2 decreases, and the electron concentration $\left(\frac{e}{a}\right)^{\frac{2}{3}}$ as well as temperature T increases, the hydrogen-absorption plateau pressure of the alloy rises, and the absolute value of the enthalpy of hydride formation decreases. The contributions of each factor to the hydrogen storage capacity of the multi-elemental alloys follow the order of $\Delta X^2 > T > \left(\frac{e}{a}\right)^{\frac{2}{3}} > Z/R$. The hydrogen storage capacity in the alloys is enhanced with an increase in the electronegativity difference (ΔX^2) and the temperature (T), as well as a decrease in electron concentration $\left(\frac{e}{a}\right)^{\frac{2}{3}}$ and the charge-radius ratio (Z/R). The absolute error bar in the enthalpy change for alloy hydride formation predicted by the model is lower than $\pm$ 5 kJ/mol H_2, and the maximum absolute error bar between the theoretical and experimental values of the hydrogen storage capacity for the alloy is $\pm$ 0.3 $w_{[H]}$ (Fig. 5.8).

Apart from adding alloying elements similar to Mg_2Ni, which can form new hydrides, there are also methods to lower the overall enthalpy change of the reaction during hydrogen desorption. This can be achieved by adding alloying elements (e.g., Al) to form intermetallic compounds. Adding substances that can react with metal hydrides during hydrogen desorption can lead to the formation of additional compounds, thereby reducing the enthalpy change of the hydrogen desorption reaction. The mechanism of altering the reaction pathway is illustrated in Fig. 5.9. If the metal hydride of AH_2 directly decomposes into A and H_2 after desorption, the high enthalpy change of this reaction will result in a lower equilibrium pressure for the decomposition of the hydride. Hence, a higher temperature is needed for hydrogen desorption. If the AH_2 hydride reacts with the additive B to form AB_x during hydrogen desorption, and the enthalpy value of AB_x formation is negative,

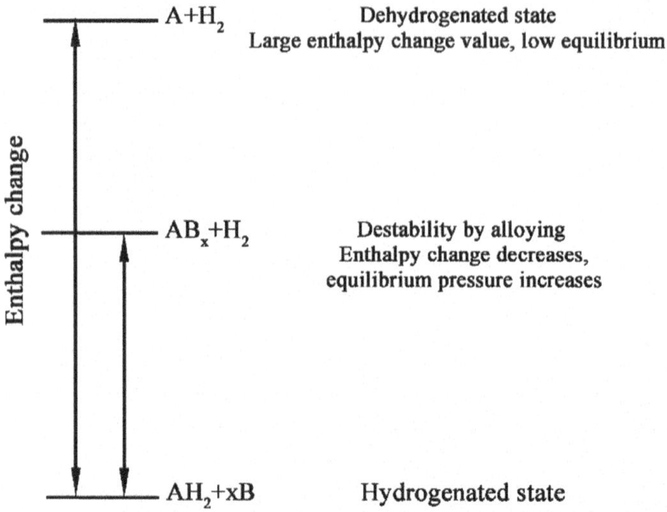

Fig. 5.9 Generalized enthalpy illustration of the reduced stability of the hydride via forming intermetallic compounds during the hydrogen desorption process. [14]

i.e., AB_x is thermodynamically more stable than A, the heat required during this reaction process will be compensated by the formation enthalpy of AB_x. This results in a decrease in the enthalpy change of the hydrogen desorption reaction, leading to an increase in hydrogen desorption equilibrium pressure and a decrease in desorption temperature. Thus, the thermodynamic stability of AH_2 during hydrogen desorption can be adjusted by altering the reaction pathway without changing the bonding strength of AH_2. The corresponding reaction is as follows:

$$AB_x + H_2 \rightarrow AH_2 + xB \tag{5.11}$$

For instance, experimental results show that Si can decrease the stability of the hydrogen desorption thermodynamics of MgH_2 in a MgH_2-Si mixture [15]. Compared with the MgH_2 ball-milled under identical conditions, the dehydrogenation equilibrium pressure of the MgH_2-Si mixture is sharply increased from 0.0165 to 0.0750 MPa at 573 K. While the ball-milled sample contains MgH_2 and Si, after dehydrogenation, the sample transforms into Mg_2Si with a trace amount of Si. The reversible hydrogen absorption/desorption reactions are as follows:

$$2MgH_2 + Si \rightarrow Mg_2Si + 2H_2 \tag{5.12}$$

The generated silicide during the dehydrogenation is more thermodynamically stable than hydride, resulting in a lower enthalpy change in MgH_2 dehydrogenation. Calculations show that the Mg_2Si compound reduces the hydrogen desorption enthalpy of MgH_2 from 75.3 kJ/mol H_2 to 36.4 kJ/mol H_2, with desorption temperature at equilibrium pressures of 0.1 MPa and 10 MPa being approximately 293 K and

423 K, respectively. Furthermore, the high hydrogen storage capacity of 5.0 wt.% is maintained after adding Si. Despite these advantages, challenges remain for the MgH_2-Si system, such as the high stability of the generated Mg_2Si, which makes the reverse reaction difficult.

Aluminum (Al), a cost-effective element neighboring Mg on the periodic table, forms stable compounds such as $Mg_{17}Al_{12}$, $Mg_{23}Al_{30}$, and Mg_2Al_3 [16]. In a previous study, three alloys with initial compositions of Mg_xAl_{100-x} (where $x = 100$, 70, and 39) were prepared using Nb_2O_5-catalyzed mechanical alloying of high-purity magnesium and aluminum powder [17]. $x = 39$ corresponds to the Mg_2Al_3 alloy, $x = 70$ corresponds to a mixture of 25 mol% Mg and 75 mol% Mg of $Mg_{17}Al_{12}$, and $x = 59$ corresponds to a single-phase $Mg_{17}Al_{12}$ alloy. The hydrogen absorption P–C–T curve of $Mg_{70}Al_{30}$ at 623 K shows three plateaus, indicating that three-phase transitions during hydrogen absorption (Fig. 5.10). Both intermetallic compounds $Mg_{17}Al_{12}$ and Mg_2Al_3 undergo disproportionation reactions with hydrogen. The hydrogenation process of Mg_2Al_3 is as follows:

$$Mg_2Al_3 + 2H_2 \rightarrow 2MgH_2 + 3Al \tag{5.13}$$

Meanwhile, $Mg_{17}Al_{12}$ reacts in two steps: first with hydrogen to form Mg_2Al_3, then followed by the formation of MgH_2 and Al:

$$Mg_{17}Al_{12} + 17H_2 \rightarrow 9MgH_2 + 4Mg_2Al_3 + 8H_2 \rightarrow 17MgH_2 + 12Al \tag{5.14}$$

The enthalpy change of these hydrogen absorption reactions is approximately −73.8 kJ/mol H_2 and −68.2 kJ/mol H_2 for Eqs. (5.13) and (5.14), respectively. The

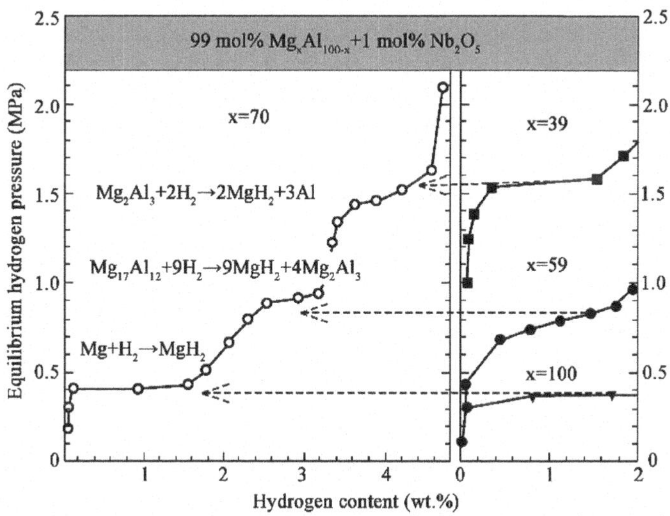

Fig. 5.10 Hydrogen absorption P–C–T curve for Mg_xAl_{100-x} alloy at 623 K [17]

hydrogen storage capacity of $Mg_{70}Al_{30}$ is around 5.74 wt.%. However, the Mg–Al system exhibits limited improvement in thermodynamic stability, (a decrease of approximately 50 K in desorption temperature and 6 kJ/mol H_2 in enthalpy). Additionally, reaction kinetics are slow below 573 K, thereby, it is seldom applied in practical applications of those compounds.

(3) Addition of coordination hydrides

The addition of alloying elements reduces the thermodynamic stability of MgH_2, but also significantly decreases the hydrogen storage capacity compared to pristine MgH_2. To counteract this and maintain the hydrogen storage capacity of the system, researchers have synthesized compounds using complex hydrides and MgH_2. Coordination hydrides are formed by group IA or VA elements (such as Al, B, and N) and hydrogen atoms with covalent bonding, followed by forming ionic bonds with metal ions. Three hydrides can be broadly categorized into three types based on the coordinating ligand: the first one is metal aluminum hydrides, containing $[AlH_4]^-$ ligands, such as $LiAlH_4$, $NaAlH_4$, and $Mg(AlH_4)_2$; the second type is metal boron hydrides, containing $[BH_4]^-$ ligands, such as $LiBH_4$, $NaBH_4$, and $Mg(BH_4)_2$; the third type is metal nitrogen hydrides, containing $[NH_2]^-$ ligands, such as $LiNH_2$ and $Mg(NH_2)_2$. Compared with traditional metal hydrides, complex hydrides exhibit higher hydrogen storage capacities. However, the combination of ionic and covalent bonds makes them possess relatively high thermodynamic stability. The compound composed of a complex hydride and MgH_2 can change the reaction path of the hydrogen desorption reaction by forming intermetallic compounds, which reduces the thermodynamic stability of the system while maintaining a high hydrogen storage capacity.

For instance, the complex hydride $LiBH_4$ and $TiCl_3$ were added into MgH_2 as an additive and catalyst under an inert atmosphere using the mechanical alloying method [18]. The results show that the addition of $LiBH_4$ results in the formation of MgB_2, which changes the hydrogen desorption pathway of MgH_2, as shown in Eq. (5.15). The enthalpy change of the hydrogen desorption reaction was measured to be 40.5 kJ/mol H_2, lower than that of pristine MgH_2 or $LiBH_4$. The thermodynamic stability of both MgH_2 and $LiBH_4$ is decreased with the formation of MgB_2.

$$MgH_2 + 2LiBH_4 \rightarrow 2LiH + MgB_2 + 4H_2 \tag{5.15}$$

The influence of varied $LiBH_4$ contents on the hydrogen desorption temperature of MgH_2 is shown in Fig. 5.11 [19]. The non-isothermal Temperature-Programmed Desorption (TPD) curves of $MgH_{2-x}LiBH_4$ ($x = 0, 5, 10, 20$ and 30 mol%) complexes are shown in Fig. 5.11a. The initial hydrogen desorption temperature of pristine MgH_2 ($x = 0$) is 620 K. With 5 mol% $LiBH_4$, the initial hydrogen desorption temperature decreases to 604 K with a maximum release of 7.3 wt.% H_2. As the $LiBH_4$ content increases to 10, 20, and 30 mol%, the initial hydrogen desorption temperatures of the composites were 598 K, 611 K, and 613 K, respectively. Figure 5.11b shows the non-isothermal hydrogen desorption Differential Scanning Calorimetry (DSC) curves for MgH_2-$x$$LiBH_4$ composites. The results further confirm the positive effect of $LiBH_4$

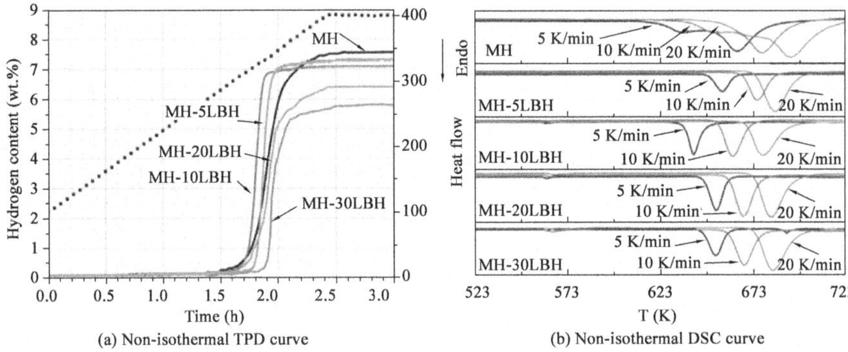

Fig. 5.11 Hydrogen desorption properties of $MgH_{2-x}LiBH_4$ ($x = 0, 5, 10, 20, 30$ mol%) complexes [19]

on hydrogen desorption for MgH_2. For the pristine MgH_2 sample, two endothermic peaks at 635.8 K and 664.7 K were observed in the dehydrogenation DSC curve with a ramping rate of 5 K/min, corresponding to the hydrogen desorption processes of γ-MgH_2 and β-MgH_2 phases, respectively. After adding $LiBH_4$, two endothermic peaks were observed in the DSC curves of the MgH_2-$LiBH_4$ samples as well. For example, the DSC curve at a ramping rate of 5 K/min has two endothermic peaks at 553 and 643 K for MgH_2-10 mol% $LiBH_4$ sample, corresponding to the dehydrogenation processes of $LiBH_4$ and MgH_2. Comparing the exothermic peaks in the DSC curves, it is evident that the dehydrogenation temperature of MgH_2 decreases initially and then increases with increasing $LiBH_4$ content. The optimal content was found to be 10 mol%, demonstrating that a suitable content of $LiBH_4$ can enhance the dehydrogenation properties of MgH_2 by forming MgH_2- $LiBH_4$ composites.

5.1.3 Nanonization

Materials with particle or grain sizes smaller than 100 nm are commonly referred to as nanomaterials. A key characteristic of nanomaterials is their extremely large surface/volume ratio, which introduces additional surface and interfacial free energies. This increases the Gibbs free energy and decreases the thermodynamic stability of the overall system. To date, nanostructured hydrogen storage materials, prepared through various methods, can generally be categorized into two types: the first type is nanoscale hydrogen storage material, where the external contour of the material reaches the nanoscale in at least one dimension; the second type is internal nanoscaled material, where the external dimensions are not the nanoscale, but the internal phases and structures are at the nanoscale. The research focus for the first type is on the surface properties of the materials, while for the second type, internal interfaces are the primary consideration. The different dimensions of nanomaterials are illustrated in Fig. 5.12 [20].

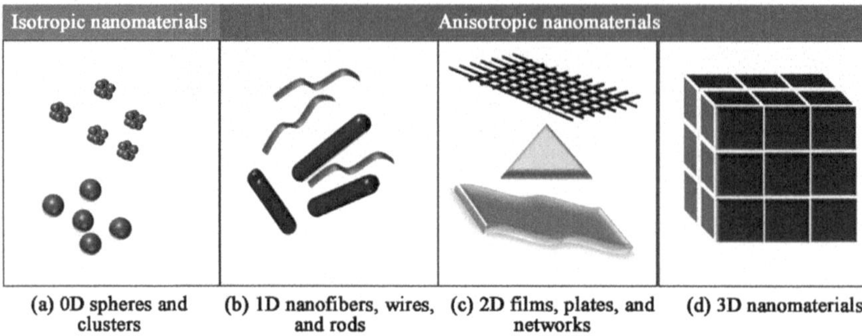

Fig. 5.12 Nanomaterials in different dimensions [20]

(1) Nanoparticles or Grains

Nanosizing enhances the specific surface area and the volume fraction of interfacial regions in the material, reducing the reaction enthalpy for hydrogen absorption/desorption. First-principles calculations show that smaller cluster particle sizes lead to lower thermodynamic stability of both Mg and MgH_2, with the stability decrease being more pronounced in MgH_2 [21]. When the grain size of MgH_2 is less than 1.3 nm, a notable decrease in the hydrogen desorption reaction enthalpy is observed. For instance, MgH_2 with a grain size of 0.9 nm exhibits a hydrogen desorption reaction enthalpy and temperature of 63 kJ/mol H_2 and 473 K, respectively. Reactive Force Field (RFF) calculations indicate no significant difference in the formation enthalpy between ca. 2 nm MgH_2 cluster and bulk Mg [22].

In nanoparticle-powder systems, reducing particle size does not infinitely lower the formation enthalpy of the system. The condition for smaller particle radius resulting in lower formation enthalpy assumes no interactions between particles (solid line in Fig. 5.13a). However, van der Waals forces create strong attraction between neighboring particles, leading to surface recombination and a decrease in specific surface area. As the particle radius decreases further, the loss of specific surface area becomes more pronounced (Fig. 5.13b). Consequently, the formation enthalpy exhibits a maximum positive shift at an optimized particle size, as indicated by the dashed line in Fig. 5.13a.

From a reaction entropy perspective, the ΔS of bulk materials at room temperature, calculated for the dehydrogenation reaction $MgH_2 \rightarrow Mg + H_2$, is $\Delta S = 130.5$ J/(mol·K) H_2, and the ΔS is typically treated as a constant. Based on the van't Hoff equation, an increased ΔS raises the equilibrium pressure of the exothermic reaction, thereby reducing hydride stability. Nanosizing reduces particle size and increases the disorder of both Mg metal and its hydrides, increasing S_{Mg} and S_{MgH_2}. MgH_2 has two elements, Mg and H. Materials with more elements tend to experience a greater increase in disorder for the same size reduction. Thus, $\Delta\ S_{MgH_2} > S_{Mg} > 0$ and $\Delta S = S_{Mg} + S_{H_2} < 130.5$ J/(mol·K) H_2. To determine the reduced ΔS caused by nanosizing, mathematical models and systematic theoretical calculations are needed.

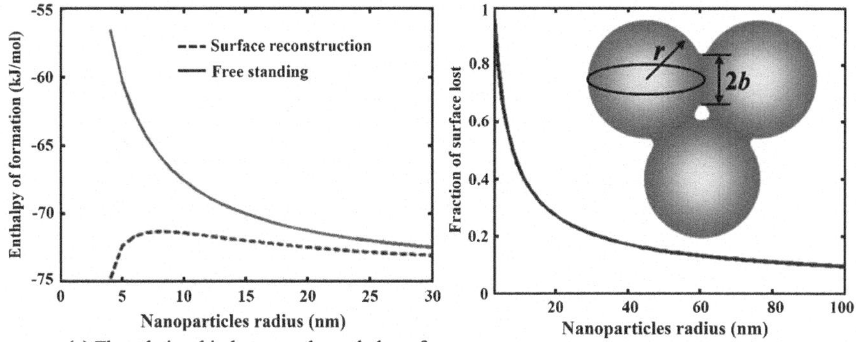

(a) The relationship between the enthalpy of MgH_2 powder formation and particle size

(b) The relationship between MgH_2 powder surface loss fraction and particle size

Fig. 5.13 Enthalpy of formation and fraction of surface loss of MgH_2 powders as a function of particle radius [23]

For nanoparticles and nanograins, atomic coordination at grain boundaries differs from that of the bulk lattice. Coordination distances increase, and coordination numbers decrease, resulting in excess volume (ΔV) at the grain boundaries. Compared with the normal lattice volume (V_0), the volume of nanomaterials can be expressed as $V' = V_0(1 + \Delta V)$, where the excess volume (ΔV) increases as the particle size decreases [24]. Statistical thermodynamics can be used to adjust for nanosizing's impact on the entropy change during hydrogen desorption from MgH_2. The decrease in the atomic coordination number at the grain boundaries is considered a reduction in density at the grain boundary. By treating the grain boundaries of nanocrystals as the normal crystals with a reduced coordination number, the nanosized S_{Mg} and S_{MgH_2} can be calculated as follows:

$$S_{Mg} = S_{Mg}^{\theta} + \Delta S_{rMg} + \Delta S_{rMg} \tag{5.16}$$

$$S_{MgH_2} = S_{MgH_2}^{\theta} + \Delta S_{rMgH_2} + \Delta S_{vMgH_2} \tag{5.17}$$

where

ΔS_r—rotational entropy increment.

ΔS_v—vibration entropy increment.

Thus, the change in entropy for the dehydrogenation reaction $\Delta S_{MgH_2 \rightarrow Mg + H_2}$ is given by:

$$\Delta S_{MgH_2 \rightarrow Mg+H_2} = (S_{Mg}^{\theta} + \Delta S_{rMg} + \Delta S_{vMg}) + S_{H_2}^{\theta} - (S_{MgH_2}^{\theta} + \Delta S_{rMgH_2} + \Delta S_{vMgH_2}) \tag{5.18}$$

The adjusted entropy of hydrogen desorption of MgH_2 is calculated as:

$$\Delta S_{MgH_2 \to Mg+H_2} = 130.5 - \frac{8}{3} R \ln\left[\frac{0.8[d^3 - (d - 0.556)^3] + d^3}{d^3}\right] \tag{5.19}$$

It can be observed that $\Delta S_{MgH_2 \to Mg+H_2}$ decreases as the particle size (d) decreases. When the particle size is smaller than 10 nm, the reduction in desorption entropy becomes more significant. Considering the influence of nanosizing on the enthalpy change of hydrogen desorption and using the van't Hoff equation, the hydrogen desorption temperature of nanostructured MgH_2 at a plateau pressure of 0.1 MPa can be expressed as:

$$T_d = \frac{\Delta H_{nano}}{\Delta S_{nano}} = \frac{75 - \frac{69.57}{d}}{130.5 - \frac{8}{3} R \ln\left[\frac{0.8[d^3 - (d - 0.556)^3] + d^3}{d^3}\right]} \tag{5.20}$$

Based on Eq. (5.20), the correlation between hydrogen desorption temperature and particle size is illustrated in Fig. 5.14. Compared to bulk materials, the hydrogen desorption temperature at 0.1 MPa hydrogen pressure decreases as the nanoparticle size is reduced, when considering only the impact of nanosizing on the entropy change of hydrogen desorption. However, when accounting for the effect of nanosizing on both the enthalpy and entropy changes, the hydrogen desorption temperature of MgH_2 nanoparticles is relatively higher. Calculations indicate that the hydrogen desorption temperature at a plateau pressure of 0.1 MPa is lower than 373 K when the particle size is smaller than 2.3 nm.

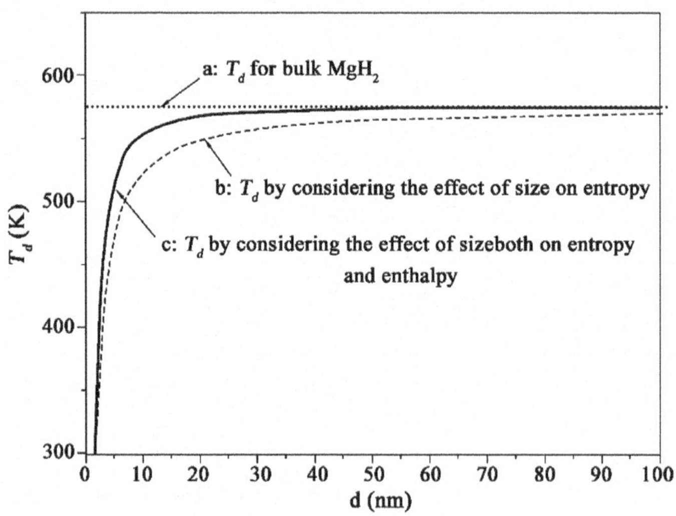

Fig. 5.14 Particle size versus hydrogen desorption temperature at an equilibrium pressure of 0.1 MPa [25]

In addition to the surfaces and grain boundaries of nanoparticles, the formation enthalpy of hydrides is influenced by the crystal orientation of magnesium alloys. Studies have shown that the thermodynamic stability of MgH_2 varies with different crystal planes, being enhanced by the (111) plane and reduced by the (200) plane. The effect of nanosizing effect on the stability of MgH_2 represents as average effect of the (111) and (200) planes. Researchers have theoretically predicted the influence of crystal planes on the hydrogen desorption enthalpy of MgH_2 (Fig. 5.15) [26]. When the particle size of MgH_2 exceeds 25 nm, the (111) plane predominantly governs the stability of MgH_2, rendering the hydrogen desorption enthalpy similar to that of bulk samples. Conversely, when the particle size of MgH_2 is smaller than 5 nm, the (200) plane gradually becomes dominant, significantly decreasing the hydrogen desorption enthalpy of MgH_2. Although there is ongoing debate about the size criterion for the nanosizing effect, most researchers agree that the particle size should be lower than 10 nm to observe significant nanoscale effects.

Theoretical predictions provide guidance for experimental studies to adjust the thermodynamics of hydrogen desorption of MgH_2 through nanosizing. There are two main approaches for preparing nanoparticles: top-down and bottom-up processing. The top-down approach involves reducing the size of bulk materials to the nanoscale, while the bottom-up approach focuses on synthesizing nanoparticles from the atomic scale via nucleation and growth processes. Top-down methods include techniques such as ball milling and lithographic etching. In contrast, bottom-up processing can be achieved by controlling nucleation and growth processes in liquid or gas phases, or through non-uniform nucleation on substrates via techniques such as vapor deposition, hydrothermal synthesis, and sol–gel processes. Additionally, a combination of the two approaches is possible. For instance, magnesium nanoparticles of various

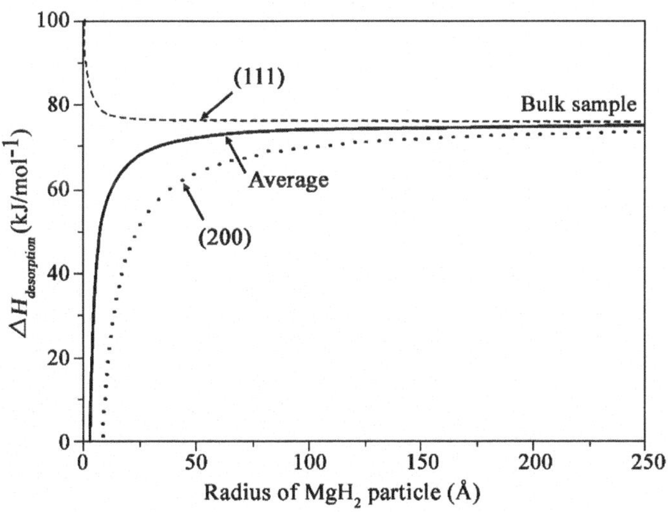

Fig. 5.15 Influence of MgH_2 particle size on the hydrogen desorption enthalpy of Mg ($\Delta H_{desorption}$) in considering the surface energy of the Mg plane of (111) and (200) [26]

sizes can be produced at room temperature using a chemical reaction ball milling method. The reaction is as follows:

$$MgCl_2 + 2LiH \rightarrow MgH_2 + 2LiCl \quad (5.21)$$

By adding an excess of LiCl to shift the equilibrium to the left, MgH_2 nanoparticles with particle sizes of 2–7 nm were synthesized via high-energy ball milling. These particles exhibited a hydrogen desorption enthalpy of 71.22 ± 0.49 kJ/mol H_2. Comparatively, magnesium nanoparticles with sizes of 12~20 nm and 15~40 nm exhibited hydrogen desorption enthalpies of (74.15 ± 0.37) kJ/mol H_2 and (74.75 ± 0.42) kJ/mol H_2, respectively [27].

In recent years, the concept of nanoconfinement has been proposed to reduce the size of alloy particles to the nanoscale while confining them to prevent particle agglomeration and growth. This enhances the stability of hydrogen absorption and desorption cycles. For instance, the reduction of graphene oxides was employed for the nanoconfinement of magnesium [28], resulting in an Mg grain size of approximately 3.2 nm. The selective permeability of graphene allows H_2 molecules to diffuse while blocking O_2 and H_2O, significantly improving its stability in air. Subsequently, nanoconfined Mg composite with a trace amount of transition metal catalysts of Ni, (Fig. 5.16) [29], achieved a hydrogen storage capacity of 6.5wt.%. The Ni-catalyzed composite exhibits reduced hydriding/dehydriding enthalpy changes of approximately 11 kJ/mol H_2 and 9 kJ/mol H_2, respectively, compared to bulk Mg. Notably, the decrease in enthalpy change with Ni catalyst was relatively small compared to the nanoconfined composite without Ni, suggesting that improved thermodynamic properties are primarily attributed to the Mg and the encapsulating action of graphene.

In summary, the thermodynamic properties of Mg-based hydrogen storage alloys are modified by nanosizing, which increases the Gibbs free energy of the system through optimization of specific surface area, crystal surface, excess volume, and

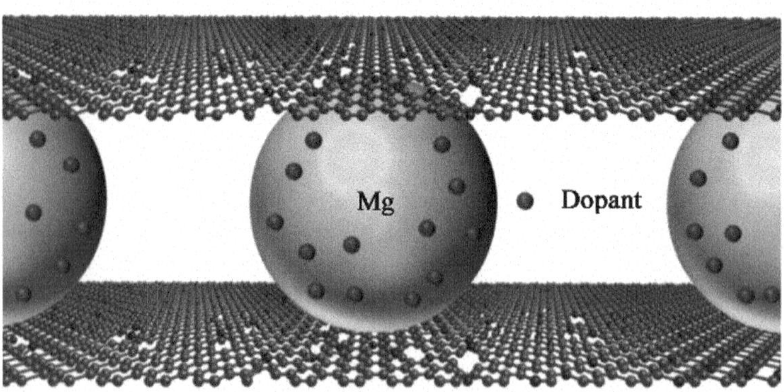

Fig. 5.16 Schematic of reduced graphene oxide (rGO) nano-confined Mg composites [29]

metastable phases. To preserve the favorable thermodynamic properties after nanosizing, it is essential to prevent nanoparticle agglomeration. Nanoconfinement emerges as an effective strategy to improve the thermodynamic performance of Mg-based hydrogen storage alloys. Specifically, nanoconfinement reduces the thermodynamic stability of MgH_2 through two mechanisms: decreasing the size of the hydride particles and preventing nanoparticle growth or agglomeration with scaffold materials.

(2) Nanofilm

Thin films are widely used in the field of Mg-based hydrogen storage materials due to their unique properties, which differ from bulk materials because of limited growth in one direction. Multilayer films with thicknesses below 100 nm are also considered as nanomaterial. These films provide spatial constraints on magnesium, reducing the hydrogen desorption temperature of MgH_2 while maintaining a relatively high equilibrium pressure. In such cases, the interlayer stress in the multilayer films plays a dominant role in nanoscale effects.

Using nanofilms to alter the hydrogen desorption properties of Mg is closely linked to the stress and strain they introduce. This is due to the abundant interfaces with interfacial elastic constraints present in the nanomaterials. For instance, in epitaxial multilayers or superlattices, lattice mismatch or differences in thermal expansion induce interfacial elastic constraints, introducing additional strain energy. This stress, strain, or strain energy can significantly alter the structure and hydrogen desorption properties of Mg-based hydrogen storage materials. Such internal stress is closely related to grain size, with larger grain sizes correlating to higher internal stress. In Pd/Mg/Pd layered thin films prepared by sputtering, the Mg layer comprises fine columnar crystals (diameters: 10–30 nm) aligned along the *c*-axis (Fig. 5.17). Elastic interactions between the nanostructured Mg and Pd layers significantly modify the thermal stability of MgH_2, lowering the adsorption temperature of MgH_2 to slightly above room temperature [30].

The effect of biaxial strain on the dehydrogenation thermodynamics of MgH_2 was investigated using the first-principles calculations [29]. Biaxial isotropic strains

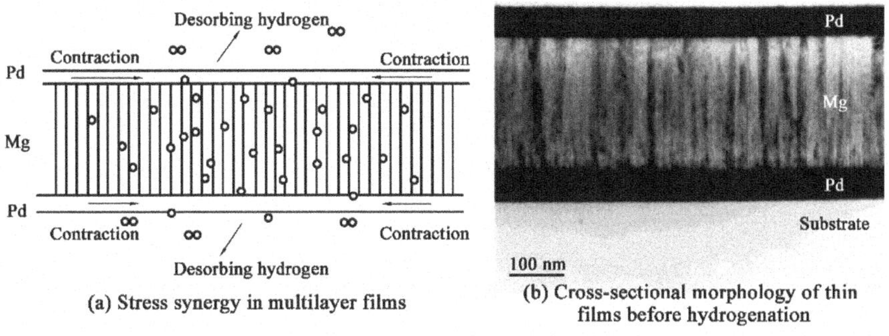

Fig. 5.17 Stress synergism phenomenon and cross-sectional morphology of Pd (50 nm)/Mg (200 nm)/Pd (50 nm) multilayer films [30]

ε_{xx} and ε_{yy} affect the MgH_2 lattice via constrained lattice constants along the x- and y-axis directions (Fig. 5.18a). The calculations showed that strained MgH_2 exhibits higher total energy and strain energy compared to unstrained MgH_2. Both tensile and compressive strains exhibit similar trends, with strain energy defined as the difference in total energy between strained and unstrained MgH_2.

Figure 5.19a shows that tensile strain has a more significant effect on increasing total and strain energies than compressive strain. The total energy, which reflects crystal structure stability, increases with tensile strain, indicating reduced stability. Thus, both biaxial tensile and compressive strains destabilize MgH_2, with tensile strain having a more pronounced effect on improving the thermodynamic property of hydrogen desorption.

The introduction of strain energy modifies the Van't Hoff equation for the hydrogen desorption reaction of MgH_2:

$$\ln\left(\frac{P_{\text{eq,strained}}}{P_0}\right) = -\frac{\Delta H^\circ}{RT} + \frac{\Delta S^\circ}{R} + \frac{\int_0^{\varepsilon_{ij}} \sigma_{ij} d\varepsilon_{ij}}{RT} \tag{5.22}$$

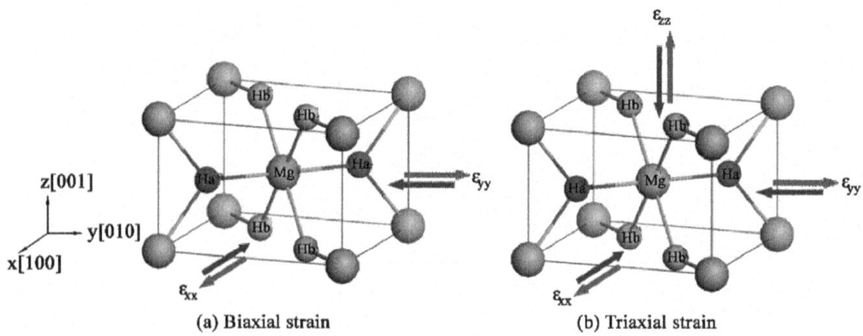

Fig. 5.18 Schematic diagram of the strained MgH_2 crystal cell models [31]

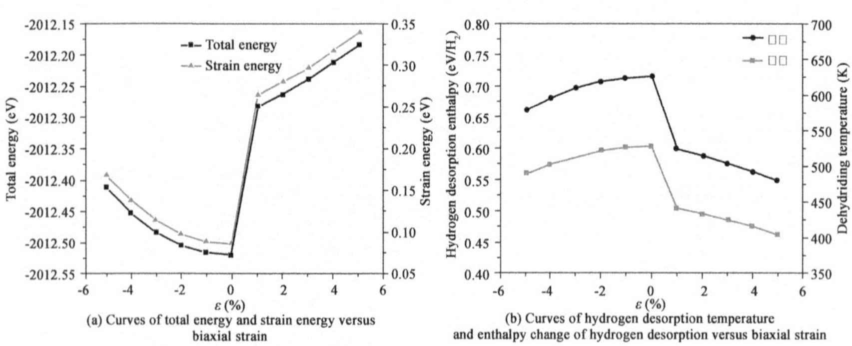

Fig. 5.19 Total biaxial strain energy, strain energy, hydrogen desorption temperature, and enthalpy change versus strain [31]

where

$P_{eq,strained}$—the equilibrium pressure of MgH_2 after strain is applied.

$\int_0^{\varepsilon_{ij}} \sigma_{ij} d\varepsilon_{ij}$—the strain energy difference between strained and unstrained MgH_2.

The effect of strain energy on hydrogen desorption temperature at an equilibrium pressure of 0.1 MPa is expressed as:

$$T_{\text{strained}} = T^0 - \frac{\int_0^{\varepsilon_{ij}} \sigma_{ij} d\varepsilon_{ij}}{\Delta S^0} \tag{5.23}$$

Calculated curves (Fig. 5.20b) show the relationship between the change in hydrogen release enthalpy and temperature with different strains. Compared to unstrained MgH_2, both enthalpy and temperature of hydrogen release decrease as the applied strain increases. These values are lower under tensile strain than under compressive strain. This indicates that due to the contribution of strain energy, biaxial tensile or compressive strains applied to MgH_2 facilitate the hydrogen release thermodynamics, with more notable improvements observed under tensile strain.

Based on the biaxial strain analysis, the effect of triaxial strain on the thermodynamics of hydrogen release from MgH_2 was explored via first-principles calculations, as shown in Fig. 5.18b. The results indicate that MgH_2 subjected to triaxial tensile strain exhibits increased total energy and reduced hydrogen release enthalpy compared to its uniaxial or biaxial tensile strained counterparts. This suggests that triaxial strain can further reduce the thermodynamic stability of MgH_2, as depicted in Fig. 5.20. Similar to the biaxial strain, the tensile strain has a more pronounced effect on the thermodynamics of hydrogen release from MgH_2.

Although alloying, compositing, and nanosizing strategies can enhance the hydrogenation thermodynamic properties of Mg-based alloys to some extent, differences exist in the methods used to improve these properties of each material, as well as in the degree of enhancement achieved. To comprehensively improve the performance

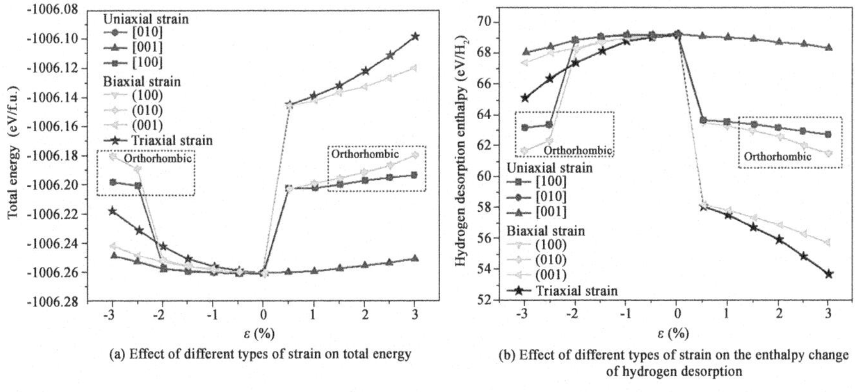

Fig. 5.20 Effect of different types of strain on the total energy and enthalpy change of hydrogen desorption [32]

of Mg-based materials, significantly lower thermodynamic temperatures, and enable mild conditions for hydrogen absorption/desorption, a multi-strategy approach has gained popularity. In particular, beyond alloying with elements, diverse preparation techniques are employed to control the particle and grain sizes of Mg or MgH_2 at the nanoscale. This approach weakens the Mg-H bonding energy, reduces the reaction enthalpy change, and minimizes the hydrogen desorption temperature to the greatest extent possible.

(3) Nanowire

In addition to Mg nanoparticles and nanofilms, magnesium nanowires can also reduce the thermodynamic stability of MgH_2. Theoretical calculations show that the hydride of Φ1 nm Mg nanowires exhibits a decomposition enthalpy of 37.55 kJ/mol H_2 and a desorption temperature of 287 K [33]. Analysis of the electronic structure and stability of Mg nanowires reveals that the Mg-H bond is ionic, with an energy barrier of 3.57 eV in bulk MgH_2. Mg nanowires exhibit discrete energy levels, with an energy barrier of ~1.36 eV in the electronic structure of the MgH_2 nanowire, indicating lower stability compared to bulk MgH_2 [33]. From the perspective of thermodynamic instability, Mg nanowires can be effectively considered as a promising hydrogen storage material.

Mg nanowires with diameters of 30–50, 80–100, and 150–170 nm were synthesized via vapor phase reaction [34] (Fig. 5.21). The hydrogenation enthalpies for these three types of Mg nanowires were approximately 65.3 kJ/mol H_2, 65.9 kJ/mol H_2, and 67.2 kJ/mol H_2, respectively. Notably, the hydrogenation enthalpy of Mg nanowires is lower than that of bulk Mg (74.0 kJ/mol H_2) and can be further decreased as the diameter of the nanowires reduces. First-principle calculations were employed to study the relationship between the thermodynamic stability of MgH_2 nanowires and their sizes [35]. The results indicate that when the diameters of the Mg nanowires were 1.24 nm, 0.85 nm, and 0.68 nm, their hydrogen release enthalpies were −61.86 kJ/mol H_2, −34.54 kJ/mol H_2, and −20.64 kJ/mol H_2, respectively. The decrease in diameter results in thermodynamic instability, allowing nanowires with a diameter of 0.85 nm to store hydrogen at room temperature. Energy and electronic structure analysis indicates that the thermodynamic properties of MgH_2 nanowires are related to the energy of both Mg nanowires and MgH_2. The reduced nanowire diameter leads to higher energy instability in both, but the instability is more pronounced in MgH_2 nanowires, causing increased thermodynamic instability as the diameter decreases. Additionally, for MgH_2 nanowires with a diameter of 0.68 nm, each hydrogen atom is single-atom coordinated. For those with a diameter of 0.85 nm, each hydrogen atom is coordinated with two Mg atoms. In Φ 1.24 nm MgH_2 nanowires, 25% of hydrogen atoms are single-atom coordinated, another 25% of hydrogen atoms are three-coordinated, and the remaining are four-coordinated. Therefore, the varying diameters of Mg nanowires result in different hydrogen coordination states, weaker Mg-H bonds, and lower hydrogen desorption temperatures [35]. In summary, decreasing the diameter of Mg nanowires can effectively influence the thermodynamic properties of MgH_2, making them promising candidates for hydrogen storage applications.

Fig. 5.21 TEM images of Mg nanowires synthesized by gas-phase reaction at different flow rates [34]

In conclusion, alloying, compositing, and nanosizing strategies can improve the thermodynamic properties of hydrogen absorption and desorption for Mg-based alloys to varying degrees. To more effectively enhance the properties of Mg-based materials, sharply lowering their thermodynamic temperature, and facilitating hydrogen absorption and desorption under mild conditions is preferred. The combination of various strategies is a key focus in future research. For example, when introducing alloying elements, different fabrication methods can be employed to control the particle and grain sizes of Mg or MgH_2 at the nanoscale. This approach substantially weakens the Mg-H bond strength, reduces reaction enthalpy, and lowers the temperature required for hydrogen desorption.

5.2 Kinetic Modification of Mg-Based Hydrogen Storage Alloys for Hydrogenation Reactions

5.2.1 The Influence Factors on Kinetics of Hydrogen Absorption/Desorption Reactions

The hydrogen absorption/desorption process of Mg-based hydrogen storage alloys involves complex multiphase gas–solid reactions, with each stage characterized by energy barriers (Fig. 5.22). Overcoming these barriers is essential for the reaction to occur. The energy required to do so is referred to as the activation energy for the specific reaction stage. As discussed in Chap. 2, the hydrogen absorption process in Mg-based alloys encompasses several steps, including physical adsorption, chemical adsorption, surface permeation, hydrogen diffusion, β-phase nucleation and growth, and interfacial chemical reaction. The reverse process constitutes hydrogen desorption. Therefore, when any of these steps becomes the rate-controlling step, strategies must be proposed to reduce its activation energy.

The root causes of sluggish kinetics in magnesium hydrogen adsorption include:

① Formation of passivation layers: The rapid formation of an oxide or hydroxide passivation layer on the surface of pristine magnesium prevents the adsorption and dissociation of hydrogen molecules on the particle surface, as well as the inward diffusion of hydrogen atoms.
② Low hydrogen diffusion rates: The diffusion rate of hydrogen in MgH_2 (1.5×10^{-16} m^2/s, 300~600 K) is significantly lower than that in Mg (4×10^{-13} m^2/s, 493~473 K). Once the MgH_2 layer forms, hydrogen atom diffusion becomes exceedingly difficult.

The main reasons for the sluggish dehydrogenation kinetics of MgH_2 are as follows:

① The diffusion rate of hydrogen is low in both MgH_2 and Mg.
② High energy is required for the nucleation of Mg on the surface of MgH_2.
③ The recombination of hydrogen atoms into hydrogen molecules on the Mg surface requires overcoming a high energy barrier.
④ Even in extremely clean laboratory environments, a MgO or $Mg(OH)_2$ layer inevitably forms on the magnesium surface, hindering the dissociation of hydrogen molecules and subsequent diffusion of hydrogen atoms.

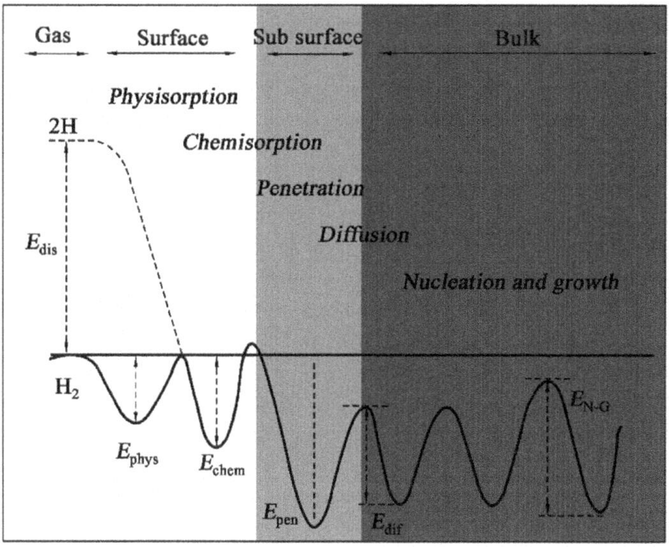

Fig. 5.22 Lennard–Jones potential mapping corresponding to the energy barrier during the Mg/ MgH_2 hydrogen absorption/desorption process [26]

To address these challenges, two primary strategies can be employed to improve the hydrogen absorption/desorption kinetics of magnesium alloys. One method is enhancing the dissociation rate of hydrogen molecules. This can be achieved by adding catalysts that facilitate the dissociation rate of hydrogen molecules on the material surface. The other method is reducing activation energy and boosting diffusion rates. This includes modifying the hydrogen desorption pathway of MgH_2 through methods such as alloying, nano-compositing, and introducing crystalline defects. Additionally, nano-compositing and the creation of structural defects can enhance hydrogen atom diffusion.

5.2.2 Catalyst Addition

Adding catalysts is an effective method to enhance the kinetics of chemical reactions. Catalysts can disrupt the metal oxide layer on the particle surfaces, facilitating the dissociation of hydrogen molecules and activating the surface for hydrogen desorption. This reduces the incubation period for hydriding/dehydriding reactions in Mg-based hydrogen storage alloys and accelerates the reaction rates. The effect of catalysts on the hydrogen storage kinetics of Mg-based alloys can be explained by the "hydrogen pump effect", "spillover effect", "electron transfer", "channel effect", etc. The hydrogen pump effect involves weakening the crystal stability of MgH_2 by doping it with transition metals such as Ti, V, Nb, or RE elements. These doping atoms form hydride clusters of T_mH_x (T_m = Ti, V, Nb, RE, etc.) with neighboring hydrogen atoms. During the dehydrogenation process, T_mH_x clusters preferentially release hydrogen, thereby improving the dehydrogenation kinetics of the MgH_2 system by providing an alternative path for hydrogen release. The spillover effect describes how hydrogen molecules are easily dissociated into hydrogen atoms on the doped catalyst surface, facilitating the migration of hydrogen atoms into the magnesium matrix. The electron transfer effect refers to the excess electrons generated during the valence changes of catalysts, which are transferred into MgH_2, promoting its decomposition and enhancing reaction kinetics. The channel effect describes the doped catalysts acting as channels for the transportation of hydrogen atoms in MgH_2, accelerating hydrogen transfer within the material and consequently enhancing the rates of hydrogen absorption and desorption. Based on these catalytic mechanisms, common catalysts for MgH_2 can be categorized into metal monomer catalysts, compound catalysts, and carbon-based catalysts.

(1) Metal Monomer Catalysts

Transition metals share a unique feature: their valence electrons fill the *d* orbitals sequentially, which makes them effective catalysts for MgH_2. For example, a powder mixture of MgH_2 and 8 mol% of a transition metal (M = Al, Ti, Fe, Ni, Cu, or Nb) was fabricated via mechanical alloying to enhance the hydrogen storage capacity of MgH_2. Upon doping with elements such as Nb, Ti, or Ni, intermediate hydrides such as TiH_2, Mg_2NiH_4, and (Nb, Mg) H_x are formed after hydrogen absorption. In contrast, elements like Fe, Ni, Cu, and Al do not form hydrides when reacting with hydrogen [31]. The hydrogen desorption capacities of these samples at 573 K are illustrated in Fig. 5.23 [35]. Both un-ball-milled and ball-milled MgH_2 exhibit relatively low hydrogen desorption capacities (approximately 1.0 wt.% and 1.7 wt.%, respectively). However, after Ni doping, the maximum hydrogen desorption capacity in mechanically alloyed mixtures reaches 7.5 wt.%, achieving equilibrium within just 15 min, demonstrating excellent kinetic properties. The catalytic effects of other transition metals follow this descending order: Al, Fe, Ti, Nb, and Cu. The corresponding hydrogen desorption capacities are 5.7, 5.3, 2.6, 2.6 and 1.7 wt.%, respectively.

(2) Compound catalysts

Compound catalysts, including oxides, halides, carbides, sulphides, and hydrides, significantly enhance the hydrogen storage performance of MgH_2. For instance, the $TiVO_{3.5}$ catalyst, synthesized by the oxidizing $(Ti_{0.5}V_{0.5})_3C_2$, exhibits a synergistic catalytic effect on MgH_2 [36]. A composite of MgH_2 with 10 wt.% $TiVO_{3.5}$ can release 5.0 wt.% H_2 within 10 min at 523 K and absorb 3.9 wt.% H_2 within 5 s at 373 K under 5 MPa hydrogen pressure. The hydrogen desorption activation energy of this system is reduced to 62.4 kJ/mol H_2, demonstrating the significant enhancement of the hydrogen desorption kinetics of MgH_2. The catalytic mechanism involves the reduction of $TiVO_{3.5}$ to metallic Ti and V during ball milling. These metals coat the MgH_2 particle surfaces, creating active sites that facilitate the dissociation and recombination of hydrogen molecules during absorption/desorption, thereby improving hydrogen storage performance. Oxide catalysts exhibit varying degrees

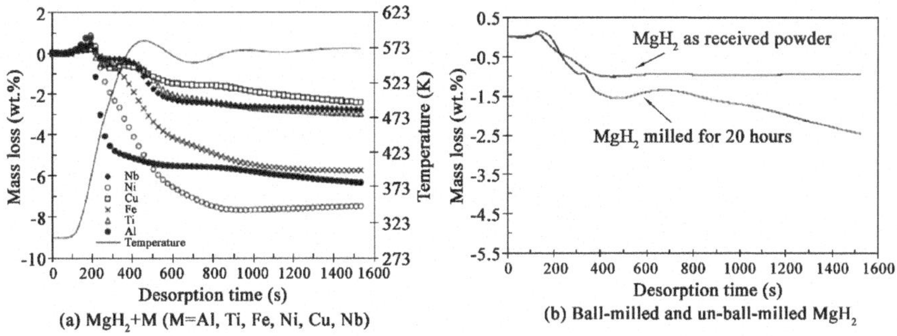

Fig. 5.23 The relationship between capacities and time of hydrogen desorption for different MgH_2-Catalyt mixtures at 573 K [35]

Fig. 5.24 The comparison of hydrogen desorption rates of MgH_2 with different oxide catalysts under vacuum at 573 K [37]

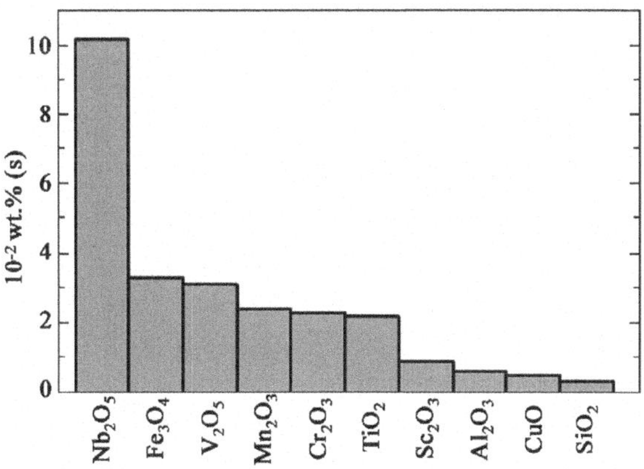

of effectiveness in enhancing hydrogen desorption rates. For example, composites of MgH_2 with oxide catalysts at 573 K under vacuum conditions show differing hydrogen desorption rates, with Nb_2O_5 demonstrating the best catalytic performance (Fig. 5.24) [37].

Polyvalent Titanium-based catalysts, such as TiH_2, TiO_2, and $TiCl_3$, have garnered significant research interest due to their outstanding catalytic properties. The strategy of using multivalent titanium-based catalysts (TiH_2, TiO_2, and $TiCl_3$) to coat MgH_2 particles [38] forms a protective layer (~10 nm thick) on the MgH_2 surface, significantly improving hydrogen desorption kinetics. This approach achieves a hydrogen desorption capacity of approximately 5 wt.% within 15 min at 523 K (Fig. 5.25a). As shown in the figure, numerous interfaces exist between MgH_2, Ti^{2+}, and $Ti^{3+/4+}$ ions, facilitating the transitions between different titanium valence states. The hydrogen absorption and desorption process of MgH_2 particles coated with multivalent Ti-based catalysts is illustrated in Fig. 5.25b, and involves the following steps:

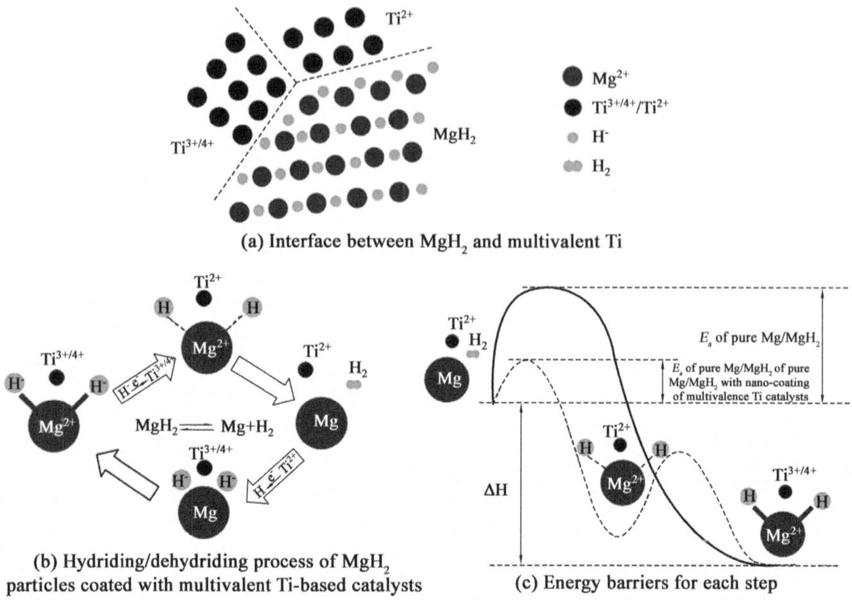

Fig. 5.25 Schematic illustration of the catalytic mechanism in the hydriding/dehydriding process of Mg-Catalyst composites [38]

① Hydrogen dissociation: H_2 molecules dissociate into H atoms on the catalyst surface and diffuse to the catalyst-Mg interface.
② Valence transition: Ti^{2+} ions spontaneously transition to $Ti^{3+/4+}$ ions, transferring electrons to H atoms, which transform into H^-.
③ Electron transfer: Electrons from Mg are transferred to $Ti^{3+/4+}$, forming Mg^{2+} ions and Mg-H bond, initiating the hydrogen absorption.
④ Nucleation and growth: MgH_2 nucleation and growth of MgH_2 complete the hydrogen absorption process.

The electron transfer process during hydrogen desorption follows the reverse steps described above. Various analyses suggest that multivalent Ti-based catalysts can act as carriers for electron transfer in the hydrogen absorption/desorption process, enhancing the electron transfer efficiency between Mg^{2+} and H^- ions. This characteristic facilitates the binding and release of two H atoms to form H_2 on the surface of the composites, thereby reducing activation energy and improving the system's kinetics (Fig. 5.25c).

Experimental studies have revealed the influence and underlying mechanisms of fluoride compounds (ZrF_4, NbF_5, TaF_5) and chloride compound ($TiCl_3$) additives on the hydrogen absorption and desorption properties of MgH_2 [39]. The results demonstrate that these halides effectively enhance the hydrogen desorption performance of MgH_2, as evidenced by the hydrogen absorption and desorption curves of the samples at 598 K shown in Fig. 5.26. Among these additives, ZrF_4 exhibits the most

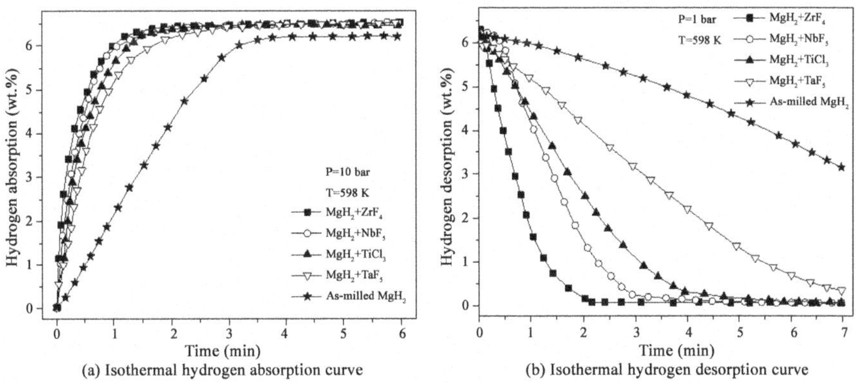

Fig. 5.26 Hydrogen absorption/desorption kinetics of MgH_2 doped with different catalysts at 598 K [39]

significant catalytic performance. For instance, the pristine MgH_2 shows hydrogen absorption and desorption rates of 0.4 wt.%/min and 2.2 wt.%/min, respectively. After the addition of ZrF_4 catalyst, these rates increase to 3.2 wt.%/min and 4 wt.%/min, respectively. The catalytic performance of the additives follows the descending order: ZrF_4, NbF_5, $TiCl_3$, and TaF_5. Notably, ZrF_4 does not participate in chemical reactions during the cycling process. Its catalytic mechanism relies on Zr and F atoms facilitating the dissociation or recombination of hydrogen molecules on the surface of ZrF_4. In contrast, NbF_5 and TaF_5 participate in the reactions during the hydrogen absorption and desorption processes, forming MgF_2. This reaction weakens the Mg-H bond and enhances the hydrogen atom diffusion rate. The catalytic mechanism of $TiCl_3$ differs from the two mentioned above. The multivalent Ti ions serve as active sites for electron transfer during the cycling process.

After doping TiC into MgH_2 using the ball milling technique, the activation energy for MgH_2 dehydrogenation decreases from 191.27 kJ/mol H_2 to 145.62 kJ/mol H_2 [40]. Furthermore, the composite achieves hydrogen desorption of 6.3 wt.% at 573 K, with the potential for further hydrogen absorption of 6.0 wt.% under 1 MPa hydrogen pressure. In addition to the improved hydrogen absorption and desorption kinetics, this system also maintains good cycling stability. After 10 absorption/desorption cycles, the capacity loss is approximately 3.3 wt.%. The underlying catalytic mechanism of TiC includes two key aspects: On one side, Ti atoms can combine with hydrogen molecules to form TiH_2. On the other side, the autocatalytic effect of TiC promotes the dissociation and adsorption of hydrogen, thereby improving the hydrogen adsorption and desorption kinetics of MgH_2. These mechanisms can be expressed by the following equations:

$$Ti + H_2 \rightarrow TiH_2 \tag{5.24}$$

$$TiC_3 + xH_2 \rightarrow 2Ti - CH_x \tag{5.25}$$

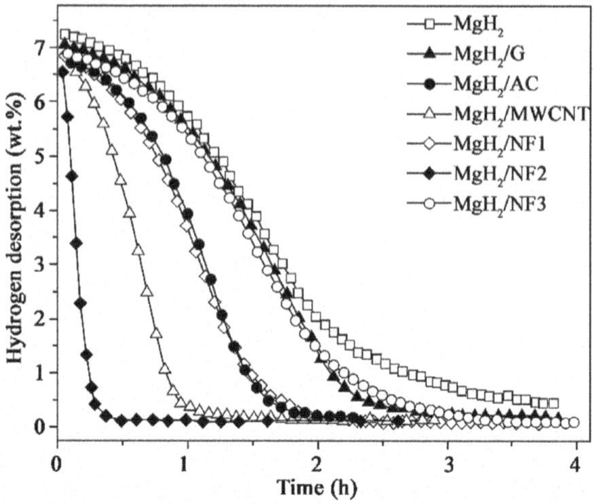

Fig. 5.27 Hydrogen desorption curves of MgH_2-carbon composites [41]

(3) Carbon-based catalysts

Carbon-based materials with different morphologies, e.g., graphite, activated carbon, carbon nanofibers, and multi-walled carbon nanotubes (MWCNTs), can enhance the hydrogen storage performance of magnesium to varying extents [41]. Previous studies indicate that carbon nanofibers and MWCNTs exhibit significantly better performance compared to graphite and activated carbon (Fig. 5.27). The composite of MgH_2-5 wt.% MWCNTs can completely release hydrogen within 20 min at 573 K, whereas at least 4 h are required without the assistance of MWCNTs. Moreover, the MgH_2-MWCNTs composite maintains rapid hydrogen absorption and desorption rates during cycling. The mechanism underlying the enhancement of hydrogen absorption and desorption kinetics can be summarized as follows:

① Carbon nanotubes serve as transport channels for hydrogen diffusion, thereby accelerating hydrogen diffusion in the Mg phase.
② The interaction between Mg and C provides abundant active sites for hydrogen dissociation and diffusion.
③ Carbon materials effectively inhibit the growth and aggregation of Mg or MgH_2 particles during cycling.
④ The lubricating effect of carbon nanotubes enhances the efficiency of mechanical alloying, facilitating the formation of smaller MgH_2 particles [42].

For different types of catalysts, the mixed addition of various catalysts demonstrates significantly improved performance. For instance, MWCNTs and TiF_3 catalysts were individually and jointly added to a melt-spun Mg-10 wt.% Ni alloy via high-energy ball milling. The activation kinetics were investigated and are shown in Fig. 5.28 [43]. The slope of the initial hydrogen absorption kinetics curves indicates a significant improvement in the activation kinetics of the Mg-10 wt.% Ni alloy after the individual addition of either MWCNTs or TiF_3. However, when MWCNTs and TiF_3

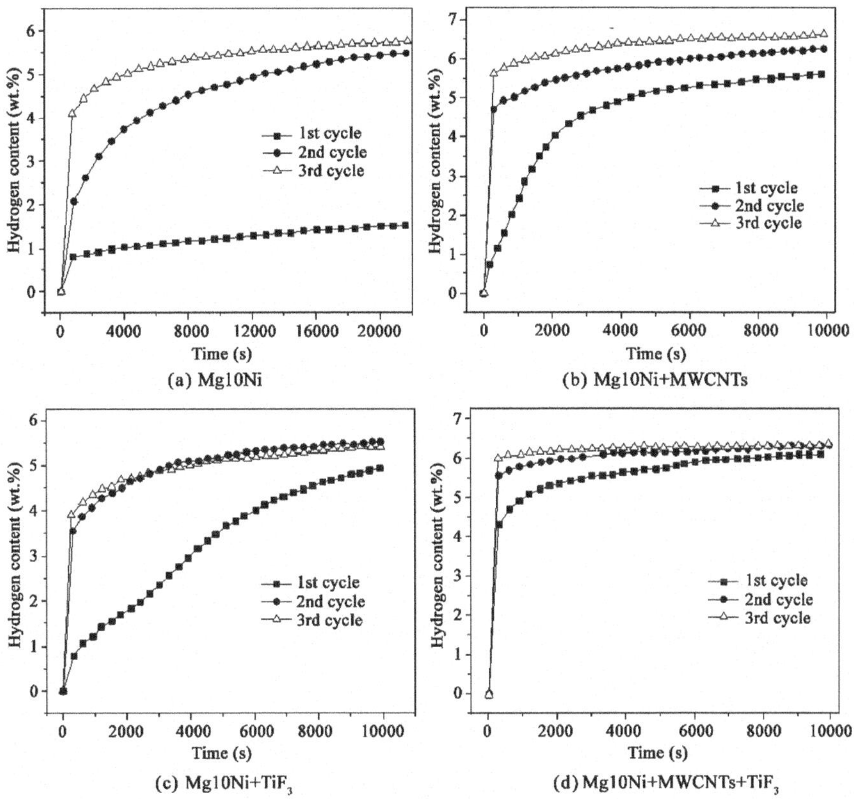

Fig. 5.28 Activation kinetics of Mg-10Ni alloys solely or synergistically catalyzed by MWCNTs and TiF_3 [43]

were co-added, the Mg-10 wt.%Ni-5 wt.%–5%MWCNTs-5 wt.%TiF_3 composite exhibited a rapid hydrogen absorption rate upon initial exposure, with a hydrogen absorption capacity reaching 6.1 wt.%. The result indicates that the MWCNTs-TiF_3 composite acts as an effective catalysis for improving the hydrogenation kinetics of magnesium-rich alloys. The excellent catalytic performance is attributed to the synergistic effect of the assisted mass transfer from MWCNTs and the heterogeneous nucleation facilitated by TiF_3.

5.2.3 *Alloying and Compounding*

Alloying and compounding are among the most effective and commonly used methods to improve the hydrogen absorption and desorption kinetics of magnesium. Introducing alloying elements into magnesium to form solid solutions can modify the lattice parameters of magnesium alloys, thereby facilitating hydrogen diffusion and reducing the activation energy required for magnesium hydrogen absorption and desorption. Furthermore, the introduction of composites or intermetallic compounds through alloying can alter the reaction pathways of hydrogen absorption and desorption, further lowering activation energy. Based on the materials formed, alloying methods can be categorized into the following: Mg-based single-phase solid solutions, intermetallic compounds, and composites involving intermetallic compounds or hydrides.

(1) Single-phase solid solution alloys

When certain elements exhibit solubility in magnesium, they can form single-phase solid solutions, as observed in the Mg(In, Y) solid solution and the intermetallic compound of Mg_2Ni. The Mg(In, Y) solid solution can be synthesized via a two-step method involving sintering and ball-milling. Its hydrogenation kinetics are shown in Fig. 5.29 [44]. The formation of Mg(In, Y) solid solution is preliminarily due to the incorporation of indium (In) in Mg, which causes lattice expansion of Mg, thereby facilitating the introduction of yttrium (Y). Upon hydrogenation, the $Mg_{90}In_5Y_5$ solid solution transforms into compounds such as MgH_2, YH_3, In_3Y, and MgIn. Notably, except from YH_3, all hydrogenation products can reversely convert back to the Mg(In, Y) solid solution after hydrogen release. Constructing the Mg(In,Y) ternary solid solution enhances the hydrogenation kinetics of MgH_2. Additionally, the element In has been introduced into Mg_2Ni through powder sintering and mechanical milling to form the $Mg_2In_{0.1}Ni$ solid solution [45]. This solid solution increases the lattice constant of Mg_2Ni, promoting hydrogen atom diffusion and improving the hydrogen desorption kinetics. Compared to Mg_2Ni, the activation energy for hydrogen desorption decreases significantly, from 80 kJ/mol H_2 to 28.9 kJ/mol H_2.

(2) Formation of Intermetallic Compounds

As mentioned in Sect. 5.1.2, magnesium (Mg) can form metal hydride alloys with various elements, and substitution on both the A-side and B-side of magnesium

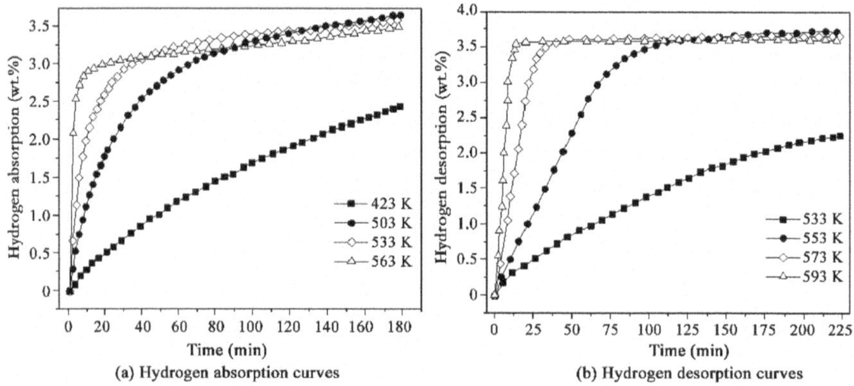

Fig. 5.29 Isothermal hydrogen absorption/desorption kinetics of $Mg_{90}In_5Y_5$ solid solutions [38]

alloys can enhance their kinetic performance. A-side substitution elements typically include rare earth (RE), titanium (Ti), vanadium (V), etc., while B-side substitution mainly involves transition metal elements. The substitution of rare earth elements in magnesium alloys forms different phases depending on the amount added, enabling substitution on both the A-side and B-side. For example, $Mg_{2-x}Nd_xNi$ ($x = 0, 0.1, 0.2, 0.3$) alloys were prepared by induction melting, where Nd mainly substitutes Mg on the A-side [46]. The results show that different amounts of Nd substitution result in varied hydrogen absorption and desorption kinetics of the alloy. Based on the slope of the initial hydrogen desorption curves (Fig. 5.30), it is evident that Nd substitution for Mg can enhance the hydrogen desorption rate compared to the pristine Mg_2Ni alloy. Moreover, as the Nd content increases, the hydrogen desorption rate continuously improves. The primary root cause is the formation of multiphase structures, including Mg_2Ni, Nd_2Ni_7, $NdMg_{12}$, and Nd_5Mg_{41}, which generate numerous phase boundaries and introduce synergistic and catalytic effects for hydrogen absorption and desorption. However, it is noteworthy that the improved kinetics after Nd substitution come at the expense of a portion of hydrogen storage capacity.

The kinetic properties of the Mg_2Ni system can also be modified via the replacement of the B-side element nickel (Ni) with yttrium (Y). The initial hydrogen absorption kinetic curves at 3 MPa and 623 K of $Mg_{67}Ni_{33-x}Y_x$ ($x = 0, 1, 3, 6$) samples, prepared via the melt quenching method, are shown in Fig. 5.31a [47]. Compared with pristine Mg_2Ni, the introduction of rare earth element Y on the B-side with varying amounts ($x = 1, 3, 6$) results in improved initial hydrogen absorption kinetics to some extent [48]. Based on the Kissinger curves and hydrogenation activation energies (E_a) (Fig. 5.31b, c) it is observed that the hydrogenation activation energy decreases to some extent after the substitution of Y on the B-side.

In conclusion, hydrogen absorption and desorption kinetics of magnesium can be significantly enhanced by incorporating transition metal and rare earth elements to form Mg-based solid solutions or intermetallic compounds. Furthermore, element

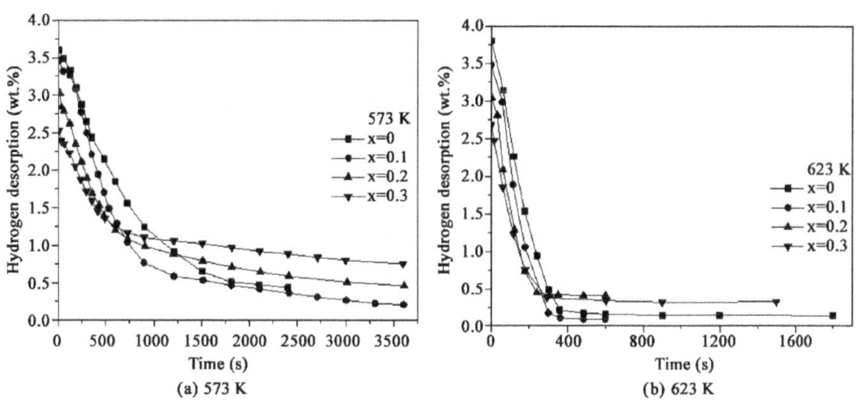

Fig. 5.30 Kinetic curves of hydrogen desorption at 0.1 MPa for $Mg_{2-x}Nd_xNi$ ($x = 0, 0.1, 0.2, 0.3$) alloys [46]

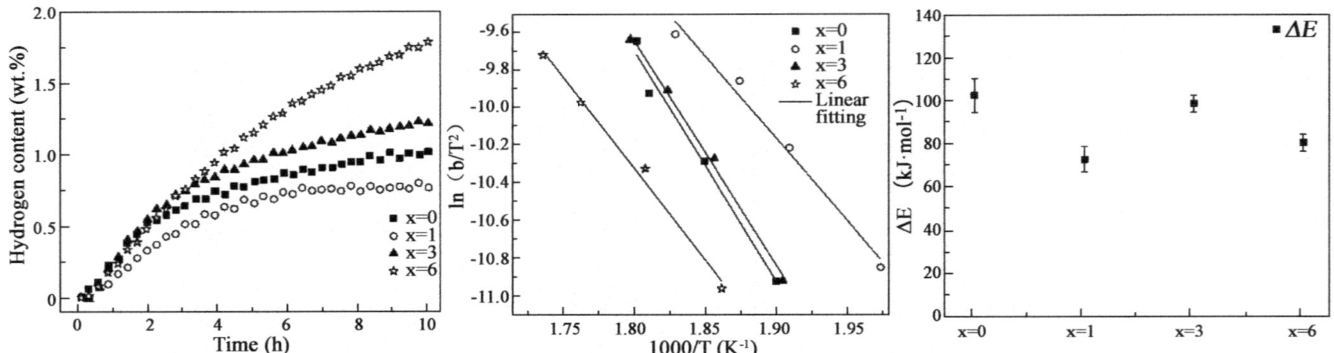

(a) The initial hydrogen absorption kinetic curves at 30 bar and 623 K

(b) Kissinger-curves

(c) The hydrogenation activation energy

Fig. 5.31 The activation energy and kinetic properties of melt quenched $Mg_{67}Ni_{33-x}Y_x$ ($x = 0, 1, 3, 6$) samples [47, 48] substitution on the A-side or B-side of intermetallic compounds further improves hydrogenation kinetics. The mechanisms behind these enhancements include:

① Lattice volume expansion and generated phase boundaries: These structural changes facilitate hydrogen mass transfer.
② Introduction of active intermediates: These intermediates catalyze hydrogen desorption from MgH_2 through synergistic effects.

However, it is essential to note that the addition of alloying elements typically reduces the hydrogen storage capacity of the system. To balance the improved kinetics with high hydrogen storage capacity, the formation of solid solutions or multiphase composites via multi-element microalloying strategies is a rational option.

(3) Complexes with intermetallic or hydrogen compounds

The hydrogen absorption/desorption kinetics of Mg can be significantly improved by compositing it with other intermetallic hydrogen storage compounds to create magnesium matrix composites. This approach leverages the kinetic advantages of metal hydride materials. For instance, combining traditional hydride materials such as AB_5, AB, AB_2 or other alloys, which possess excellent hydrogen absorption and desorption kinetics, with Mg to form composites, especially in the form of nano-material composites, greatly enhances the performance of Mg. For example, Mg or MgH_2 were mixed with 30 wt.% $LaNi_5$ using the ball-milling technique, resulting in composites with significantly enhanced hydrogen absorption and desorption kinetics (Fig. 5.32) [49]. This improvement is attributed to the transformation of the composite into a mixture of $MgH_2 + LaH_3 + Mg_2NiH_4$ after hydrogenation. Notably, LaH_3 positively influences hydrogen absorption but has a negligible effect during desorption. Furthermore, Mg_2Ni exhibits a more pronounced catalytic effect than LaH_3 at high temperatures. Thus, the phases formed by adding $LaNi_5$ play a synergistic role in the hydrogen absorption/desorption processes of Mg.

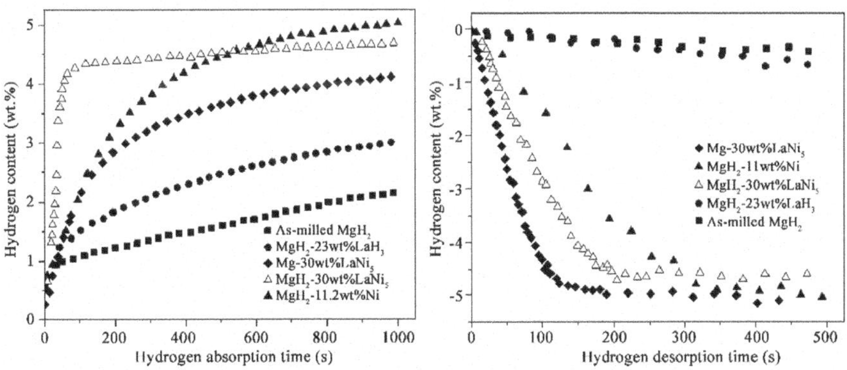

(a) Hydrogen absorption kinetic curves at 423 K and 1 MPa hydrogen pressure
(b) Hydrogen desorption kinetic curves at 423 K

Fig. 5.32 Hydrogen absorption/desorption kinetics for the Mg-$LaNi_5$ and MgH_2-$LaNi_5$ systems [49]

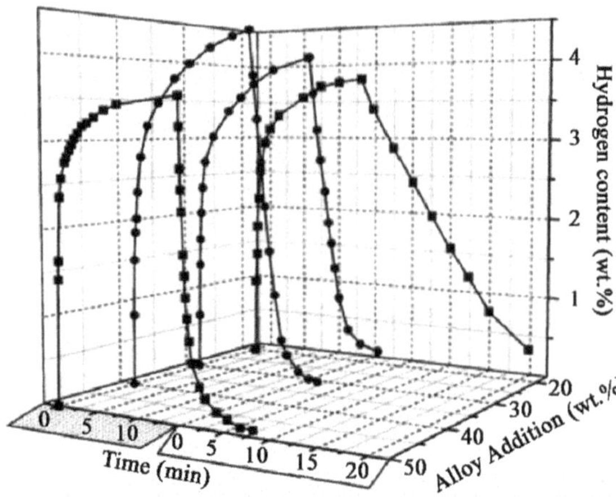

Fig. 5.33 Hydrogen absorption/desorption kinetics of Mg-x wt.% $ZrFe_{1.4}Cr_{0.6}$ at 623 K [50]

Complexes can also be formed between Mg and AB intermetallic compounds. For instance, the hydrogen absorption and desorption kinetics of Mg-Mg-$ZrFe_{1.4}Cr_{0.6}$ complexes are shown in Fig. 5.33 [50]. The Mg-40 wt.% $ZrFe_{1.4}Cr_{0.6}$ system exhibits the best hydrogen absorption and desorption kinetics, with a hydrogen storage capacity of approximately 4.25 wt.%. Over 80% of the hydrogen can be absorbed within one minute at 623 K and 2 MPa hydrogen pressure, while 80% of the hydrogen is desorbed within 5 min at 623 K and 0.1 MPa. The exceptional kinetic properties of this system arise from the catalytic effect of the $ZrFe_{1.4}Cr_{0.6}$ alloy and the nanostructure of magnesium. Unlike typical catalytic methods, where the catalyst (usually less than 5 wt.%) contributes little or negatively to hydrogen storage capacity, the catalytic phases, such as AB_5, AB, and AB_2, directly participate in hydrogen absorption and desorption. Consequently, their maximum content in the composite can reach 30–50%.

For coordination hydrides, systems like MgH_2-$LiBH_4$ significantly reduce reaction enthalpy through the formation of MgB_2. However, these systems require more reversible hydrogen absorption conditions. Additionally, the H–H exchange between $LiBH_4$ and MgH_2 partially hinders the decomposition of MgH_2, increasing the dehydrogenation activation energy from 77 kJ/ mol H_2 to 125 kJ/mol H_2 after the addition of Nb_2O_5. Therefore, when designing composites using hydrogen-containing hydrides and MgH_2, the effect of H–H exchange on hydrogen desorption performance must be carefully considered.

In summary, the hydrogen absorption and desorption kinetics of Mg-based composites can be improved by incorporating intermetallic compounds and coordination hydrides. The enhancements in hydrogen absorption and desorption kinetics of Mg-based metal hydrides can be attributed to the following factors:

① Phase boundaries and grain boundaries: A high density of phase boundaries and grain boundaries provides numerous nucleation sites for new phases, accelerating the hydrogen absorption reactions.

② Hydrogen transport channels: These high-density phase boundaries and grain boundaries act as fast transportation pathways for hydrogen, effectively reducing the diffusion distance at the nanoscale.
③ Catalytic effects: Catalysts within the composite facilitate hydrogen absorption and desorption.
④ Synergistic effects: The interaction between different hydrogen storage phases enhances overall hydrogen adsorption and desorption properties.

5.2.4 *Nano-crystallization*

Nano-crystallization not only enhances the thermodynamics but also significantly improves the kinetics of hydrogen absorption/desorption reactions. Nano-, or non-crystallization is the key method for improving the kinetics of hydrogen absorption/desorption by refining the structures, increasing diffusion channels, and shortening diffusion paths. For instance, in hydride PdH_x, the nanostructure increases the solubility of H_2 at nanograin boundaries and enhances the diffusion rate of H atoms [51], as hydrogen mainly diffuses along Mg/MgH_2 interfaces, grain boundaries or defects. The most important factor for the improved hydrogen absorption/desorption kinetics with nanocrystalline structure is the shortened diffusion distance and increased growth rate of hydrides. Furthermore, the increased solid solubility of hydrogen atoms in nanoparticles reduces the energy barrier of hydride formation, which reduces the overall activation energy of the entire hydrogen absorption/desorption process.

(1) Nanoparticles/Grains

Compared with the bulk polycrystalline Mg, nanocrystalline Mg produced via high-energy ball milling exhibits fast hydrogen absorption and release rate without the activation process [52]. Nanocrystalline Mg can be completely hydrogenated within 6.5 min at 573 K and dehydrogenated within 1 h [53]. Usually, polycrystalline Mg cannot desorb hydrogen at 573 K, and the activation energy of nanocrystalline Mg is reduced by ca. 40–60 kJ/mol H_2 as well, this progress is a significant improvement originating from the nanosizing effect. Some researchers attribute the improvement to the reduced particle size from micrometers to nanometers after ball-milling, which exposes more fresh active reaction interfaces previously covered by the oxidized MgO layer. Additionally, ball milling generates numerous defects and grain boundaries as well as increases the specific surface area by one order of magnitude versus polycrystalline Mg. These factors promote hydrogen dissociation and diffusion and finally improve the overall kinetic performance. However, some literature also reported that nanocrystalline Mg prepared by the high-energy ball milling process is easily recrystallization during the cycling process of hydrogen absorption/desorption, leading to increased grain size from a few nanometers to more than 100 nm [54]. Furthermore, the defects also disappear after the hydrogen absorption/desorption cycles. However, the decay in kinetic performance was not so obvious, implying that the improved kinetic performance by the ball-milling process is relatively less related to generated defects and stress [55]. it is mainly related to the reduced particle and grain size, which shortens the diffusion path and increases the active reaction interfaces due to the breakdown of the surface oxide layer.

Inspired by the kinetic enhancement of Mg/MgH_2 by the particle down-sizing effect via ball milling, researchers previously endeavored to synthesize nanoscale Mg (nanoparticles, one-dimensional nanowires, and two-dimensional nanofilms) to further improve the hydrogen absorption/desorption kinetics. For instance, a polymer-based Mg nanocomposite with the function of isolating water and oxygen was developed by adding poly methyl methacrylate (PMMA) into naphthalene-based lithium-reduced magnesium dichalcogenide ($MgCp_2$) [56]. The average grain size of the nanocrystalline Mg is ca. 4.9 nm. The resulting composite can reach the maximum 4 wt.% hydrogen absorption within 30 min under 473 K and 3 MPa hydrogen pressure. Additionally, magnesium nanoparticles with sizes of 25, 32, and 38 nm were synthesized by reducing $MgCp_2$ with different reductants (i.e., potassium biphenyl, potassium phenanthrene, and potassium naphthalene) [57]. The hydrogen absorption/desorption curves of the three samples at different temperatures are shown in Fig. 5.34. We can see that the magnesium nanoparticle with size of 25 nm exhibits much faster hydrogen absorption and desorption rates versus the larger one, demonstrating that the decreased particle size improves kinetics.

Nanoparticle/grain magnesium alloys exhibit remarkable kinetic behavior in hydrogen absorption/desorption, but the agglomeration of nanoparticles and growth of nanograins is harmful to the reaction kinetics and cycling stability. Hence, the nanoconfinement technique is proposed by researchers to inhibit the agglomeration and growth of nanoparticles. For instance, $MgBu_2$ is introduced into mesoporous carbon aerogels by permeation, and MgH_2 structure and n-butane (HBu) can be obtained by the hydrogenation of $MgBu_2$ and carbon aerogel complex. After the evaporation of n-butane (HBu), the nanoconfined MgH_2 structure can be obtained [58]. The resulting nanoconfined MgH_2 releases ca. 2 wt.% hydrogen within ca. 8 h at 525 K, which is five times faster than the ball-milling activated MgH_2 (1.5 wt.%

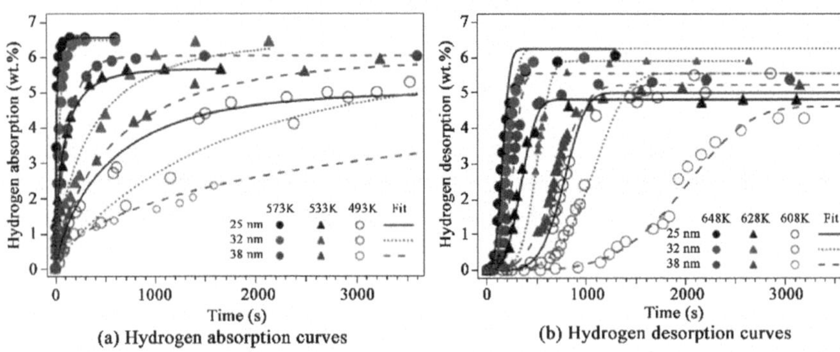

Fig. 5.34 Kinetic properties of magnesium nanoparticles with different particle sizes [57]

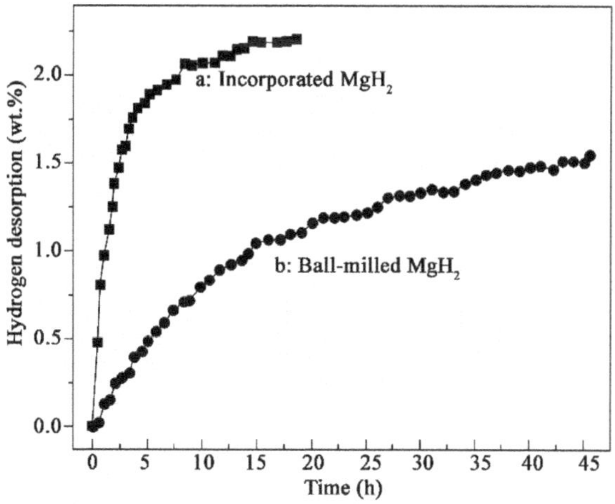

Fig. 5.35 Kinetic curves of hydrogen desorption of MgH_2 prepared by ball-milling and domain restriction methods [58]

hydrogen within 44 h, Fig. 5.35). The common sense is that the enhanced kinetics is primarily attributed to the reduced particle size.

The nanoconfined magnesium composites were prepared by reducing graphene oxide, and the grain size of Mg in the composite is ca. 3.2 nm. Furthermore, by adding a trace amount of Ni to the composite, Mg-Ni nano-alloys can be formed, the resulting sample exhibits a hydrogenation capacity of ca. 6.5 wt.% and achieves 90% hydrogen absorption within 2.5 min at 473 K and 1.5 MPa H_2. Furthermore, 90% hydrogen release in 4.6 min at 573 K and 0.001 MPa H_2 could be achieved (Fig. 5.36a) [29]. To the best of our knowledge, this is the best Mg-based hydrogen storage material reported so far. The enhancement lies in the synergistic effect of magnesium nanosizing, graphene confinement, and the catalytic effect of Ni. Thus, the nano-restricted-domain combined catalyst approach represents a novel strategy to simultaneously enhance the thermodynamic and kinetic properties of Mg-based hydrogen storage alloys.

(2) Nanofilms

Mg nanofilms with different thicknesses obtained by vapor deposition method can absorb hydrogen near room temperature but the hydrogen desorption temperature is still high. Moreover, after several hydrogen absorption and desorption cycles, the changed film structure leads to reduced kinetic performance [59]. The hydrogenation reaction kinetics of sandwiched Pd-Mg-Pd film with different thicknesses of Mg layer at room temperature was investigated (Fig. 5.37) [60]. The results show that the hydrogen absorption/desorption kinetics of the Pd-Mg-Pd film were highly dependent on the thickness of the Mg layer. Particularly, when the thickness of Mg layer is lower than 60 nm, MgH_2 is formed immediately after the sample is exposed to H_2 atmosphere at room temperature. Meanwhile, Mg layer is formed rapidly when a hydrogen desorption reaction happens in the ambient environment. Optical observation and electric impedance spectrum reveal that the hydrogen diffusion process is a critical step for hydrogen absorption/desorption.

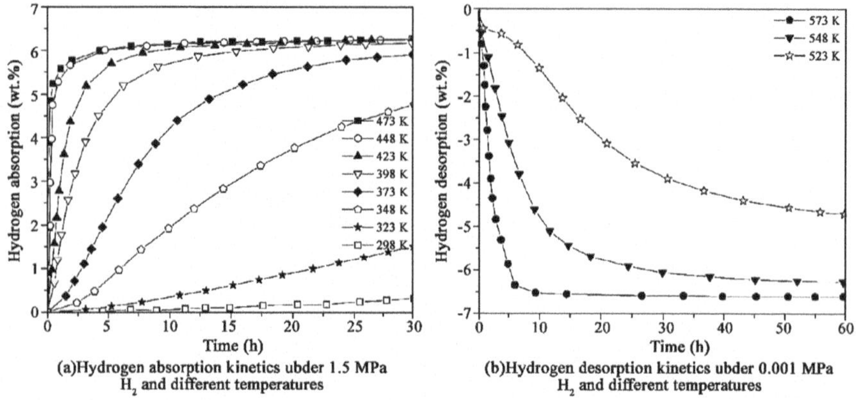

Fig. 5.36 Nanoconfined and Ni-doped magnesium composite prepared by graphene oxide reduction [29]

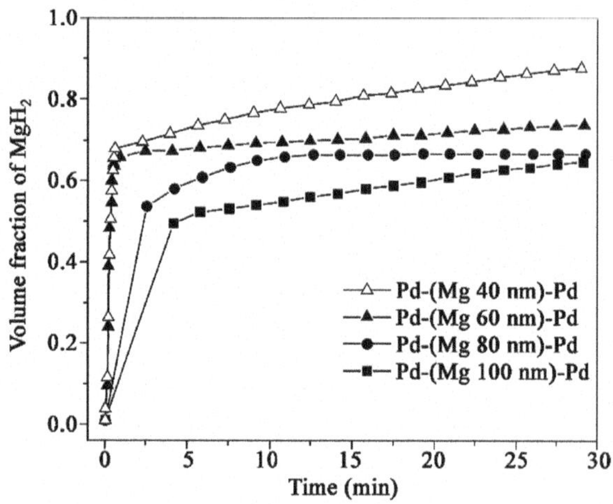

Fig. 5.37 The relationship between the volume fraction of MgH_2 and hydrogenation time of Pd-Mg-Pd film (under 4×10^3 Pa H_2 and at 298 K) [60]

(3) Nanowires

Mg nanowires with different diameters can be obtained by vapor deposition. The hydrogen absorption and desorption kinetics of the nanowires were experimentally investigated. The results revealed that nanowires with diameters ranging from 30 to 50 nm exhibited hydrogen absorption and desorption activation energy of 33.5 and 38.8 kJ/mol H_2, respectively. Moreover, the absorption and desorption activation energy decreases as the nanowire diameter decreases. The absorption/desorption curves of Mg nanowires with different diameters at various temperatures are shown in Fig. 5.38 [34].

Considering that the adsorption or desorption process involves six steps on the surface or inside of the material: (1) physical adsorption of H_2 molecule on Mg

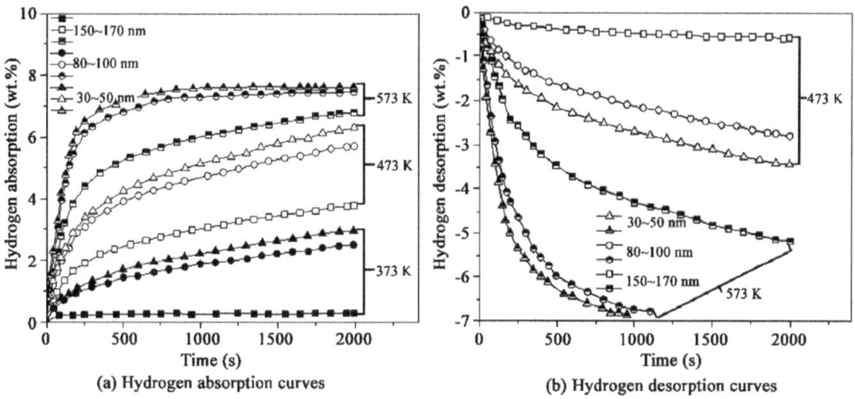

Fig. 5.38 Hydrogen absorption and desorption kinetic curves of Mg nanowires with different diameters [34]

surface; (2) H_2 molecule dissociation on Mg surface; (3) chemical adsorption of the dissociated H atom on Mg; (4) H atom permeation on the surface of Mg; (5) H atom diffusion in Mg, and (6) phase transformation ($\alpha \rightarrow \beta$). The strategies, including adding catalysts, alloying, nanosizing, compositing, etc. all improved hydrogenation kinetics to some extent. Although these strategies focus on different aspects (e.g., crystal structure, polymeric structure, micro/nanostructure, surface structure, etc.), each of them has certain limitations. Hence, to boost the overall performance of hydrogenation kinetics of Mg-based materials, the complex methods combining the strategies mentioned above are the research focus in the future.

5.3 The Application of Mg-Based Hydrogen Storage Materials

Hydrogen storage plays a crucial role in the application of hydrogen technologies. Magnesium-based solid-state hydrogen storage systems are regarded as a promising solution for advanced hydrogen storage techniques. To date, magnesium-based hydrogen storage materials have been widely used in many fields, such as large-scale safe hydrogen storage and transportation, hydrogen sources for fuel cell-powered vehicles and ships, phase-change heat storage, and energy storage. Depending on the energy requirements in different application scenarios, the conversion between chemical energy, electrical energy, thermal energy, and mechanical energy can be achieved through the reversible reactions between hydrogen and Mg-based hydrogen storage materials. This enables the development of a wide range of extended applications. The energy conversion mechanism and application areas of Mg-based hydrogen storage materials are illustrated in Fig. 5.39. Due to space limitations, this section focuses on classifying the application fields based on the principles and scales of

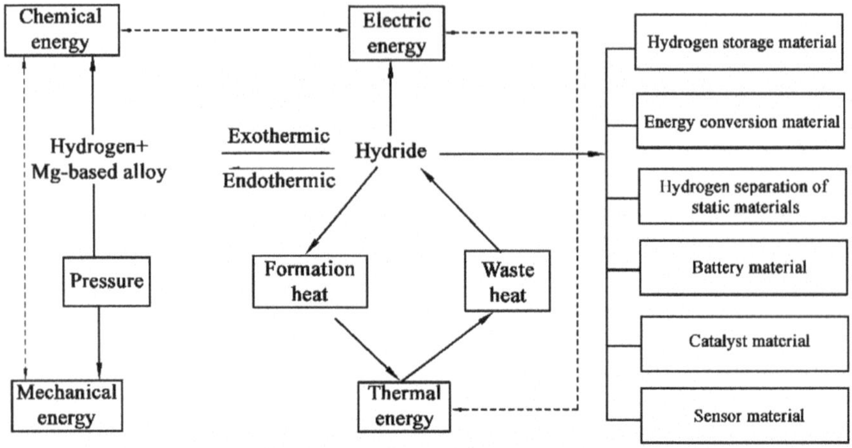

Fig. 5.39 Energy conversion mechanisms and related application fields of Mg-based hydrogen storage materials

Mg-based hydrogen storage materials, with a particular emphasis on their utilization in hydrogen storage, heat storage, energy storage, and chromogenic film.

5.3.1 Hydrogen Storage

In the hydrogen energy industry chain, high-density hydrogen storage and transportation are critical for the development of hydrogen energy and represent a bottleneck in the advancement of hydrogen energy infrastructure in China. Developing a safe, economical, efficient, and practical transportation method is essential for optimizing the full life cycle of hydrogen energy. Currently, hydrogen storage technologies are primarily categorized into three types: low-temperature liquid hydrogen storage, high-pressure gaseous hydrogen storage, and solid-state ambient-pressure hydrogen storage. Among these, solid-state hydrogen storage can be further classified into physical adsorption and chemical hydride absorption. In particular, hydrogen storage alloys react with hydrogen to form metal hydrides, offering significant advantages in safety, efficiency, high hydrogen storage density, and convenience. As a result, the chemical hydride absorption method has emerged as a research hotspot, representing the cutting-edge and the most promising solution in the field of hydrogen storage. Since the discovery of hydrogen storage alloys in 1968, researchers have focused on four major categories of hydrogen storage materials: lanthanum (rare earth)-based, titanium-iron-based, Mg-based, and zirconium (vanadium)-based materials. Among these materials, Mg-based hydrogen storage materials are considered the most promising materials due to their abundant geological reserves, high theoretical value of hydrogen storage capacity, moderate hydrogen absorption and desorption properties, and good cycling stability. Table 5.1 summarizes the hydrogen storage performance of some typical hydrogen storage alloys.

(1) Hydrogen storage and transportation media

Compared to the traditional high-pressure and liquid hydrogen storage methods, Mg-based solid-state hydrogen storage materials offer several advantages:

① High mass/volume hydrogen storage density: For instance, the theoretical value of MgH_2 is 7.6 wt.%.
② Enhanced safety: Mg-based hydrogen storage materials release hydrogen at specific temperatures and pressures, increasing the safety of hydrogen storage systems.
③ Low energy consumption: Combined with the abundant geological reserves of magnesium, this makes the method highly practical.

Additionally, hydrogen storage and transportation media must meet the following requirements:

① Easy activation with the minimum heat generation.
② Minimal equilibrium hydrogen pressure difference between hydrogen absorption and desorption.
③ Maintaining a constant platform pressure, with a near room-temperature decomposed pressure between 0.2 and 0.3 MPa.
④ Fast hydrogen absorption/desorption rates.
⑤ Minimal degradation and good cycling life.

The application of Mg-based hydrogen storage materials relies on the development of Mg-based metal hydride reactors. Generally, these reactors must exhibit excellent heat transfer properties, a reasonable filling structure and density, and good sealing properties. Most Mg-based metal hydride reactors are constructed using stainless steel in a cylindrical shell design. Research teams worldwide have conducted extensive fundamental studies on Mg-based metal hydride reactors across different scales.

Table 5.1 Hydrogen storage capacity, equilibrium pressure, and operating temperature of typical hydrogen storage alloys [61, 62]

Types	Alloys	Hydrides	Mass hydrogen storage capacity/wt.%	Equilibrium pressure/MPa	Operating temperature/K
Pristine metal	Mg	MgH_2	7.6	0.1	573
A_2B	Mg_2Ni	Mg_2NiH_4	3.6	0.1	528
AB_2	ZrV_2	$ZrV_2H_{5.5}$	3.0	0.1	443
AB	TiFe	$TiFeH_2$	1.9	0.5	303
AB_5	$LaNi_5$	$LaNiH_6$	1.4	0.2	298
BCC	TiV_2	TiV_2H_6	2.6	1.0	313

For example, the Garrier team [63] in France designed a medium-sized magnesium metal hydride reactor filled with 1.8 kg of 5 wt.% expandable graphite and Ti-V-Cr alloy-doped MgH_2 powder briquettes. The reactor demonstrated a hydrogen storage capacity of approximately 1.2 Nm^3 and a mass density of 0.8 wt.%. Similarly, the Delhomme team [64] in France designed a large-scale Mg-based metal hydride reactor capable of accommodating 10.0 kg of 10 wt.% expandable graphite and Ti-V-Cr alloy-doped MgH_2 powder briquettes. This reactor achieved a hydrogen storage capacity of 6.7 Nm^3 and an average hydrogen desorption flow rate of 40 NL/min.

Based on the application scenarios, Mg-based metal hydride reactors are primarily categorized into stationary and mobile hydrogen storage systems. Stationary hydrogen storage is typically employed in hydrogen stations or small- to medium-size hydrogen storage instruments. These systems generally have relatively lower requirements for gravimetric hydrogen storage density but higher demands for volumetric hydrogen storage density. Hydrexia Inc., based in the United States, has conducted various tests on Mg-based solid-state hydrogen storage and transport equipment. Economic and technical feasibility analyses confirm that the Mg-based hydrogen storage technique can be commercially viable on a large scale [65]. A notable stationary application of Mg-based hydrogen storage in China is the showcase hydrogen station built by Hydrogen Maple Energy in 2020 in Jining, Shandong Province, which utilizes MgH_2. The station has a hydrogen refueling capacity is 550 kg/day, sufficient to support two bus lines. The cost of Mg-based hydrogen stations is approximately half that of traditional technologies. Moreover, these stations require less space, offer high refueling capacities without the need for large pressure tanks, and are adaptable to multi-function station configurations, such as combined hydrogen/gasoline stations, hydrogen/natural gas stations, and mobile hydrogen stations [66].

Mobile hydrogen storage devices should possess key features such as lightweight design, compactness, and high hydrogen storage density. As early as 1981, the Institute of Chemical Technology of Japan developed a mobile Mg-based hydrogen storage device filled with 3.5 kg of Mg-10Ni alloy, achieving a hydrogen storage capacity of approximately 2.9 Nm^3 [67]. In 2019, the Hydrogen Science Center of Shanghai Jiao Tong University and Shanghai Hydrogen Maple Energy Technology Co., Ltd., collaboratively launched a prototype of large-scale Mg-based metal hydride reactor. This reactor can accommodate more than 1,000 kg of Mg-based hydrogen storage alloys in a single tank and is capable of absorbing or releasing approximately 60 kg of hydrogen within 8 h. Furthermore, by connecting multiple storage tanks in parallel, the system addresses the challenges of low-energy consumption and controllable hydrogen release through high-temperature heat transfer oil heating. This innovation led to the successful development of a ton-level Mg-based solid-state hydrogen storage and transport system, with a capacity of up to 1,000 kg of hydrogen. This system can transport three to four times the volume of hydrogen compared to the current mainstream high-pressure pipe trailers and operates at ambient temperature and pressure, significantly enhancing safety. It is well-suited for applications

Fig. 5.40 Mg-based solid-state hydrogen storage trailer developed by Shanghai Hydrogen Storage Co. Ltd. [68]

in hydrogen transportation, hydrogen metallurgy and chemical processing, seasonal energy storage, and distributed energy storage systems.

Hydrogen Storage (Shanghai) Ltd. has developed a Mg-based solid-state hydrogen storage trailer [68] (Fig. 5.40), capable of carrying up to 1.2 tons of hydrogen in a single trip, three times more than a standard pipe trailer. This advancement enables the efficient transportation of large quantities of Mg-based solid-state hydrogen. The Mg-based solid-state trailer operates at near ambient temperature and pressure, eliminating the need for additional apparatus at hydrogen refueling stations. Moreover, the optimized range for high-pressure gaseous hydrogen storage in a traditional pipe trailer is approximately 100 to 150 km. In contrast, the Mg-based solid-state hydrogen storage trailer, developed by Hydrogen Storage Co. Ltd. in collaboration with Shanghai Jiao Tong University, can quadruple the hydrogen storage capacity within the same volume and extend the operating range to over 400 km. This innovation effectively addresses the challenge of limited hydrogen refueling station distribution and significantly facilitates the advancement of hydrogen energy infrastructure.

Figure 5.41 illustrates the application scenario of a Mg-based solid-state hydrogen storage trailer [69]. Clean energy sources such as wind and solar power are converted into hydrogen energy and stored in the Mg-based solid-state hydrogen storage trailer. The stored hydrogen can be safely transported over long distances at room temperature and ambient pressure, supplying Mg-based solid-state hydrogen to refueling stations for hydrogen fuel cell vehicles. This integration system enables renewable energy to achieve production, storage, and transportation. Moreover, the Mg-based solid-state hydrogen storage trailer can be combined with solid oxide fuel cells (SOFC) for distributed power generation. Specifically, the trailer supplies high-quality hydrogen to an SOFC power station and the heat generated from the SOFC facilitates the hydrogen desorption reaction of the Mg-based hydrogen storage material.

In 2021, the Hydrogen Science Center of Shanghai Jiaotong University, in collaboration with Baowu Clean Energy Co., Ltd, developed a skid-mounted system with "hydrogen generation from solar power—Mg-based solid hydrogen storage—high-pressure hydrogen refueling". This equipment is designed for refueling hydrogen

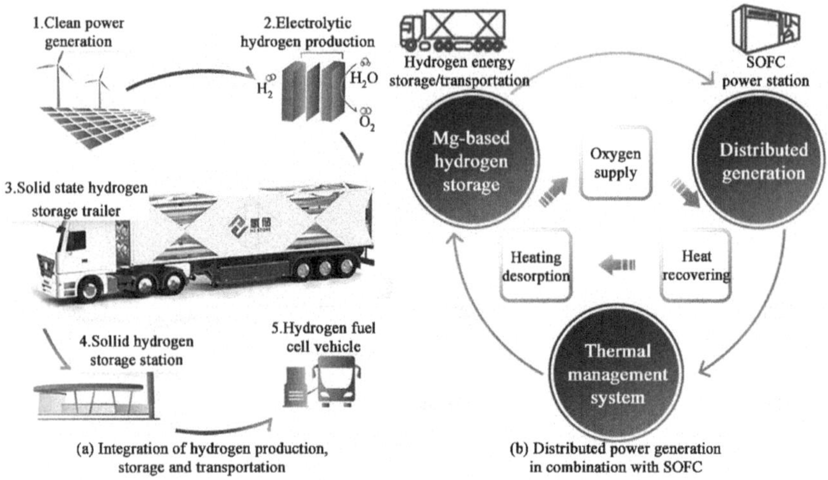

Fig. 5.41 Application scenarios for Mg-based solid-state hydrogen storage trailers [69]

fuel cell forklift trucks within industrial parks. It is currently being demonstrated at Baowu Company and operates efficiently, showcasing the feasibility of coupling Mg-based solid-state hydrogen storage and transportation technology with renewable energy power generation and hydrogen production technology.

(2) Hydrogen source for fuel cells

The hydrogen fuel cell is an innovative power generation device characterized by high energy conversion efficiency, "zero" emission, and low noise. With support from local and national policies and successful demonstration projects, hydrogen fuel cells are now widely utilized in both civil and military sectors, including transportation, distributed power stations, unmanned aerial vehicles, portable power sources, aerospace, and submarines. In practical applications of hydrogen fuel cell power systems, efficient, safe, and cost-effective hydrogen storage technology is of paramount importance, as hydrogen storage is a critical component of fuel cell technologies. With advancements in hydrogen storage materials, metal hydrides have emerged as dominant hydrogen sources for fuel cells.

The U.S. Department of Energy (DOE) has set a 2025 target for hydrogen storage systems for light-duty vehicles: a volumetric density of 50 g/L, a gravimetric density of 5.5 wt.%, an operating temperature range of 233 to 358 K, a pressure range of 0.5 and 1.2 MPa, and a hydrogen release rate of 0.02 g/(s·kW), with a cycle life exceeding 1,000 cycles. Mg-based hydrides are considered the most promising hydrogen source for fuel cells due to their significant hydrogen storage capacity, fast kinetics and outstanding cycling life. Additionally, Mg-based materials offer several advantages in hydrogen fuel cell applications including high specific energy, safety, convenience, low cost, a broad operational temperature range, and low environmental impact.

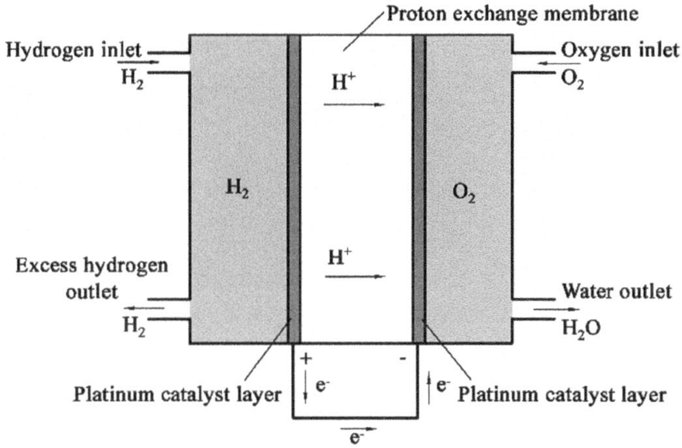

Fig. 5.42 Reaction mechanism of proton exchange membrane fuel cell

To date, fuel cell technologies are primarily classified into two approaches. The first is proton exchange membrane fuel cells (PEMFCs), which are the most commonly employed as mobile power sources, such as in fuel cell electric vehicles. In China, hydrogen fuel cells are primarily used for large vehicles, including city buses, heavy-duty trucks, vans, and forklifts. For these vehicles, the key parameter is volumetric hydrogen storage density rather than the mass density of the systems. The second approach is solid oxide fuel cells (SOFCs), which are mainly used as stationary power sources, such as in power stations. Currently, PEMFCs represent the most mature, rapidly advancing, and widely adopted hydrogen fuel cell technology across various applications (Fig. 5.42).

The hydrogen fuel cell system primarily consists of hydrogen storage tanks and a fuel cell stack. In these systems, Mg-based powder material is used in the storage tanks, serving as the hydrogen supply source for the fuel cells. The fuel cell stack functions as the power generator in hydrogen vehicles. Once hydrogen is transferred from the storage tanks to the stack, hydrogen oxidation and oxygen reduction reactions occur at the anode and cathode, respectively, converting hydrogen energy into electrical power to drive the vehicle. Integrating Mg-based solid-state hydrogen storage devices into fuel cell systems allows the heat generated by the fuel cell stack to facilitate hydrogen release from the storage tanks, thereby enhancing the energy efficiency of the fuel cell power system. Figure 5.43 illustrates the integrated system structure of a Mg-based solid-state hydrogen storage device and fuel cell stack developed by Shanghai Magnesium Power Company [70]. The Mg-based solid-state hydrogen storage-fuel cell vehicle offers the following advantages:

① The hydrogen storage and power generator components form an all-solid structure, avoiding side effects caused by liquid electrolytes and enabling long-term operation.

② The only by-products are heat and water, achieving true "zero" emission.

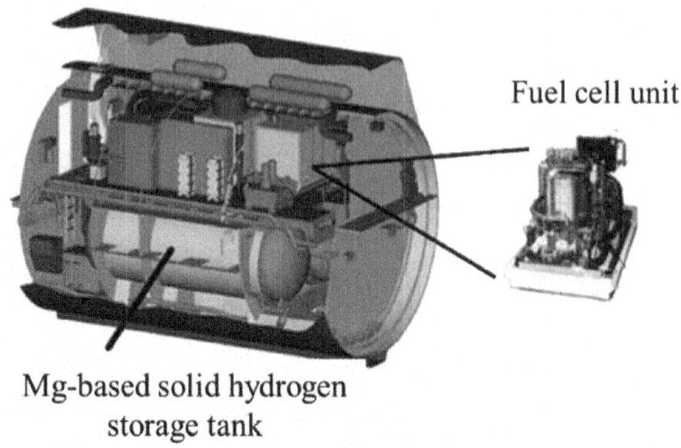

Fig. 5.43 Mg-based solid-state hydrogen storage device integrated with fuel cell stack [70]

③ The fuel cell stacks typically operate at approximately 353 K, allowing the generated heat to promote the hydrogen release of Mg-based hydrides.
④ Solid-state hydrogen storage and fuel cell stack are compact and space-saving.

As a hydrogen source for fuel cell stacks, Mg-based hydrogen storage materials must meet the following requirements:

① High reversible hydrogen storage capacity (mass density > 3.0 wt.%).
② Fast hydrogen absorption and desorption rates, with hydrogen absorption and desorption time < 10 min.
③ Operation at moderate temperatures, with pressures typically ranging from 1.0 to 5.0 MPa at room temperature.
④ High cycling stability, with less than 5% capacity loss after 500 cycles and less than 10% capacity loss after 1,000 cycles.

Chengdu Xinkeli Chemical Science and Technology Company has designed and fabricated a layer-ordered Mg-based composite hydrogen storage material for fuel cells [71]. This material uses an amorphous carbon network (20–60 μm) as a support material, embedding Magnesium-transition metal-magnesium alloy particles (Φ 40–60 nm, with a layer thickness of 5–10 nm). The resulting magnesium-carbon composite possesses a reversible hydrogen storage capacity of 5.42 wt.%.

Antai Chuangming Company has developed a hydrogen fuel cell-powered electric two-wheeled vehicle utilizing a Mg-based solid-state hydrogen storage system as the hydrogen source for the fuel cell stack [72]. The hydrogen refueling pressure of the vehicle is below 0.1 MPa, and the tank replacement takes less than one minute. Equipped with two Mg-based hydrogen storage tanks, the vehicle has an operational range exceeding 120 km. The nano Mg-based hydrogen storage material in the tank has a capacity of approximately 5.5 wt.%, with excellent kinetic and cycling performance, maintaining a reversible capacity of 4.5 wt.% after 300 cycles at 548 K. The heat required for the hydrogen release comes from the exhaust heat of the fuel cell stack.

A new type of magnesium hydride fuel cell system has been developed, utilizing the hydrolysis reaction of magnesium hydride to generate hydrogen for the fuel cell stack. Since part of the hydrogen atoms in water molecule can be reduced during the hydrolysis reaction, the hydrogen release of this hydride reaches 15.2 wt.%, significantly surpassing the traditional high-pressure gaseous hydrogen storage systems (~3.0 wt.% at 35 MPa). Hydrogen Science Center of Shanghai Jiaotong University developed a 200 W fuel cell backup power system and a high-energy–density MgH_2 hydrolysis hydrogen production system for Shenhua Group, capable of operating continuously for 96 h. In 2020, the same group, in collaboration with Shanghai Yuji Power Limited Company, further developed a magnesium hydride fuel cell backup power system with an energy density exceeding 600 Wh/kg. This system can start at temperatures between 233 and 328 K and continuously provide a field power supply of 60 W for 50 h. Hydrolysis-based hydrogen fuel cell technology using MgH_2 can be applicable in fields such as backup power supplies, unmanned aerial and underwater vehicles. The controllable hydrolysis-based hydrogen production-fuel cell technology was jointly developed by Shanghai Magnesium Power Company and Shanghai Jiao Tong University using MgH_2. This technique can continuously provide hydrogen to the fuel cell stack with a power output of 1000 W. The hydrogen supply rate is adjustable, the hydrolysis degree of the material exceeds 99%, and water usage approaches the theoretical value of the reaction. The magnesium hydride hydrolysis fuel cell system is illustrated in Fig. 5.44 [70], hydrogen is produced by the hydrolysis of magnesium hydride, which can power fuel cells from watt-level to kilowatt-level. Its energy density far exceeds that of the conventional lithium batteries, making it suitable for electronics, marine/submarine applications, aerospace, and clean-energy vehicles, and remote emergency backup power underwater equipment. Furthermore, Shanghai Mg Power Technology Co., Ltd., in cooperation with Singapore's HES company, designed a hydrogen-based unmanned aerial vehicle using Mg-based hydrogen storage composites as a hydrogen source for the fuel cell stack. This unmanned aerial vehicle achieved a record flight duration of 6 h and a range of 300 km.

Mg-based hydrogen storage materials can also be used in stationary fuel cell power systems. Mg-based materials can store hydrogen produced by renewable energy sources and, in combination with Solid Oxide Fuel Cells (SOFCs), generate electric power when needed. In SOFCs, a solid oxide serves as electrolyte and the operating temperature ranges between 1,073 to 1,273 K. The heat by-product generated by the SOFCs provides the necessary thermal energy for the desorption of hydrogen from the Mg-based storage materials. The released hydrogen, acting as fuel, reacts with an oxidant (O_2/air) to convert chemical energy into electric energy. Fuel cell power stations typically have lower requirements for gravimetric hydrogen storage density, but the volumetric hydrogen storage density remains a critical criterion [73]. Currently, most global demonstrations of fuel cell systems use rare-earth-element-based $LaNi_5$-type and TiFe-based hydrogen storage alloys, while Mg-based hydrogen storage materials are rarely reported.

Shanghai Mg Power Technology Co., Ltd. has designed and proposed a power station system based on Mg-based hydrogen storage materials [74]. This system consists of a hydrogen generation device, a hydrogen storage device, and a hydrogen fuel cell stack. The hydrogen storage device is filled with Mg-based hydrogen storage

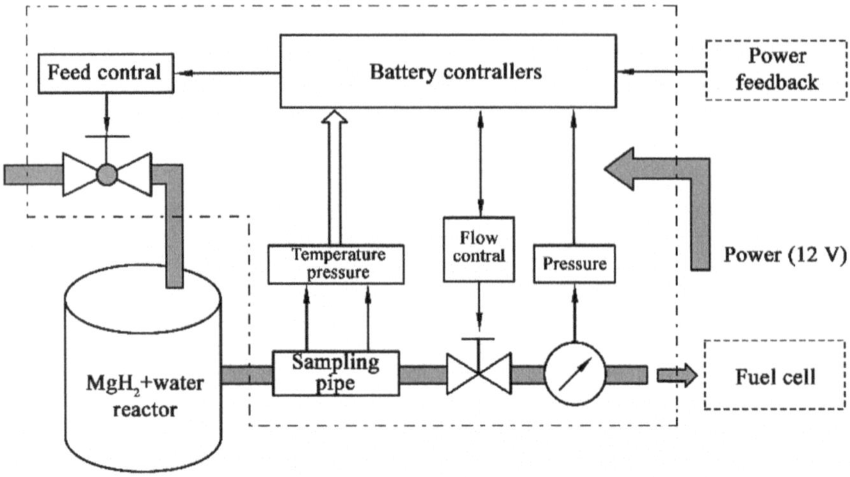

Fig. 5.44 High energy density magnesium hydrogenation fuel cell system [70]

materials, potentially a combination of several materials such as Mg-graphite, Mg-Fe, Mg-carbon nanotube, Fe-graphene, Mg-Cu, and Mg-Ni. The Mg content ranges from 70 to 99 wt.%, with a reversible hydrogen storage capacity of approximately 4.6 wt.%. Furthermore, the power station incorporates a phase transition thermal storage layer and a thermal insulation layer for the storage of the heat generated during the hydrogenation of Mg-based materials. This stored heat is preserved for the dehydrogenation process, enhancing overall energy utilization efficiency.

To date, high-pressure gaseous hydrogen storage technology remains the mainstream for supplying hydrogen to fuel cells, while significant progress has also been made in low-temperature liquid hydrogen storage techniques. The application of Mg-based hydrogen storage materials in hydrogen fuel cell systems is still primarily at the lab-scale research or medium-scale production testing stage. For large-scale commercial applications, Mg-based materials continue to face challenges such as high hydrogen absorption and desorption temperatures and limited cycling life. As a result, the development of dual-purpose tanks capable of both high-pressure gaseous storage and ambient-pressure solid-state storage in fuel cell systems remains a prominent research focus in fuel cell systems.

(3) Hydrogen source for internal combustion vehicles

In the context of carbon peak and carbon neutrality, hydrogen internal combustion vehicles, which use hydrogen as a substitution for gasoline, have received significant attention due to the clean exhaust consisting of heat and water vapor. Notably, in terms of volume density, metal hydride exhibits 4 to 6 times the density of high-pressure hydrogen and offers much greater safety. Table 5.2 presents a comparative analysis of the weight and volume of fuel tanks for various fuel sources.

The hydrogen internal combustion vehicles primarily consist of an engine, a metal hydride reactor, a hydrogen supply system, and a heating component for the metal

Table 5.2 The comparison of weight and volume of fuel tanks [75]

Fuel source	Weight		Capacity	
	kg	Ratio	L	Ratio
Petrol	61	1	78	1
MgH_2	328	5.4	257	3.5
Mg_2NiH_4	649	10.7	544	7.0
Liquid hydrogen (20 K)	160	2.6	289	3.7
High-pressure hydrogen (14.0 MPa)	1020	18.2	1870	23.9

hydride reactor. The operating conditions for the hydrogen stored in the metal hydride reactor are as follows:

① Low heat absorption, with the hydrogen desorption pressure platform lower than 1.0 MPa.
② High hydrogen storage density (in mass%).
③ Fast hydrogen desorption kinetics.
④ Low cost and long cycling life.

The application of Mg-based hydrides in hydrogen engine vehicles is primarily based on their properties of high-temperature/low-pressure hydrogen desorption, which easily provides high-quality hydrogen fuel for vehicles. For a typical gasoline vehicle, only minor modifications are needed to convert it into a hydrogen vehicle. The heat from the vehicle's exhaust gases is used to heat the Mg-based hydride, promoting hydrogen desorption. Additionally, when a mixture of hydrogen and gasoline is used as fuel, the thermal efficiency is much higher than when using gasoline alone.

As early as 1969, the Brookhaven National Laboratory in the U.S. proposed a hydrogen supply system for combustion engines using Mg-based and TiFe-based metal hydrides [75]. In this system, TiFe metal hydrides are low-temperature hydrides, and the engine's cooling water can be used as the heating source for hydrogen desorption. Even in low-temperature regions, TiFe-based metal hydrides can provide sufficient hydrogen pressure without an additional heating source. In contrast, Mg-based metal hydrides (such as Mg, Mg_2Ni, and Mg_2Cu) are high-temperature hydrides, and their hydrogen desorption (above 573 K) requires a high-temperature heating source to start. The engine exhaust heat provides high-quality heat for hydrogen desorption. To date, most hydrogen combustion vehicles employ improved TiFe-based metal hydrides and Mg-based metal hydrogenation reactors. The combustion engine requires hydrogen pressure between 0.1 and 0.2 MPa. Both Mg and Mg_2Ni hydrides are much lighter than TiFe, which helps improve the mass and volume density of hydrogen.

Among the various developed metal hydrides, Mg-based hydrides have a high hydrogen capacity (maximum of approximately 7.6 wt.%), but their drawback is the high operating temperature, which requires an external heating source. In contrast, the hydrogen capacity of TiFe-based hydrides is only 1.8 wt.%, resulting in an increased reactor weight. To overcome these challenges, Mercedes-Benz integrates

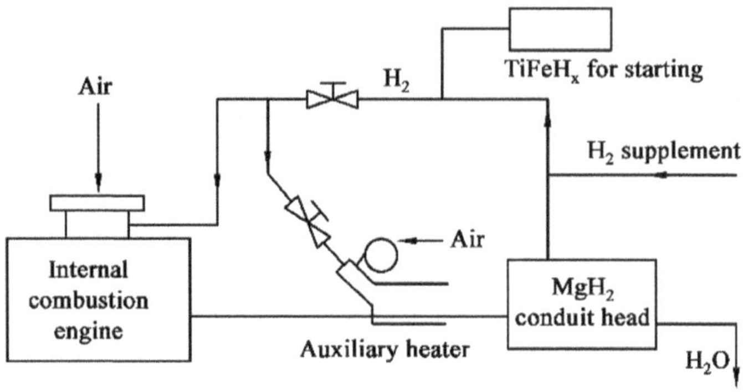

Fig. 5.45 Automotive hydrogen supply system with a combination of high-temperature and low-temperature metal hydrides [75]

TiFe-based hydride reactors with Mg-Ni series hydride reactors, achieving success in the lightweight aspect of reactor (Fig. 5.45). Specifically, the vehicle is equipped with three separated metal hydride reactors: the first is a Mg-Ni series metal hydride reactor that is directly heated by the engine's exhaust heat; the second is a TiFe-based metal hydride reactor that utilizes the residual exhaust heat; the third is a TiFe hydride fuel tank equipped with a liquid heat exchanger from the vehicle's air conditioner. This hydrogen supply system has a total weight of 340 kg and can store 5.4 kg of hydrogen (equivalent to 20.8 L of gasoline), with a road test range of more than 100 km [75].

Although the working range of hydrogen combustion vehicles with Mg-based hydrogen storage materials cannot yet compare to that of gasoline vehicles, they offer the environmental benefit of being eco-friendly, making them suitable for urban or short-range transportation. However, the hydrogen desorption from Mg-based hydride materials requires high temperatures, which reduces the engine efficiency. Therefore, reducing the enthalpy of hydride dehydrogenation by altering thermodynamic properties to achieve a milder dehydrogenation temperature will be a key focus of future research in this field. Additionally, exploring the technique for hydrogen-gasoline hybrid fuel is another direction for future development. The hybrid fuel vehicle uses gasoline to replace part of the hydrogen, solving the problem of the heating source when Mg-based hydride materials are solely used as the hydrogen source, and improving fuel utilization efficiency. Compared to traditional gasoline vehicles, this method not only reduces air pollution but also addresses the problem of insufficient engine power.

(4) Hydrogen Separation and Purification

Hydrogen separation and purification typically involve extracting hydrogen from other waste gases to obtain higher-purity hydrogen. In industries such as the chemical industry, petroleum refining, and metallurgy, exhaust gases often contain significant amounts of hydrogen, sometimes reaching 50–60%. Recycling and utilizing this part of hydrogen can greatly enhance economic efficiency. Moreover, ultra-high purity hydrogen is essential for producing electronic materials, such as integrated circuits, semiconductor devices, and optical fibers. If hydrogen can be effectively separated and purified from by-products of coal chemical production, the economic and social benefits of the coal chemical industry will be improved significantly.

The commonly used hydrogen separation methods primarily include variable pressure adsorption, deep-freezing separation, and membrane separation. Among them, variable pressure adsorption and deep-freezing separation suffer from high energy consumption, complicated processes, and relatively low hydrogen purity. Membrane separation, on the other hand, is hindered by high costs and low permeation flux. As a cutting-edge technology, recycling and purifying hydrogen using metal hydrides has attracted worldwide attention. Mg-based hydrogen storage systems, when combined with other equipment for hydrogen separation and purification, offer advantages such as abundant raw material, low cost, high hydrogen recovery rates, and high hydrogen purity [76].

The application of Mg-based hydrogen storage materials in hydrogen purification and separation is based on two fundamental principles: (1) The reaction of Mg-based storage materials with hydrogen to form metal hydrides, followed by the reversible desorption of hydrogen upon heating; (2) The unique affinity of Mg-based materials with hydrogen atoms allows for the selective absorption of hydrogen while repelling impurity gases such as Ar, N_2, CH_4, CO, CO_2, and NH_3. This method of hydrogen absorption and separation contrasts with other separation methods, which typically use adsorbents to capture impurities. The hydrogen purification and separation process employing Mg-based hydrogen storage materials is illustrated in Fig. 5.46. The purifier, filled with Mg-based hydrogen storage material, integrates both purification and compression functions to produce high-purity hydrogen from crude hydrogen. The application of Mg-based hydrogen storage materials in hydrogen fine-refining technology is primarily suited for gases with high hydrogen content and relatively simple gas compositions. However, industrial hydrogen often contains various impurities such as N_2, O_2, and CO_2, in varying proportions.

Zhou et al. [77] from Central South University developed a Mg-based hydrogen storage alloy nano-powder for hydrogen separation. Figure 5.47 illustrates the schematic of the hydrogen purification and separation device. This device operates at low pressure, and achieves high hydrogen purity, low energy consumption, a simple process, safety, and reliability. Impurities such as O_2, CO, H_2O, and H_2S, which can

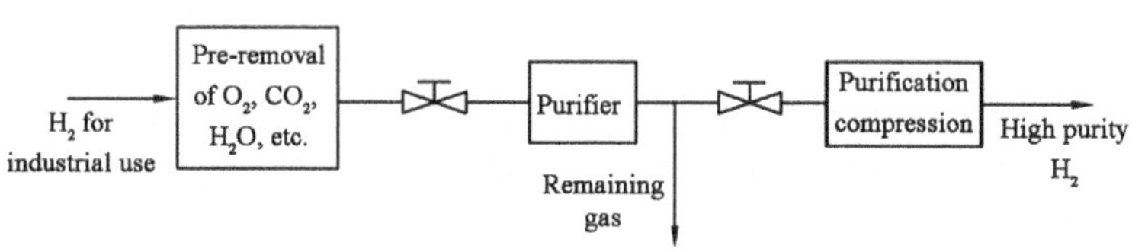

Fig. 5.46 The schematic illustration of hydrogen separation and purification process

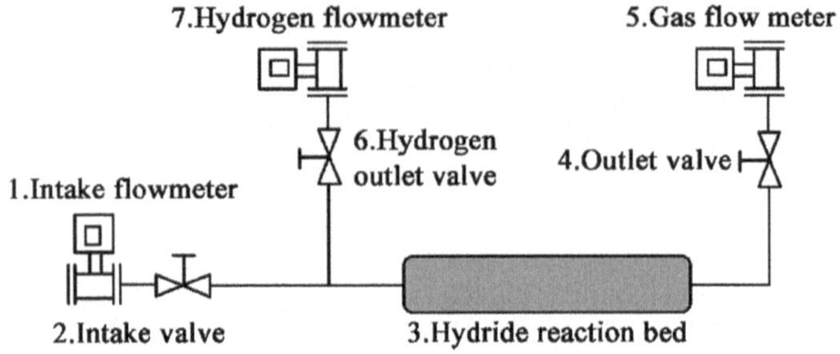

Fig. 5.47 Schematic illustration of a hydrogen purification and separation device with Mg-based hydrogen storage materials [77]

cause oxidation and contamination in mixed gases, are effectively removed. Furthermore, the use of Mg-based hydrogen storage material for hydrogen purification eliminates the need for the precious metal Pd, offering significant economic value and broad application potential [77]. However, Mg-based materials require high hydrogen purity in the feed gas (approximately 99.99%), and brittleness and powdering tend to occur after multiple cycles. These limitations hinder their large-scale applications, particularly in gases containing ammonia and carbon monoxide.

To date, the commonly used hydrogen storage materials in hydrogen separation and purification applications are rare-earth-based, TiFe-based, and TiV-based alloys. In contrast, reports on Mg-based hydrogen storage materials are relatively scarce. This is primarily due to the high temperatures required for hydrogen absorption and desorption in Mg-based materials. Additionally, most of the techniques developed for hydrogen purification and separation using Mg-based hydrogen storage materials are limited to mixtures with high hydrogen content and relatively simple gas compositions. There are no reports involving complex gas mixtures with low-light components, such as coke oven gases.

The primary barrier to the application of Mg-based hydrogen storage materials in this field is the poisoning effect of alloys/hydrides during the absorption/desorption processes caused by various non-hydrogen impurities. For example, the presence of CO can hinder hydrogen release, while impurities such as O_2 and water can lead to hydride failure, thus deteriorating cyclic stability. Furthermore, the anti-poisoning ability of alloys/hydrides against impurities is closely related to their intrinsic properties of the materials. Therefore, the ultimate solution for this issue lies in developing novel Mg-based hydrogen storage materials. Currently, the effects of the chemical composition and content of Mg-based hydrogen storage materials on hydrogenation/dehydrogenation properties remain unclear, resulting in a lack of theoretical guidance. Developing Mg-based hydrogen storage materials with high recovery rates, high hydrogen purity, and excellent anti-poisoning properties is crucial for achieving efficient hydrogen recovery.

(5) Hydrogen isotope separation

Hydrogen isotope separation refers to the use of different formation rates of hydrogen isotope compounds with hydrides during the hydrogen absorption reaction of hydrogen storage alloys. In the nuclear industry, heavy water serves as both a coolant and a neutron moderator in atomic fission reactors, while tritium acts as the primary nuclear fuel for fusion reactions. Recycling radioactive tritium from nuclear fusion waste is essential for minimizing the emission into the atmosphere. Therefore, hydrogen isotope separation holds great significance in the nuclear sector [78].

Metal hydrides commonly exhibit hydrogen isotope effects, meaning the equilibrium pressure and adsorption capacity of metals or alloys differ when absorbing hydrogen, deuterium, and tritium. The diffusion and absorption rates of these isotopes in alloys also vary. The former phenomenon is referred to as the thermodynamic isotope effect, while the latter is the kinetic isotope effect [7]. These characteristics allow for the effective separation of hydrogen (H_2), deuterium (D_2), and tritium (T_2).

Currently, the most commonly used metals for hydrogen isotope separation at the laboratory scale include palladium, uranium, zirconium, and TiNi alloy. Palladium, in particular, is highly effective for isotope separation but is prohibitively expensive. To develop cost-effective alternatives, significant research efforts have been dedicated to investigating alloys such as Mg_2Ni, TiFe, $LaNi_{4.25}Al_{0.75}$. With advancements of Mg-based hydrogen storage materials, research in this field is steadily progressing. Future studies may focus on preparing Mg-Pd multilayer nanofilms, which offer the advantages such as dimensional stability, high permeability, and isotope selectivity under optimal temperature and pressure conditions.

5.3.2 Thermal Storage

Thermal energy conduction is one of the most common methods for energy conversion and transfer. The development of thermal storage technology enhances energy utilization efficiency, protects the environment, and addresses the mismatch between heat supply and demand. This technology has broad applications, including solar thermal power generation, industrial waste heat utilization, and peak-shifting in electrical demand.

Thermal storage materials are substances that undergo phase transitions at specific temperatures, accompanied by heat absorption or release. These materials can be used for temperature regulation or thermal energy storage. Based on the heat storage method, thermal storage materials can be categorized into three types: sensible heat storage materials, latent heat storage materials, and thermochemical storage materials. Among these, thermochemical storage materials have gained significant attention due to their high energy density, minimal heat loss, and lack of need for heat preservation. In solar thermal power generation, heat storage technology based on the decomposition reaction of metal hydrides offers several advantages, including good reaction reversibility, high reaction heat, low corrosion, and easy control, making it a promising option with wide application prospects [79]. Additionally, metal hydrides are also suitable for use in refrigeration, air conditioning, and heat pumps.

For heat storage applications using hydrogen storage materials, the requirements differ from those for hydrogen storage. Specifically, the enthalpy of hydride formation should be as large as possible, and the materials should exhibit good thermal conductivity. Due to the significant exothermic and endothermic heat generated by magnesium-based hydrogen storage materials during hydrogen absorption and desorption, these materials play a crucial role in thermal energy utilization. Currently, Mg-based hydrogen storage materials are primarily utilized in thermal energy applications in two ways [80, 81]:

① Recycling industrial waste heat: By utilizing the exothermic reaction during hydrogen absorption and the endothermic reaction during hydrogen desorption, Mg-based hydrogen storage materials enable the recycling of industrial waste heat and residual heat, facilitating energy conservation through heat storage.
② Heat pump development: Hydrogen storage materials with compatible hydrogen storage properties can be used to develop heat pumps, converting low-quality heat sources into high-quality ones while simultaneously achieving refrigeration effect.

(1) Thermal storage materials in thermal storage systems

Mg-based hydrogen storage materials can release heat during hydrogen absorption and absorb heat during hydrogen desorption. These properties allow for thermal energy storage or release through the reversible enthalpy change, enabling applications such as storing industrial waste heat, geothermal energy, and solar heat. Based on operating temperature, thermal storage metal hydrides are classified into high-temperature, medium-temperature, and low-temperature types. Mg-based hydride materials are categorized as medium-temperature types, typically operating between 473 to 823 K [73]. However, Mg-based hydride materials can be further subdivided based on specific temperature ranges. Table 5.3 presents the thermodynamic properties and thermal storage performance of selected Mg-based hydride materials used for thermal storage.

In 1982, Kawamura et al. [82] in Japan experimentally studied the thermal storage performance of a high-temperature Mg-based hydride system. This system primarily consists of a thermal storage reactor and a hydrogen storage tank, effectively utilizing waste heat in the range of 573–773 K. The reactor, featuring a single-tube construction filled with 6.27 kg of Mg_2Ni alloy, achieved a total thermal storage capacity of 8 MJ. Experimental results indicated an overall heat transfer coefficient of approximately 11.61 W/(m^2·K). In 1989, Bogdanovic et al. [83] in Germany developed a high-temperature metal hydride thermal storage system. This system used Ni-doped MgH_2 in the thermal storage reactor and $Ti_{0.98}Zr_{0.02}V_{0.43}Fe_{0.09}Cr_{0.05}Mn_{1.2}$ alloy in the hydrogen storage reactor. Experimental results showed that the total hydrogen release amount, maximum release rate, reabsorption capacity, and total thermal storage capacity of the thermal storage reactor decreased as the temperature of the hydrogen storage reactor increased.

Based on the cases discussed above, the configuration of high-temperature Mg-based hydride thermal storage systems can be divided into two types: "thermal

Table 5.3 Thermodynamic properties and thermal storage properties of some Mg-based hydrides [84]

Materials	Hydrogen storage capacity	Reaction enthalpy / ($kJ \cdot mol^{-1} H_2$)	Temperature/ K	Equilibrium pressure/ MPa	Heat storage density/ [$MJ \cdot kg^{-1}$]
Low-temperature Mg hydride	–	–	–	–	–
Mg_2Ni/Mg_2NiH_4	3	62.2	523~623	0.1~1.1	0.916
Mg/MgH_2 + 2%Ni	6	74	563~693	0.1~2.5	2.257
Medium-temperature Mg hydride	–	–	–	–	–
Mg/MgH_2	7.6	74	623~723	0.6~4.4	1.837
NaH-Mg/$NaMgH_3$	3.27	86.6	655~766	0.1~1.0	1.415
High-temperature Mg hydride	–	–	–	–	–
Mg-Co/Mg_2CoH_5	3.5	76	723~823	1.1~4.8	1.260
Mg-Fe/Mg_2FeH_6	5	77.2	723~826	2.8~13.8	1.817
Mg-Co/Mg_6CoH_{11}	3.5	89	–	–	1.472

storage reactor + hydrogen storage reactor" and "thermal storage reactor + hydrogen storage tank" [85]. Figure 5.48 illustrates the configuration of a high-temperature thermal storage system, including a high-temperature thermal storage reactor, a low-temperature hydrogen storage reactor, valves, and other components. The thermal storage reactor and the hydrogen storage reactor are filled with high-temperature metal hydrides and low-temperature hydrogen storage alloys, respectively. The working principle of this system is as follows: The thermal storage reactor is heated until the high-temperature metal hydrides decompose. The released hydrogen gas reacts with the hydrogen storage alloy in the hydrogen storage reactor, and the heat released is absorbed by the cooling water, achieving thermal storage. The heat release process works as follows: When high-temperature heat is needed, a low-temperature heat source (e.g., solar energy, factory waste heat, geothermal energy, etc.) is used to heat the hydrogen storage reactor. The released hydrogen reacts with the hydrogen storage alloy, providing heat. The storage and release of heat can be conveniently controlled by the valves between the two reactors.

The "thermal storage reactor + hydrogen storage reactor" configuration is widely used for its large heat storage per unit volume and compatibility with high-grade low-temperature heat sources. However, the low-temperature hydrogen storage alloys (Ti-based) used are expensive, raising system costs. On the other hand, the "thermal storage reactor + hydrogen storage tank" configuration is more cost-effective but has slower heat storage and release rates. Additionally, pressure-enhanced systems, employing hydrogen compressors to improve the dynamic performance have been developed. The usage of the compressor can reduce the pressure in the thermal storage

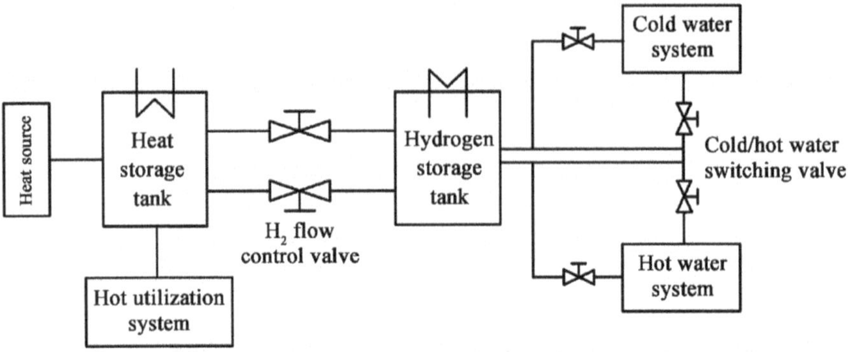

Fig. 5.48 The configuration of a high-temperature thermal storage system with Mg-based hydride

reactor and increase the pressure within the hydrogen storage reactor (or hydrogen storage tank), accelerating the hydrogen absorption and desorption reaction rates as well as the flow velocity of hydrogen. However, the addition of a compressor significantly increases system cost, energy consumption from electrical energy during operation, and the overall complexity of the operation [79].

Mg-based hydrogen storage materials and their hydrides are the most extensively studied heat storage materials, with well-understood reaction kinetics, thermodynamic properties, and cyclic stability. As shown in Table 5.3, the thermal storage temperatures of Mg-based hydrides developed to date range from 523 to 823 K, which align well with the operating temperature of trough solar thermal power stations. However, a major hurdle for Mg-based hydride materials in thermal storage applications is their extremely high equilibrium pressure at high temperatures. For instance, at 723 K, the equilibrium hydrogen pressure of Mg_2FeH_6 and MgH_2 reach 6.6 MPa and 9.2 MPa, respectively. Recently, NaMgH was proposed for high-temperature thermal storage applications due to its lower hydrogen equilibrium pressure at elevated temperatures. For example, at 723 K, the hydrogen equilibrium pressure is only 1.1 MPa, significantly lower than that of Mg_2FeH_6 and MgH_2.

To date, the Mg-based hydrides suggested for high-temperature thermal storage include MgH_2, Mg_2NiH_4, Mg_2FeH_6, Mg_2CoH_5, and $NaMgH_3$. Among these, MgH_2 is the most extensively investigated. For example, Bogdanovic et al. from the Max Planck Institute in Germany have conducted long-term studies on MgH_2. As early as 1975, Alefeld et al. in Germany proposed MgH_2 as a high-temperature thermal storage material. However, researchers from Siemens found that the low thermal conductivity and poor reaction kinetics limit the practical application of MgH_2 in thermal storage systems [86]. Bogdanovic et al. [86] discovered that the uniformly distributed transition metal compounds on Mg or MgH_2 surfaces enhance reaction kinetics. In 1989, the Max Planck Institute, Bomin Solar Ltd, and the University of Stuttgart collaborated to develop a small-scale solar thermal power station, which included a solar collector, a heat pipe heat transfer system, a Stirling engine, a generator, an MgH_2/Mg thermal storage device, and a hydrogen storage tank (Fig. 5.49).

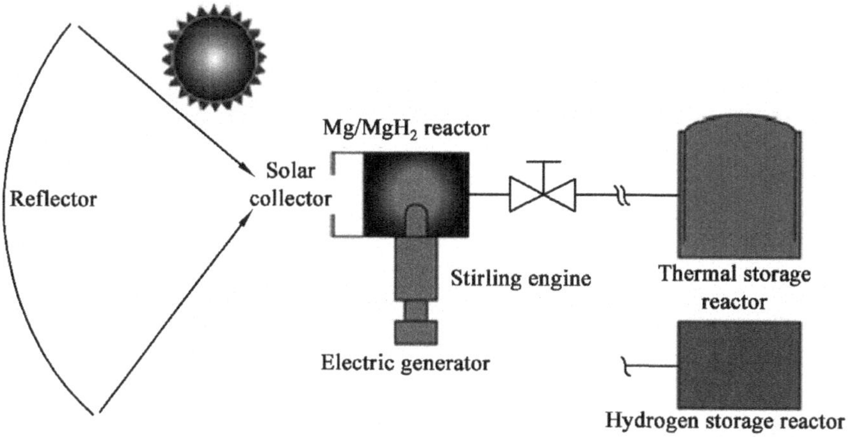

Fig. 5.49 Schematic illustration of a small-scale solar thermal power station based on Mg-based hydrides [87, 88]

The thermal storage device comprised 14 cylindrical reactors filled with approximately 24 kg of MgH_2. The operating temperature ranged from 573 and 753 K, and the total thermal storage capacity reached 12 kW·h, enabling continuous operation of a Stirling engine for two hours [87, 88]. Reiser [84] et al. systematically investigated the fundamental properties of hydrogen storage materials such as Mg-Ni/Mg_2NiH_4, Mg-Fe/Mg_2FeH_6, and Mg–Co–H systems. They found that the temperature range for storing thermal energy was between 523 to 823 K, with an energy density of up to 2,257 kJ/mol. In 2017, Urbanczyk et al. [89] in Germany developed a thermal storage system based on Mg_2FeH_6, using molten salt as a heat-conducting medium at temperatures above 573 K. This system achieved 1.5 kW·h of thermal energy storage, which can be used for industrial waste heat recovery. In China, researchers such as Jiang from the Beijing Non-Ferrous Metal Research Institute developed a demonstration solar thermal power generation and thermal storage system, using 600 kg of Mg-Ni-based alloy as the thermal storage material and $MmNi_{5.5}Mn_{0.5}$ alloy for hydrogen storage [90]. Tests showed that the system could store up to 1,000 MJ of thermal energy, release heat at a rate of $\geq$ 6 MJ/min, and operate continuously for three hours at a hydrogen absorption rate of $\geq$ 2,000 L/min. This represents the first integration of a Mg-based hydride thermal storage system with a solar thermal collection system in China.

Researchers have previously focused on developing Mg-based hydride materials as high-temperature thermal storage materials, and significant progress has been made, albeit at a slower pace in recent years. Currently, the commercialization of the thermal storage technique has not yet been achieved, with most of the projects limited to small-scale demonstration units. Challenges remain in reaction kinetics, cycle stability, thermal conductivity, and hydrogen pressure, among other factors. To further promote the development and application of Mg-based hydrides in thermal storage, it is essential to develop more advanced high-temperature Mg-based hydride materials. Additionally, achieving compatibility between the high-temperature Mg-based hydrides and the low-temperature hydrogen storage alloys used in the thermal and hydrogen storage reactors is crucial. In the future, research on Mg-based hydride materials for thermal storage should focus on developing improved hydride materials, ensuring their compatibility with low-temperature hydrogen storage alloys, simulating and optimizing reactor design, and conducting dynamic analysis of thermal storage systems. These efforts aim to enhance performance, reduce costs, and gradually promote the commercialization of high-temperature Mg-based hydrides for thermal storage applications [79].

(2) The applications of metal hydrides in heat pumps and refrigeration

As core materials in air conditioners, heat pumps and refrigerators, fluorocarbons exhibit outstanding performance. However, due to their severe environmental impact, such as ozone depletion and the greenhouse effect, fluorocarbons should be phased out in industrial applications. In response, researchers worldwide have made significant efforts to develop the alternatives, resulting in notable progress.

Metal hydrides, developed over the past two decades, are novel functional materials capable of reversibly absorbing and desorbing hydrogen with significant thermal effects. A heat pump can be constructed using two hydrogen storage materials with different equilibrium hydrogen pressures, one placed on the high-temperature side and the other on the low-temperature side. Hydrogen gas serves as the working medium in the hydrogen absorption/desorption cycles. By selecting suitable hydrogen storage alloys, the functions of cooling, warming, or heating in an air conditioner system can be achieved, depending on the application. Metal hydride heat pumps represent an environmentally friendly cooling/heating solution, as they neither pollute the environment nor contribute to global warming. Additionally, they can utilize waste heat, significantly reducing conventional energy consumption. As a result, they are currently a major focus of research worldwide [91].

The application of Mg-based hydrogen storage materials in refrigeration, air conditioning, heat pumps, and thermal storage systems shares a common foundation: hydrogen as the working medium and hydrogen storage alloys as energy conversion carriers. However, the selection of materials and operations varies depending on the specific objectives and functions. Mg hydrogen storage material-based heat pumps utilize hydrogen as the working medium, and hydrogen storage alloys as energy conversion carriers. Typically, the two hydrogen storage alloys with a significant difference in equilibrium pressure, referred to as M1 and M2, are paired to form a thermodynamic cycling system. The heat pump device generally consists of two interconnected reactors (Fig. 5.50), with valves controlling the flow of hydrogen between them. The function of cooling/heating is realized by the generation or loss of heat, which is transferred by the heat-conducting fluid [92]. Based on the purpose and characteristics of the heat transfer cycles, Mg hydrogen storage material-based heat pumps can be categorized into three groups: heating type, temperature-enhancing type, and refrigeration type.

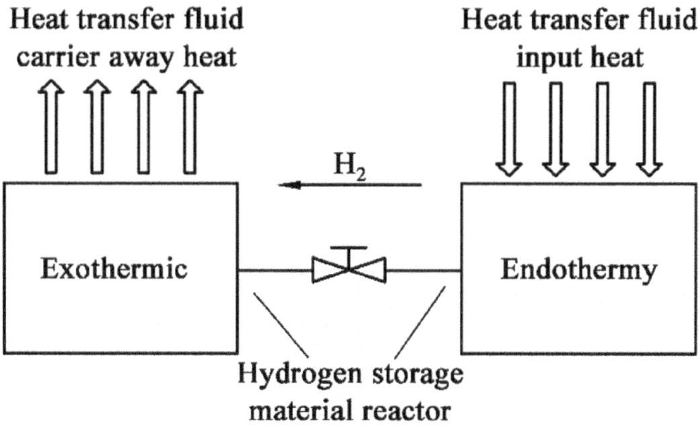

Fig. 5.50 The schematic illustration of heat pump process

Table 5.4 presents the properties and applications of heat pumps using Mg-based hydrogen storage materials. The mainstream and widely used materials are rare-earth-based, Ti-based, Zr-based, and V-based solid solution alloys, while Mg-based materials are less frequently reported. In 1995, a systemically study was conducted by Bogdanovic et al. [93] from Germany on a solar cooker/ice-making system equipped with a metal hydride thermal storage device. The thermal storage reactor of the system, which was connected to the cooking appliance, was filled with 5.4 kg of Ni-doped MgH_2, and the total thermal storage capacity reached 3 kW·h. The hydrogen storage reactor, connected to the cold storage and water tank, was filled with 30 kg of $MmNi_{5.22}Fe_{0.78}$ alloy. The system could run continuously for 5 to 6 h, with the cooker reaching 573 K and the hydrogen storage reactor cooling to 263 K, achieving a cooling capacity of 0.9 kWh.

To date, the major hurdle in the research and development of Mg hydrogen storage material-based heat pump lies in enhancing the performance of Mg-based hydrogen storage alloys and their compatibility in alloy pairs, particularly addressing the hysteresis loop and plateau pressure differences between hydrogenation and dehydrogenation process. A large hysteresis loop reduces the driving force of the cycles, while a low ramping rate of the plateau pressure impacts hydrogen transfer efficiency. Another challenge affecting the performance of Mg hydride-based heat pumps is the sluggish heat and mass transfer within the heat pump and air conditioning system. Future efforts should focus on developing Mg-based hydrogen storage materials with enhanced properties, including efficient and highly reversible hydrogen absorption and desorption capacities, fast hydrogenation and dehydrogenation rates, large enthalpy changes, minimal hysteresis, fast ramping rate of plateau pressure, high thermal conductivity, easy activation, good resistance to poisoning and pulverization, and strong cycling stability. Additionally, more attention should be given to improving the compatibility of alloy pairs and addressing the sluggish heat and mass transfer challenges.

Table 5.4 Heat pump used Mg-based hydrogen storage materials [91]

Developer	Heat pump performance						System features	Output power/kW
	Cycle type	Hydrogen storage alloys	Temperature/K			Cycle time/min		
			T_H	T_M	T_L			
Daimler BenzAG, Germany	Heating	$Ti_{0.8}Zr_{0.2}CrMn$/Mg $Mg_{0.95}Ni_{0.05}$/Ti	623 ~1042	532~952	288~296	376~250	Vehicle air conditioning	–
Bogdanovic, Germany	Heating/cooling	MgH_2/ $MmNi_{5.22}Fe_{0.78}$	–	–	–	360~300	Solar cooker/ refrigeration	0.9

5.3.3 *Energy Storage*

The application of Mg-based hydrogen storage materials in the field of energy storage primarily utilizes the reversible reactions between Mg and hydrogen to facilitate the conversion between chemical and electrical energy. For instance, Mg-based hydrogen storage materials serve as energetic materials to enhance the energy content. They have also served as anode materials in Ni–H batteries and lithium-ion batteries to enable energy conversion.

1. Energetic Materials

Energetic materials are characterized by their ability to react and release significant energy without assistance from other substances. These materials typically store chemical energy and are widely used in military and aerospace industries, where advancements significantly impact the energy efficiency of weapons and aerospace propulsion systems. Traditionally, the improvement of energetic materials has relied on organic nitro compounds [94]. Hydrogen, with its high power density, clean combustion, and abundant geological reserves, is a promising fuel. It possesses a combustion value of 121,061 kJ/kg, surpassing that of methane, gasoline, ethanol, and methanol. A propellant composed of liquid oxygen and hydrogen can achieve a specific impulse of 3822 N·s/kg, demonstrating hydrogen's suitability as an energetic material. However, the storage, transportation, and utilization of hydrogen require specialized conditions, such as high pressure and low temperature, which pose challenges in terms of cost and safety. Therefore, introducing hydrogen energy into solid propellants, explosives, and other energetic materials through safe, reliable, and high-capacity hydrogen storage materials presents a feasible solution [95].

Metal hydrides exhibit higher combustion heat compared to pristine metals. In energetic materials, the application principle of Mg-based hydrogen storage materials lies in storing hydrogen in a solid state and desorbing it within a certain pressure range. The desorbed hydrogen serves as a high-energy component in propellants and explosives. The decomposition of Mg-based material hydrides produces active magnesium metal, which has a combustion heat of approximately 24,773 kJ/kg. This property makes magnesium a widely used metallic fuel in solid propellants, significantly enhancing their combustion heat. Additionally, the magnesium and other metals generated from the decomposition of Mg-based hydrides can promote the thermal decomposition of other high-energy components in solid propellants and explosives, thereby boosting their power and efficiency.

In the field of energetic materials, Mg-based hydrogen storage materials offer several advantages over other metals:

① High hydrogen storage capacity: Examples include 7.6 wt.% of MgH_2, 3.6 wt.% of Mg_2NiH_4, and an impressive 14.8 wt.% of $Mg(BH_4)_2$ [96, 97].

② Abundance and Cost-Effectiveness: Magnesium is an earth-abundant and cost-effective resource.

③ Efficient Hydrogen Release: A significant amount of hydrogen can be released within a certain pressure range, ensuring high hydrogen utilization efficiency.

④ Environmental Benefits: Low air pollution results from combustion.

⑤ Chemical and Thermal Stability: Compared with other metal hydrides, Mg-based hydrides exhibit superior chemical stability and thermal stability. This reduces their predisposition to decomposition during production and storage, enhancing safety. Based on the chemical structure, Mg-based hydrogen storage materials used in energetic materials can be categorized into magnesium hydrides (MgH_2), Mg-based alloy hydrides, and Mg-based coordination hydrides.

(1) MgH_2

MgH_2, as a single light metal hydride, has a density of 1.45 g/cm^3 and a mass hydrogen storage density of 7.6 wt.%, which is significantly higher than that of Mg-based alloy and other metal hydrides. Additionally, the enthalpy change of MgH_2 in the hydrogen desorption reaction is 74.5 kJ/mol H_2, and the hydrogen desorption temperature is approximately 573 K. These two parameters demonstrate the thermal stability of MgH_2 and make it suitable for use as a component of solid propellants and explosives. To date, MgH_2 can desorb hydrogen via thermal decomposition and hydrolysis, both reaction paths, as shown in Eqs. (5.26) and (5.27), are suitable for the energetic materials sector:

$$MgH_2 \rightarrow Mg + H_2 \uparrow \tag{5.26}$$

$$MgH_2 + 2H_2O \rightarrow Mg(OH)_2 + 2H_2 \uparrow \tag{5.27}$$

MgH_2 is recognized as a high-energy constituent of next-generation solid propellants and explosives. Substitute metal fuel components in solid propellants with MgH_2 can enhance energy output, while its use as a high-energy additive can remarkably increase destructive power during explosions. Hradel [95] found that MgH_2 can improve the work output of various organic explosives, such as trinitrotoluene (TNT), 2,4,6-trinitrobenzenemethanitramine (Tetryl), and C-4. Liu et al. [98] investigated the impact of MgH_2 on the performance of ammonium perchlorate/aluminum powder/hydroxyl-terminated polybutadiene (AP/Al/HTPB) composite solid propellants. The results showed that the burning speed increased by 13.9% after adding 1.3 wt.% MgH_2, outperforming Mg_2NiH_4 and Mg_2Cu-H. Liu et al. [99, 100] fabricated a mixture of aluminum and MgH_2 powder using a mechanical alloying method. The physical parameters of the resulting product are as follows: mass hydrogen storage density of 5.8 wt.%, density of 2.476 g/cm^3, net combustion heat of –30,359 kJ/kg, and combustion efficiency of 94%. This aluminum-MgH_2 mixture demonstrates significantly higher combustion heat and efficiency than the pristine aluminum with the same granularity. Furthermore, replacing aluminum powder with an aluminum-MgH_2 mixture in the solid propellant of ammonium perchlorate/aluminum powder/hydroxyl-terminated polybutadiene composite significantly enhances the theoretical energy capacity of the propellant. When the content of the aluminum-MgH_2 mixture reaches 22 wt.%, the specific impulse (theoretical value) of the propellant reaches its maximum. Compared with pristine aluminum, the theoretical maximum specific impulse increases by 51.04 N·s/kg (approximately 1.96%), and the characteristic velocity increases by 1.8% (28.7 m/s). Zhang et al. [101] investigated the ignition delay time and flame spread speed of a mixture of MgH_2 with five energetic materials (e.g., RDX) using laser-delayed ignition and high-speed photographic visualization technologies. For energetic materials, the addition of MgH_2 enhances their ignition and combustion properties. The mechanism for this improvement is as follows: ① After hydrogen desorption, the burning of the desorbed hydrogen promotes an increase in surface temperature and the melting of the condensed phase in the energetic compounds. ② The metallic magnesium produced by the decomposition of MgH_2 acts as a heat conductor during the transition of energetic compounds from the condensed phase to the gaseous phase. This, in turn, accelerates the phase transition, temperature rise, flow rate, and diffusion of the gaseous phase, and ultimately speeding up the ignition.

However, an in-depth study is lacking in the following aspects: ① the compatibility and stability of MgH_2 with the various components of solid propellants and explosives. ② the sensitivity of these materials after MgH_2 is added. Furthermore, considering the poor moisture and oxidation resistance of MgH_2, surface modification is needed to enhance stability during storage, reduce maintenance costs, and prolong the service life of MgH_2 [95].

(2) Mg-based Alloy Hydrides

Mg-based alloy hydrides have relatively high hydrogen storage capacities (3.6 wt.% hydrogen for Mg_2NiH_4) and good thermal stability (dehydrogenation temperature: 523 K). Similar to MgH_2, the hydrogen released from Mg-based alloy hydrides not only provides high combustion heat but also promotes the combustion of solid propellants. To date, Mg-based alloy hydrides have been extensively investigated and applied in the fields of solid propellants and explosives, achieving significant progress and wide applications.

Liu et al. [102] prepared two types of Mg-based hydrogen storage alloy hydrides, namely Mg_2Cu-H and Mg_2NiH_4. They found that both Mg_2Cu-H and Mg_2NiH_4 significantly promote the thermal decomposition process of ammonium perchlorate (AP). After a 5 wt.% addition of Mg_2Cu-H or Mg_2NiH_4, the decomposition heat of AP increased from 436 kJ/kg to 1,250 kJ/kg and 1,293 kJ/kg, respectively, with the enhancement proportional to the amount of Mg-based alloy hydrides added. Furthermore, the addition of Mg_2Cu-H and Mg_2NiH_4 also significantly promotes the thermal decomposition of solid propellants composed of ammonium perchlorate/aluminum powder/hydroxyl-terminated polybutadiene (AP/Al/HTPB). With a 1.3 wt.% addition of Mg_2Cu-H or Mg_2NiH_4 addition, the temperature of exothermic peak of the propellants decreased by 41.7 K and 10.4 K, and the decomposition heat increased from 1,940 kJ/kg to 3,540 kJ/kg and 3,860 kJ/kg, respectively. This led to an increase in the burning rate of AP/Al/HTPB by 14.4% and 3.5%.

The use of magnesium-nickel or magnesium-copper alloy hydrides as an additive or substitution for metallic compounds in solid propellants is a major application of Mg-based hydrogen storage alloy hydrides in the field of energetic material. To further increase the energy level of propellants and explosives, comprehensive studies are needed on the compatibility and stability of Mg-based alloy hydrides with propellant and explosive components. The influence of Mg-based alloy hydrides on the burning rate, energy capacity, and physicochemical properties of propellants and explosives also requires more attention. Only through these studies can we gain a deep understanding of the application and value of Mg-based alloy hydrides in propellants and explosives [95].

(3) Mg-based Coordination Hydrides

Mg-based coordination hydrides have the highest mass hydrogen storage capacity among Mg-based hydrogen storage materials, a merit that has made them gradually become dominant in energetic materials. There are three major classes of Mg-based coordination hydrides: magnesium borohydrides (e.g., $Mg(BH_4)_2$), magnesium aluminohydrides (e.g., $Mg(AlH_4)_2$), and magnesium amides (e.g., $Mg(NH_2)_2$). During the combustion of solid propellants, $Mg(NH_2)_2$ struggles to release hydrogen and easily decomposes to generate ammonia; therefore, $Mg(NH_2)_2$ is not suitable as a hydrogen source of propellants.

$Mg(BH_4)_2$ and $Mg(AlH_4)_2$ are typically synthesized through ion-exchange reactions between sodium- and lithium-based coordination metal hydrides and $MgCl_2$ [103]. The mass hydrogen storage capacity of $Mg(BH_4)_2$ is approximately 14.9 wt.%, with a hydrogen desorption enthalpy of around 37 kJ/mol H_2, both of which are significantly lower than those of Mg-based alloy hydrides and MgH_2. In the early 1980s, Fifer et al. [104] proposed that metal boron hydrides could enhance the ignitability and burning rate of propellants. However, results show that $Mg(BH_4)_2$ not only increases the decomposition heat of explosives but also contributes to their stability [105, 106]. Compared to $Mg(BH_4)_2$, $Mg(AlH_4)_2$ exhibits a mass hydrogen storage capacity of around 9.34 wt.%. Unlike alkali-metal-coordinated aluminum hydrides, $Mg(AlH_4)_2$ does not release the intermediate of $[AlH_6]^{3-}$ when decomposing to form MgH_2. Furthermore, when the temperature reaches 560 K, the generated MgH_2 further decomposes to form Mg. Dehydrogenation is an exothermic process and thermodynamically irreversible. While the mass hydrogen storage capacity of $Mg(AlH_4)_2$ is much lower than $Mg(BH_4)_2$, it far surpasses Mg-based hydrogen storage alloy hydrides and MgH_2. Therefore, $Mg(AlH_4)_2$ is still more suitable for enhancing the power of solid propellants and explosives.

The mass hydrogen capacity of Mg-based coordination hydrides ranks first among hydrogen storage materials. Therefore, a deeper understanding of these materials will reveal their potential in energetic applications. They may improve the energy density of energetic materials and promote the development of this field. However, research on Mg-based coordination hydrides began relatively late, mature preparation methods are lacking, and the hydrogen absorption and desorption mechanisms remain unclear. Under these circumstances, their applications in solid propellants and explosives still have a long way to go.

2. Nickel-Metal Hydride Battery

Due to the rapid development of portable electronic devices and the increasing demand for sustainable and low-energy transportation, electrochemical energy

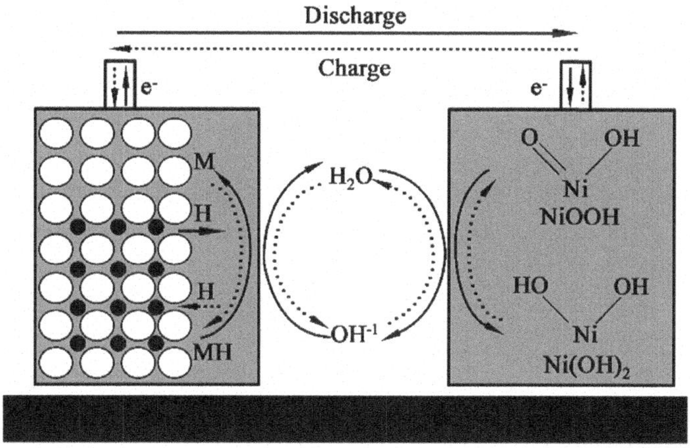

Fig. 5.51 Electrochemical charge and discharge processes of Ni/MH batteries [107]

storage and conversion systems have drawn significant attention. Commonly used secondary batteries include lead-acid batteries, nickel–cadmium (Ni/Cd) batteries, nickel-metal hydride (Ni/MH) batteries, lithium-ion (Li-ion) batteries, and lithium-ion polymer batteries. Nickel-metal hydride batteries, which feature a nickel hydroxide electrode as a cathode, a hydrogen storage alloy electrode as an anode, and a potassium hydroxide solution as an electrolyte, are considered a new type of green, rechargeable battery. Figure 5.51 illustrates the charge and discharge processes of a nickel-metal hydride battery. During the charging process, hydrogen atoms are released from $Ni(OH)_2$ at the cathode and absorbed by the hydrogen storage alloy at the anode to form a metal hydride. When the battery discharges, hydrogen is desorbed from the metal hydride at the anode and reacts at the cathode to regenerate $Ni(OH)_2$. Therefore, the charge and discharge mechanism of nickel-metal hydride batteries is based on the movement of hydrogen between the metal hydride electrode and the nickel hydroxide electrode. The overall electrochemical reaction can be described as follows:

$$\mathrm{xNi(OH)_2} + \mathrm{M} \underset{\text{Discharge}}{\overset{\text{Charge}}{\longrightarrow}} \mathrm{xNiOOH} + \mathrm{MH_x} \tag{5.28}$$

Nickel-metal hydride (Ni-MH) batteries are widely applied due to their superior energy density, long cycling life, high-rate capacity, resistance to overcharging, and good environmental compatibility. These characteristics have enabled them to rapidly replace nickel–cadmium batteries in portable electronic applications. Today, Ni-MH batteries have become the preferred power source for hybrids and pure electric vehicles (HEVs) and are also being experimentally used in electric power tools. The capacity, cycling life, and discharge behavior, of Ni-MH batteries largely depend on the intrinsic properties of the electrode materials, particularly the hydrogen storage alloys used in the anode [107].

As electrode material of Ni-MH secondary batteries, the following requirements must be met:

① High hydrogen absorption and desorption capacity between the temperatures of 253–313 K, ensuring robust charge and discharge properties.
② Good cycling stability in alkaline electrolyte (KOH solution).
③ Excellent activation properties.
④ Rational particle size, suitable for electrode plates. Numerous hydrogen storage alloys have been developed as anode materials for Ni-MH batteries, including AB_5-type rare-earth alloys, AB_3- or A_2B_7-type rare-earth magnesium alloys, AB_2-type multicomponent alloys, Mg-based alloys, and TiV-based multiphase alloys.

Mg-based hydrogen storage alloys, serving as anode material in nickel-metal hydride (Ni-MH) batteries, have drawn considerable interest due to their good reversible hydrogen storage capacity and high theoretical capacity. High-capacity Mg-based hydrogen storage materials are particularly promising for developing high-energy–density Ni-MH batteries. However, significant challenges remain. For example, Mg-based hydrogen storage alloys are easily oxidized and corroded in alkaline electrolytes, leading to decreased capacity and reduced cycling life. Although several strategies, such as elemental substitution, surface modification, and compositing, have been proposed to address these issues, they often result in a reduction of the initial discharge capacity (maximum discharge capacity), limiting their overall effectiveness. Currently, the application of Mg-based hydrogen storage alloy electrodes in Ni-MH batteries faces two major challenges:

① Improving capacity: The initial capacity of Mg-based hydrogen storage alloy electrodes can reach 700–800 mA·h/g, which is still lower than their theoretical value.
② Enhancing cycling life: After 20 charge–discharge cycles, the stable remanent discharge capacity is approximately 300–400 mA·h/g, which remains significantly below their theoretical value.

These challenges continue to hinder the widespread application of Mg-based hydrogen storage alloy electrodes in Ni-MH batteries [108].

For the first challenge, the kinetic properties of Mg-based alloy electrodes should be further improved, especially the hydrogen absorption and desorption kinetics near room temperature. Liu et al. [109] found that the maximum charge capacity of Mg_2Ni remains unchanged, but the discharge rate is significantly enhanced after introducing 1.0% La, suggesting that a trace amount of La addition can improve the discharge kinetics of Mg_2Ni. Cui et al. [110] partially substitute Mg and Ni with V and Al to obtain the alloy of $Mg_{1.9}V_{0.1}Ni_{0.8}Al_{0.2}$. They found that the hydrogen diffusion rate is significantly increased, which benefits the hydrogen absorption and desorption kinetics of the electrodes.

The second challenge is mainly caused by the high chemical activity of magnesium. During the electrochemical reaction process of Mg-based hydrogen storage electrodes in KOH electrolyte, a passivation layer of $Mg(OH)_2$ quickly forms on the electrode surface, blocking the hydrogen absorption/desorption process. Additionally, the pulverization and dissolution of the electrode in the electrolyte also contribute to the performance decay of the battery.

In recent years, strategies of compositing transition metal oxides with Mg-based hydrogen storage alloys via high-energy ball milling have been proposed to enhance the catalytic performance of Mg-based hydrogen storage alloys, such as Sc_2O_3, TiO_2, V_2O_5, Cr_2O_3, SiO_2. These methods have shown mild progress in improving hydrogen absorption and desorption kinetics and the overall storage capacity of electrode materials. Additionally, surface modifications have improved the cycling life of the batteries, such as fluorination, and copper/nickel deposition via plating. The combination of these two strategies (alloying element and surface modification) has also been reported in the literature. Results indicate significant performance improvements compared to the single strategy, but these are still far from meeting the requirements of practical applications. Xu [111] pioneered the use of unalloyed but TiF_3-catalyzed magnesium as a negative electrode in NiMH batteries (Fig. 5.52). By introducing 5 mol% TiF_3 into nano-magnesium, hydrogen adsorption kinetics were improved. They also explored several strategies to prevent magnesium passivation, such as electrode encapsulation, and using room/high-temperature ionic liquids or alkaline polymer films as the electrolyte. The developed Mg-TiF_3 composite exhibited good electrochemical performance, and further improvements are possible. Although the strategies mentioned above help mitigate the capacity decay of Mg-based hydrogen storage materials, they often come at the expense of maximum discharge capacity. Moreover, those methods frequently require large amounts of transition metals, increasing costs and hindering practical applications.

With the development of advanced electronics and other technologies, there is an increasing demand for higher power and energy capacities in batteries. For Ni-MH batteries, in particular, high-capacity hydrogen storage alloy anodes are of critical importance. Mg-based hydrogen storage alloys are among the most promising candidates for Ni-MH battery anodes. Fully utilizing and developing Mg-based hydrogen storage alloy electrodes is of great significance for building a sustainable society. The performance of Mg-based hydrogen storage electrodes is influenced by factors such as intrinsic properties and electrode fabrication technique. Therefore, future efforts should focus on enhancing the performance of Mg-based hydrogen storage alloys and advancing electrode preparation methods.

3. Lithium-Ion Battery

The commercialized lithium-ion battery (LIB) uses graphite-based carbon materials as an anode to form LiC_6 intercalation compounds during charging. However, the low specific capacity (theoretical value: 372 mA·h/g) of the graphite-based material could not meet the requirements of cutting-edge LIBs. Therefore, Mg-based metal hydrides have emerged as a focal point in battery research.

The application of Mg-based hydrogen storage materials in LIBs mainly relies on the lithium storage characteristics of Mg-based hydrides, allowing them to serve as anode materials and facilitate the conversion between chemical energy and electrical energy. The theoretical value of specific capacity magnesium hydride (MgH_2)-based

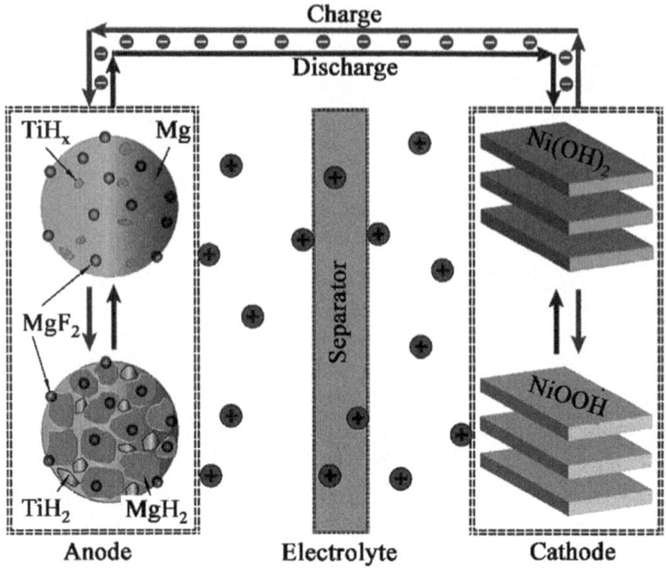

Fig. 5.52 Schematic illustration of TiF_3 catalyzed Mg as negative electrode for NiMH battery [101]

LIBs reaches 2,038 mA·h/g, with a reversible capacity of 1,480 mA·h/g in practical applications [112]. Moreover, MgH_2 possesses an average potential difference versus anode materials (0.5 V vs. Li^+/Li^0) and smaller charge–discharge hysteresis (<0.25 V) compared to other energy conversion materials. Other Mg-based hydrogen storage alloys, such as Mg_2NiH_4, Mg_2CoH_5, and Mg_2FeH_6, exhibit similar characteristics in LIBs. These findings pave the way for the application of Mg-based hydrogen storage materials in LIBs and hold significant promise.

Although Mg hydrides offer many advantages in LIBs applications, challenges remain, including slow kinetics, poor reversibility, and short cycling life. To address these issues, efforts should focus on enhancing the electrochemical performance of MgH_2 through nanosizing or forming composites with functional materials. For example, Huang et al. [113] synthesized a $0.7MgH_2$–$0.3TiH_2$ nanocomposite under a hydrogen atmosphere via reactive ball milling. Experimental results showed that the composite electrode exhibits good reversibility (reversible capacity: 1,540 mA·h/g) and reduced initial polarization. A conversion reaction between MgH_2 and Li^+ occurs at 0.52 V, while δ-$TiH_{1.5}$ reacts with Li^+ at 0.21 V to produce α-Ti and LiH. However, the capacity of the electrode decays rapidly, with only 530 mA·h/g of discharge capacity remaining after seven cycles. Despite these challenges, MgH_2 still presents better commercialization prospects compared to traditional graphite anode materials. Currently, research efforts are focused on addressing volume expansion during charge/discharge cycles and improving the poor conductivity of LiH [73].

5.3.4 Hydrogen Chromatic Film

Hydrogen-induced chromic films are novel functional materials. These films can switch their status between transparent and reflective through hydrogenation and dehydrogenation, enabling intelligent control in both the infrared and visible bands. Such materials can serve as smart window coatings to regulate the intensity of the incident light, providing thermal insulation and energy-saving for buildings, vehicles, spacecraft, and optical devices (e.g., selective light filters). [114].

At the end of the twentieth century, Huiberts et al. [115] reported the phenomenon of "hydrogen-induced transition," referring to a type of functional thin film whose optical transmittance changes during hydrogenation. Moreover, the optical transmittance of these films can return to their original state by altering the environmental atmosphere. Throughout this reversible process, the optical properties of the film continuously change, and the displayed color varies depending on the material component. As a result, photochromic fiber optic sensors and smart glass based on hydrogen-induced chromic films have developed rapidly in recent years [116–118].

The first-generation hydrogen chromatic films are primarily based on rare earth metals, such as lanthanum. However, due to the poor stability of their hydrides, low visible light transmittance, and limited optical control band, their application is quite limited. Consequently, research has focused on developing film materials with higher visible light transmittance during hydrogenation and better reflectivity during the dehydrogenation process. In 1997, Philips discovered that magnesium-rare earth alloy films exhibit excellent visible light transmittance, and Mg-based hydride films have significantly expanded the application fields of hydrogen-induced chromic functional films. Moreover, due to its abundance on earth- and high hydrogen storage capacity (7.6 wt.%) [119], magnesium has become an important material in the hydrogen-induced chromic film family, with significant progress made in Mg-based hydrogen-induced chromic film over the past two decades.

The principle of hydrogen-induced chromic films is based on reversible hydrogen adsorption, and the mechanism involves the interaction between Mg-based films and hydrogen gas, facilitated by catalysts, to form magnesium hydrides with broader optical bands. This interaction changes the photochromic layer from a reflective to a transparent state. Furthermore, when inert gas is introduced (i.e., the partial pressure of hydrogen decreases), the magnesium hydrides release hydrogen. As the content of hydrogen atoms in the photochromic layer decreases, it transitions from a transparent state back to a reflective state, thereby regulating the light transmittance of the film. Mg-based films offer advantages such as simplicity in structure, quick response, wide optical band, and good sustainability. They are currently the primary component of hydrogen-induced chromic devices [119].

(1) Hydrochromic hydrogen-induced Mg-based dimming films

Chromogenic windows for energy-saving, commonly known as smart glass, are widely used in transportation, vehicles, buildings, and display technologies. These windows consist of transparent substrates (e.g., glass) and a tinting film, and they can be classified into thermochromic, electrochromic, gasochromic, and photochromic materials based on their working mechanism. Hydrogen-induced tinting film is a kind of gasochromic film that can reflect sunlight with a specific wavelength in the presence of hydrogen gas. This property allows it to adjust the intensity and heat load of light entering the interior space, thereby reducing air conditioning energy consumption and achieving energy savings. Since the reaction process only involves hydrogen and oxygen, with water as the only byproduct, hydrogen-induced chromic films are considered environmentally friendly materials. Mg-based films, as significant members of the hydrogen-induced chromic tinting film family, have been studied for nearly 20 years, yielding significant progress. Current research primarily focuses on Mg-based systems complex with other elements [119].

The second-generation metal hydride chromic materials mainly consist of magnesium-rare earth metal films, which utilize a composite of magnesium and rare earth metals as the hydrogen-induced transition layer. The hydrides are synthesized and decomposed with the assistance of an outer palladium catalyst layer, which promotes the dissociation of hydrogen gas and the recombination of hydrogen atoms. When magnesium is added to rare earth metals (e.g., Y or La), the reflectivity of the film increases due to the metallic state of Mg. Upon hydrogenation, the resulting MgH_2 enhances the transmittance and band gap of the film. In recent years, the National Institute of Advanced Industrial Science and Technology (AIST) in Japan has focused on developing durable magnesium-based dimming films. Yamada et al. [120] developed Mg-Y alloy films using direct current (DC) magnetron sputtering. By adjusting the magnesium-yttrium ratio, the optimized film achieved 10,000 hydrogen absorption and desorption cycles. Multiple experimental tests demonstrated that the films containing 0.27–0.70 yttrium exhibited approximately 35% transmittance for visible light in the hydrogenated state and demonstrated good stability. In subsequent research, a tantalum layer was introduced between the palladium and Mg-Y layers to prevent interdiffusion between the two layers. This strategy not only significantly enhanced the durability of the films but also increased the transmittance in the hydrogenated state to 45%, while the transmittance difference between the metallic and hydrogenated states improved to 35%, further enhancing the light-modulating performance of the films [121]. Although magnesium-rare earth alloy films have been extensively studied, the scarcity of rare earth elements results in high costs, limiting their large-scale applications in devices such as architectural glass and vehicle windows.

The third-generation metal hydride chromogenic materials are Mg-transition metal (Ni, Co, Mn, Fe, Ti) hydrides [117]. A milestone achievement in this field was the development of Mg-Ni alloy film by Richardson et al. [122] in 2001. Subsequently, Griessen et al. [123] discovered that Mg-Ni film exhibits a silver-white reflection on both sides before hydrogenation. During the slow hydrogenation process, the substrate side of the film gradually transforms to a dark absorbing state, while the palladium-coated side remains unchanged. As hydrogenation proceeds, the film gradually becomes transparent and colorless. Similar behavior has also been observed in Mg-Co and Mg-Fe films. Alloy films composed of magnesium and other transition metals (e.g., manganese, titanium, vanadium, and niobium) also exhibit hydrogen-induced chromic effects. Specifically, Mg-Nb alloy films, produced by magnetron sputtering, how a peak transmittance of 45% for visible light in the hydrogenated state and a peak reflectance of 60% in the dehydrogenated state, surpassing Mg-Ti alloy films. Additionally, the full hydrogenation and hydrogenation times of Mg-Ti films are approximately 20 and 250 s, respectively. After 150 cycles, the modulation decay in the light band is negligible, which benefits from the stable structure of Mg-Ti films after repeated cycles [124]. Similar to tantalum, titanium, vanadium, and niobium can also serve as intermediate layers to prevent the diffusion between the magnesium and palladium layers. Compared with the first two generations of chromic materials, the third generation offers several advantages: a much wider range of adjustable optical transmittance, light modulation over a broader wavelength range, lower cost, and improved anti-oxidation properties. However, the chromic performance of these materials heavily depends on the alloy composition ratio. With different element ratios, Mg-transition metal alloy films exhibit varied chromic effects. For example, the color of Mg-Ni alloy films changes from crimson to pale yellow as the Ni content decreases, which limits their practical application in smart glass. Other magnesium-transition metal systems, such as Mg-Ti alloys, although nearly colorless in the transparent state, have a visible light transmittance below 40%, resulting in poor light modulation performance.

The fourth-generation metal hydride chromic materials consist of Mg-alkaline earth metal (Ca, Ba, Sr) hydrides. Yamada et al. [125] pioneered a groundbreaking advancement in Mg-alkaline-earth-metal dimmable films. Palladium-coated Mg-Ca alloy films were directly fabricated on quartz substrates using the DC magnetron sputtering method. A maximum visible light transmittance of 46% can be achieved at a calcium content of 0.06 at.%. Subsequently, strontium and barium were selected to substitute calcium, forming Mg-Sr and Mg-Ba alloy films. The visible light transmittance of the resulting $Mg_{0.8}Sr_{0.2}$ and $Mg_{0.45}Ba_{0.55}$ films reached 45% and 50% under hydrogenated states, respectively. The chromic performance of these materials also depends on the ratio of alloy components. For instance, $MgCa_x$ alloys exhibit optimal chromic properties only within the range of $0.035 < x < 0.075$.

In previously reported hydrogen-induced chromic Mg-based smart films, Mg-based alloy films have been the primary research focus. Most cutting-edge systems were developed based on hydrogen storage materials with similar components and adapted for film applications. Considering the superior performance of Mg-transition metal oxide composites in the field of Mg-based hydrogen energy materials, the fifth generation of chromic film materials, magnesium-transition metal oxide films, has been developed. Among them, Mg-TiO_2 composite films are particularly notable. TiO_2 can catalyze the reaction between Mg and hydrogen, significantly enhancing the hydrogen absorption and desorption kinetics of Mg. As a result, Mg-TiO_2 composite films exhibit excellent chromic properties. Liu et al. [126] from Shanghai Jiao Tong University prepared Pd/Mg-TiO_2 dimmable thin films with different Mg to TiO_2 ratios using magnetron sputtering. Specifically, Mg and TiO_2 were co-deposited to form a composite film layer, which was then covered by a palladium layer. The study revealed that Mg-10 mol% TiO_2 film demonstrated a 23.6% improvement in hydrogen-induced chromic performance compared to pristine Pd/Mg film, with the reflectance of the Mg-10 mol% TiO_2 film under the dehydrogenated state approaching 70%.

China has achieved measurable success in the field of hydrogen-induced chromic Mg-based films. However, research remains in the early stages, focusing primarily on fundamental studies. There is still significant progress to be made before these materials can be commercialized. Since the magnetron sputtering technique is well-suitable for synthesizing Mg-TiO_2 films, future efforts should focus on developing Mg-transition metal films and advancing their application in hydrogen-induced chromic dimmable glass. Key objectives include improving processing techniques, optimizing performance, and promoting the widespread application of Mg-transition metal films.

(2) Hydrogen-induced Mg-based chromic sensing films for fiber optics

Hydrogen, as a clean energy carrier, plays a significant role in building a sustainable society for the future. However, when the hydrogen content (volume fraction) in air reaches 4–75%, an explosion can occur when exposed to an open flame. Therefore, a highly sensitive detection device is essential for hydrogen storage, transportation, and usage processes. Traditional hydrogen detection is typically achieved using catalytic resistance detectors or electrochemical devices, which still have inherent shortcomings. As a result, optical fiber sensors have garnered significant attention. Changes in optical signals can be generated when a palladium or metal oxide layer deposited at the end of the optical fiber is hydrogenated. These changes can be used to detect hydrogen concentration. The signal collector of the optical fiber can be detached from the sensing point. Due to the small footprint, multiple fibers can be connected to a single detector, ensuring the reliability of the collected signals. Therefore, fiber films with significant optical contrast before and after hydrogenation hold great potential for practical applications. Research into the application of hydrogen-induced chromic Mg-based films in the field of fiber optic sensors has been actively pursued.

The overall design of the fiber optic hydrogen sensor is illustrated in Fig. 5.53 [127]. The core of the optical fiber is typically composed of silica, similar to the quartz-based material in dimming glass. A Mg-based film is deposited at the end of the fiber core, followed by a palladium layer acting as a catalyst, and finally covered by a protective coating layer. Slaman et al. [128] fabricated a fiber optic sensor using a Mg-Ti alloy film as a dimming layer. The resulting product achieved a reflectivity deduction from 61 to 7% within 8 to 20 s when the hydrogen concentration in the ambient environment increased from 0 to 1%. However, as the oxygen concentration in the atmosphere increased, the reaction time also increased. Although this type of optical fiber film exhibited good reproducibility and strong adhesion to the fiber surface, the time required for hydrogen desorption increased from 130 to 300 s after 50 cycles. To further optimize performance, a novel Pd/Mg/Ti optical fiber film was fabricated using a dual-layer structure of magnesium and titanium. By monitoring changes in transmittance or reflectance, the transition between metallic and insulating states of the film could be observed, enabling hydrogen detection in the pressure range of 200 Pa to 4000 Pa.

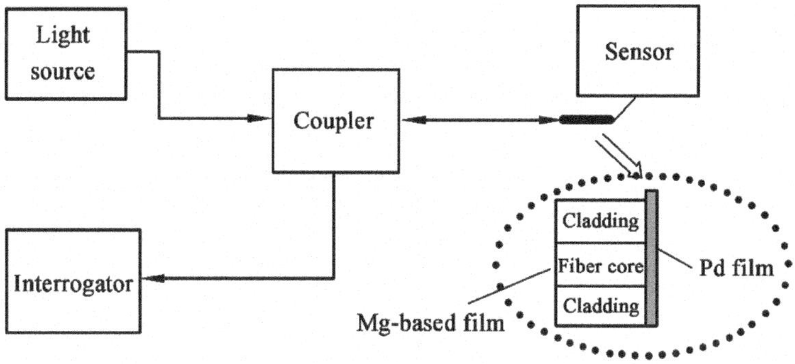

Fig. 5.53 Schematic illustration of the overall design of the optical fiber hydrogen sensor [127]

To meet the requirements for sensor response speeds, Zhao et al. [129] prepared amorphous and crystalline Pd/Mg and Pd/Mg-Ni films, respectively. They observed that the amorphous films exhibit faster optical transition properties compared to their crystalline counterparts. The amorphous structure of the Mg-based layer prevents the formation of harmful hydride layers, facilitating the diffusion of hydrogen atoms within the magnesium-based layer. This observation enhances the response properties of Mg-based hydrogen-induced chromic films.

Films containing rare earth elements exhibit significant color differences before and after hydrogen-induced transformations, making them highly suitable for optical fiber hydrogen sensors from a signal quantity perspective. However, their susceptibility to oxidation limits their application in the air. Song et al. [130] deposited a Pd-Y alloy film onto a quartz substrate via co-sputtering. Although the film suffers severe degradation after one month of exposure to air at 293 K, annealing at 473 K regenerates its fast response to hydrogen. This can be explained by the stress relaxation mechanism (Fig. 5.54). The response value of the films increases linearly with the increase of hydrogen concentration between 0.1 and 2% (volume fraction). Notably, the Pd/Mg-Y15% dimmable film maintains favorable optical conversion properties even after 10,000 cycles of hydrogen absorption and desorption, highlighting the significant application potential of Mg-based hydrogen-induced chromic films in the field of fiber-optic hydrogen sensors. Although China has made measured progress in the field of hydrogen-induced chromic Mg-based functional films, it remains at a fundamental research stage. Future efforts should focus on the development of hydrogen-induced chromic film materials and their applications in industries.

Although significant progress has been made in the application of magnesium hydride in chromogenic films, commercialization remains a distant goal. The fabrication process for Mg-based hydride color-changing materials is complex and costly, requiring expensive catalysts such as precious Pd or Pt. While improvements have been made in film durability, further enhancements are still needed, and the hydrogen release reaction of the film remains temperature-sensitive. Therefore, the development of Mg-based hydrogen storage materials in the field of chromogenic films

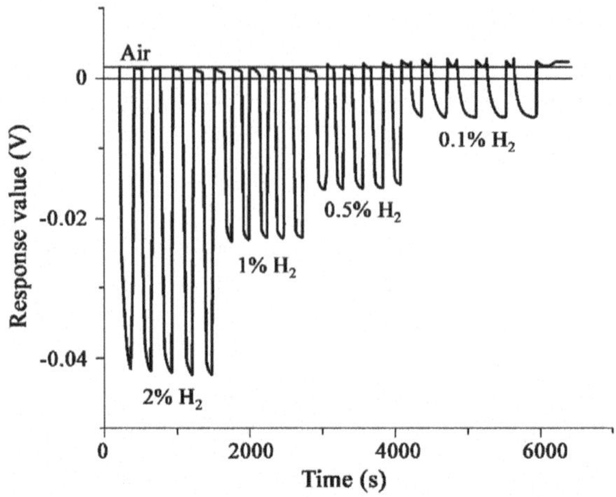

Fig. 5.54 The response value of the film as a function of increased hydrogen concentration [130]

should focus on the following two aspects: ① Leveraging the strength of Mg-based hydride color-change materials, improving their color-change properties, and reducing production cost through innovative processing techniques, such as doping Zr, Mn, Y, and V elements to alter their color properties or developing more cost-effective fabrication methods (e.g., electroplating and sol–gel processes). ② Developing advanced Mg-based hydride color-change materials with novel hydrogen storage layers in composite films that are transparent at the nanoscale and temperature sensitive.

5.3.5 *Other Applications*

(1) Catalysis for Chemical Synthesis

Intermetallic compounds, such as $LaNi_5$, Mg_2Ni, Zr_2Ni, and TiFe, exhibit reversible hydrogen absorption and desorption properties. Specifically, hydrogen is first decomposed and then absorbed as atoms on the compounds' surface, demonstrating the significant surface activity of these materials in decomposing hydrogen. Therefore, these materials can be used as catalysts, particularly for reactions involving hydrogen.

Mg-based hydrogen storage materials, such as Mg_2Ni and Mg_2Cu, have also been explored as catalysts. Imai et al. [131] investigated the catalytic performance of hydrogen storage alloys (FeTi, $(Fe_{0.9}Mn_{0.1})Ti$, $CaNi_5$, $LaNi_{4.7}Al_{0.3}$, and Mg_2Cu) in the hydrogen evolution reaction of methanol and ethanol. In 1990, Tomohico et al. [6] studied the catalytic activity of Mg_2Cu alloy in C_{18} alcohol synthesis using a pulse reactor under 0.1 MPa hydrogen atmosphere at 503 K. The results show that the Mg_2Cu pre-oxidized at 773–973 K achieved 100% reaction efficiency after the second or third pulses, with a specific area activity surpassing that of the CuCrO catalyst. For hydrogenation reactions of organic compounds, Mg-based hydrides not only serve as catalysts but also exhibit high activity. The dissolved atomic hydrogen in the hydride effectively hydrogenates unsaturated organic compounds, acting as "hydrogen source." In most hydrogenation processes of organic materials, Mg-based materials simultaneously function as catalysts and hydrogen sources.

(2) Medical Field

Hydrogen is considered an ideal medicinal gas that can work at the cellular level. Therefore, it is one of the novel treatment strategies against cardiovascular, cancer, metabolic, and respiratory diseases. As early as 1975, Dole et al. [132] reported in *Science* that malignant skin tumors in animals could be cured by continuously exposing them to a mixture of 97.5% hydrogen and 2.5% oxygen under 0.8 MPa for 14 days. The anti-oxidation effect induced by hydrogen is considered one of the primary mechanisms of its anti-cancer therapy. In July 2007, Ohsawa et al. [133] reported that 2% hydrogen inhalation significantly improves brain ischemia–reperfusion injury in animals by selectively reducing free radicals and peroxynitrite anions, both of which are important mediators of oxidative damage. Hence, the selected anti-oxidation effect of hydrogen is considered the mechanism for curing brain ischemia–reperfusion injury as well. Over the decades, hydrogen medicine in China has also made significant progress, with over a hundred research institutes conducting studies in different branches of this field. Currently, a variety of hydrogen-related products have been developed and popularized, such as hydrogen inhalers, hydrogen–oxygen breathing machines, hydrogen-enriched water cups, and hydrogen-infused foods.

The medical value of Mg-based solid-state hydrogen storage materials is primarily attributed to their two decomposition products: magnesium ions and hydrogen gas. First, magnesium is an essential element for the human body and Mg-based alloys can corrode and degrade in physiological environments generating Mg ions that are non-toxic and vital for human health. Magnesium ions participate in various physiological activities in the human body by acting as co-factors for enzymes. The functions of almost 350 enzymes, the integrity of mitochondrial cells, and all membrane structures depend on Mg ions. For example, magnesium ions help stabilize potassium intracellularly by activating the membrane Na^+-K^+-ATP enzyme. Thus, magnesium plays an important role in maintaining normal physiological activities of the myocardium, nerves, and muscles, with a particular effect on bone formation [134]. Second, extensive studies indicate that hydrogen gas has antioxidant, anti-inflammation, and anti-apoptosic functions. It has shown therapeutic effects for more than 70 chronic diseases in humans, including cardiovascular diseases, diabetes, Alzheimer's disease, and cancer, demonstrating significant potential in the fields of biomedicine.

The properties of molecular hydrogen make it inconvenient to acquire and apply in daily life. However, this problem can be effectively addressed using the products developed with Mg-based solid-state hydrogen storage materials. The hydrolysis reaction of Mg-based solid-state hydrogen storage materials can occur at room temperature under mild reaction conditions, generating approximately 15.2 wt.% hydrogen, which is twice that of their thermal decomposition [135]. Shanghai Magnesium Power Science & Technology Co., Ltd. has developed various health and

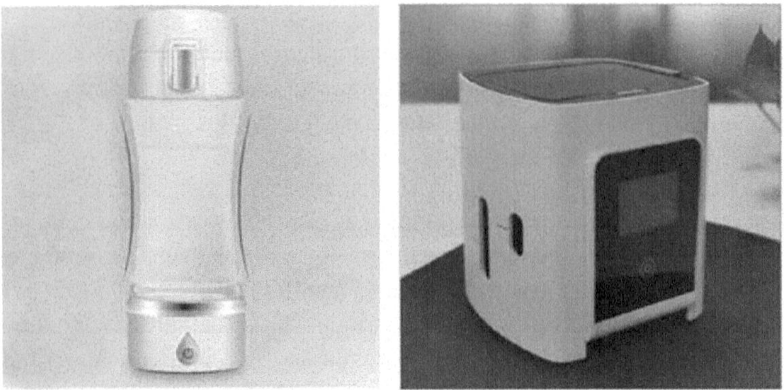

(a) Hydrogen enriched cup (b) Hydrogen productor

Fig. 5.55 Hydrogen health products

medical products based on hydrogen storage materials, including hydrogen inhalers, hydrogen-rich cups, and hydrogen-rich tablets for personal care. Unlike the conventional hydrogen inhalers and hydrogen-rich cups, these Mg-based products do not require electricity power, the produced hydrogen is of high purity, and free of residual chlorine or ozone. Figure 5.55 illustrates the hydrogen-rich cup and inhaler developed by Shanghai Magnesium Power Company. The usage of hydrogen-rich cups is not limited by the type of beverage or its temperature, making it suitable for the Chinese preference for hot drinks. Additionally, the hydrogen inhaler is compact and convenient to use.

Mg-based solid-state hydrogen storage materials exhibit promising prospects in the field of biomedicine, attracting attention in both fundamental research and clinical applications. Currently, Shanghai Magnesium Power Technology Co., Ltd., in collaboration with Huashan Hospital and the Hydrogen Science Centre of Shanghai Jiaotong University, has preliminarily validated the safety and efficacy of magnesium hydride in treating inflammation, tumors, immune disorders, and bacterial infections. Furthermore, they have partnered with enterprises from China and Japan, utilizing the sustainable released hydrogen from magnesium hydride hydrolysis to reduce oxidative free radicals, thereby providing significant health benefits. In the future, magnesium hydride is expected to be widely used in the fields of hydrogen health and medical hydrogen.

As a significant sector in the innovation and development of hydrogen science and technology, hydrogen medicine has achieved notable progress in both academic and industrial fields in recent years. With the key issue of integrated surface modification of degradable Mg-based medical materials resolved for "controlled degradation-biocompatibility", Mg-based hydrogen storage materials are emerging as a new type of medical materials. The development of Mg-based hydrogen storage medical materials has also made great contributions to promoting the integration of industry, academia, and research in the field of hydrogen science. The future direction for Mg-based hydrogen storage materials in medicine is to achieve controllable hydrogen release, enhance tissue repair functions and elucidate the related molecular mechanisms.

(3) Agricultural Sector

Hydrogen is currently a widely used clean energy source in the industrial sector and is also recognized as a biologically functional gasotransmitter with high biosafety. In 2003, Dong et al. [136] proposed the "hydrogen fertilizer theory" based on their study of hydrogen desorption in nitrogen fixation and analysis of inter-root microorganisms. The hydrogen released during the nitrogen fixation process by rhizobia lacking hydrogenase can promote the growth of hydrogen-oxidizing bacteria, thereby further enhancing plant growth. Previous studies indicate that hydrogen can effectively improve soil conditions and regulate plant growth. It can enhance the stress resistance and antioxidant capacity of plants, reduce the respiration of fruits and vegetables, and extend their shelf life [137]. Consequently, some studies suggest that hydrogen could be a potential alternative to chemical fertilizers. However, the direct use of hydrogen in large-scale field production is evidently impractical. Some researchers attempted to utilize hydrogen sticks or electrolysis to obtain hydrogen-rich water for watering plants. However, the widespread adoption of this method is limited by costs and practical conditions. Additionally, the low solubility and short retention time of hydrogen in water present o clear obstacles for its agricultural application.

The development and investigation of Mg-based hydrogen storage materials provide technical support for exploring stable and reliable hydrogen supply methods for agriculture. The application of Mg-based hydrogen storage materials in agriculture and animal husbandry relies on the combined action of magnesium and hydrogen. Mg-based hydrogen storage materials, as controlled-release hydrogen fertilizers, play an important role in regulating plant hormones, enhancing stress resistance, promoting nutrient absorption, optimizing soil microorganism composition, and supporting plant growth. Magnesium ions generated by Mg-hydrides decomposition are essential for chlorophyll synthesis and necessary for plant photosynthesis. The hydrogen generated can be directly absorbed by plant roots or released to the air-earth interface to form a hydrogen-rich atmosphere absorbed by other plant parts [138]. Adding magnesium-based hydrogen sustain-release materials to soil fertilizers, water-soluble fertilizers, or foliar fertilizers represents a feasible method for applying hydrogen in agriculture, with a positive influence on agricultural productivity.

Magnesium hydrides not only serve as an efficient hydrogen storage material but also function as safe biological hydrogen supply material. However, the alkaline nature of magnesium hydride solutions hinders plant growth, limiting their practical applications. The research team led by Academician Ding and Prof. Zou from the Hydrogen Science Center of Shanghai Jiao Tong University utilized a citric acid buffer to enhance its hydrogen production efficiency and adjust the pH value of high-purity MgH_2, making it suitable for plants. Figure 5.56 shows photos of carnation-cut flowers watered with different magnesium hydride-citric acid buffer solutions or hydrogen-rich water solutions. The experimental results reveal that, compared

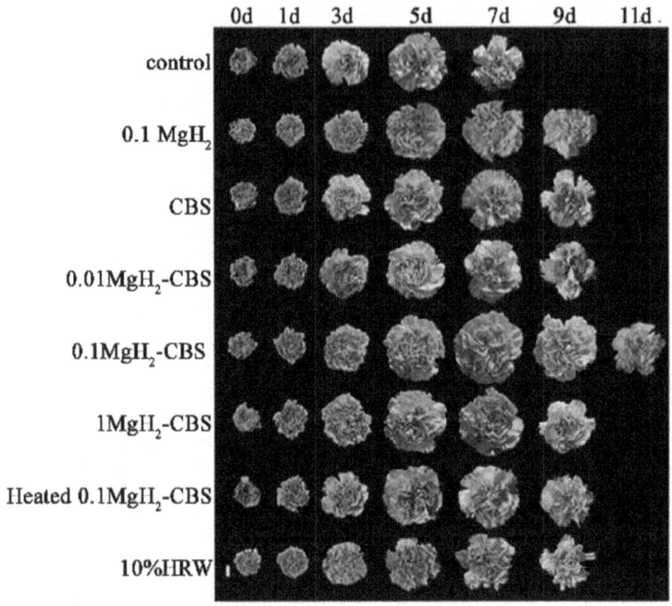

Fig. 5.56 Photographs of carnation after different days feed by various solutions of magnesium hydride-citrate buffer and hydrogen-rich water [138]

with the control group fed with distilled water, the magnesium hydride-citric acid buffer solution can extend the flowering life of carnation by approximately 52%, outperforming both magnesium hydride or electrolytically produced hydrogen-rich water used alone [138].

Using Mg-based solid-state hydrogen storage materials as hydrogen fertilizers has the following advantages: ① Low costs, low dosage, and suitability for large-scale application. ② No pollution, minimal residue, and environmentally friendly. ③ Wide applicability, including drylands, paddy fields, greenhouses, soilless cultivation, potted plants, and floral arrangements.

The hydrogen produced from magnesium hydride powder used as fertilizers can effectively regulate plant growth, enhance photosynthesis, and increase yield per acre. Field experiments with wheat demonstrate that wheat yields can be increased by approximately 15% when irrigated with MgH_2-containing solutions. Therefore, magnesium hydrides, as a fertilizer additive, can boost crop yields, presenting promising prospects for future agricultural applications and markets. Additionally, other research demonstrates that feeding cattle or sheep with hydrogen can improve their physical condition, reduce antibiotic usage, and significantly increase the protein content in milk. Moreover, Mg-based solid hydrogen storage materials can not only provide hydrogen supplementation to animals but also provide essential magnesium ions, making them excellent candidates as feed additives.

Magnesium deficiency in soil and plants in some regions highlights the great potential of magnesium hydride in agriculture. The future scope of Mg-based hydrogen storage materials will expand from industry to agriculture, further unlocking the potential of magnesium hydrides and promoting overall economic development.

References

1. LI J, LI B, SHAO H, et al. Catalysis and downsizing in Mg-based hydrogen storage materials [J]. Catalysts, 2018, 8(2): 89.
2. CHEN X, LI C, GRAETZEL M, et al. Nanomaterials for renewable energy production and storage [J]. Chemical Society Reviews, 2012, 41(23): 7909–7937.
3. ZHONG H, WANG H, LIU J, et al. Altered desorption enthalpy of MgH_2 by the reversible formation of Mg(In) solid solution [J]. Scripta Materialia, 2011, 65(4): 285–287.
4. WANG H, ZHONG H, OUYANG L, et al. Fully reversible de/hydriding of Mg base solid solutions with reduced reaction enthalpy and enhanced kinetics [J]. The Journal of Physical Chemistry C, 2014, 118(23): 12087–12096.
5. REILLY J J, WISWALL R H. Reaction of hydrogen with alloys of magnesium and nickel and the formation of Mg_2NiH_4 [J]. Inorganic Chemistry, 1968, 7(11): 2254–2256.
6. HU Z. Hydrogen storage materials [M]. Beijing: Chemical Industry Press, 2002: 147–148.
7. GUO J, HUANG C, YANG K, et al. The effects of Sn element on hydrogen storage characteristics of $Mg_{2-x}Sn_xNi$ (x= 0, 0.05, 0.1, 0.15, 0.2) alloys [J]. MRS Online Proceedings Library (OPL), 2006, 971.
8. LAN Z-Q, XIAO X, SU X, et al. Effect of doping with aluminium on the electronic structure and hydrogen storage properties of Mg_2Ni alloy [J]. Acta Physico-Chimica Sinica, 2012, 28(8): 1877–1884.
9. YANG H, YUAN H, JI J, et al. Characteristics of Mg2Ni0. 75M0. 25 (M= Ti, Cr, Mn, Fe, Co, Ni, Cu and Zn) alloys after surface treatment [J]. Journal of Alloys and compounds, 2002, 330: 640–644.
10. LI Q, JIANG L-J, LIN Q, et al. Influence of Ag and Al on hydrogen storage properties of Mg 2 Ni alloy [J]. Chinese Journal of Nonferrous Metals, 2003, 13(4): 864–870.
11. TAKAHASHI Y, YUKAWA H, MORINAGA M. Alloying effects on the electronic structure of Mg_2Ni intermetallic hydride [J]. Journal of Alloys and compounds, 1996, 242(1–2): 98–107.
12. MEYER M, MENDOZA-ZéLIS L. Mechanically alloyed Mg–Ni–Ti and Mg–Fe–Ti powders as hydrogen storage materials [J]. International Journal of Hydrogen Energy, 2012, 37(19): 14864–14869.
13. LI Q, LIN Q, CHOU K, et al. A mathematical calculation of the hydriding characteristics of $Mg_{2-x}A_xNi_{1-y}B_y$ alloy systems [J]. Journal of Alloys and Compounds, 2005, 397(1–2): 68–73.
14. VAJO J J, SALGUERO T T, GROSS A F, et al. Thermodynamic destabilization and reaction kinetics in light metal hydride systems [J]. Journal of Alloys and Compounds, 2007, 446: 409–414.
15. VAJO J J, MERTENS F, AHN C C, et al. Altering hydrogen storage properties by hydride destabilization through alloy formation: LiH and MgH_2 destabilized with Si [J]. The Journal of Physical Chemistry B, 2004, 108(37): 13977–13983.
16. ZHONG Y, YANG M, LIU Z-K. Contribution of first-principles energetics to Al–Mg thermodynamic modeling [J]. Calphad, 2005, 29(4): 303–311.
17. CRIVELLO J-C, NOBUKI T, KUJI T. Improvement of Mg–Al alloys for hydrogen storage applications [J]. International Journal of Hydrogen Energy, 2009, 34(4): 1937–1943.
18. VAJO J J, SKEITH S L, MERTENS F. Reversible storage of hydrogen in destabilized $LiBH_4$ [J]. The Journal of Physical Chemistry B, 2005, 109(9): 3719–3722.
19. PAN Y, LENG H, JIA W, et al. Effect of $LiBH_4$ on hydrogen storage property of MgH_2 [J]. International Journal of Hydrogen Energy, 2013, 38(25): 10461–10469.

20. SAJANLAL P R, SREEPRASAD T S, SAMAL A K, et al. Anisotropic nanomaterials: structure, growth, assembly, and functions [J]. Nano Reviews, 2011, 2(1): 5883.
21. WAGEMANS R W, VAN LENTHE J H, DE JONGH P E, et al. Hydrogen storage in magnesium clusters: quantum chemical study [J]. Journal of the American Chemical Society, 2005, 127(47): 16675–16680.
22. CHEUNG S, DENG W-Q, VAN DUIN A C, et al. ReaxFF$_{MgH}$ reactive force field for magnesium hydride systems [J]. The Journal of Physical Chemistry A, 2005, 109(5): 851–859.
23. BERUBE V, CHEN G, DRESSELHAUS M. Impact of nanostructuring on the enthalpy of formation of metal hydrides [J]. International Journal of Hydrogen Energy, 2008, 33(15): 4122–4131.
24. LU K. Interfacial thermodynamic properties of metal nanocrystals [J]. Journal of Physics, 1995, (09): 1454–1460.
25. CUI J, OUYANG L, WANG H, et al. On the hydrogen desorption entropy change of modified MgH_2 [J]. Journal of Alloys and Compounds, 2018, 737: 427–432.
26. AGUEY ZINSOU K F, ARES FERNáNDEZ J R. Hydrogen in magnesium: new perspectives toward functional stores [J]. Energy & Environmental Science, 2010, 3(5): 526–543.
27. PASKEVICIUS M, SHEPPARD D A, BUCKLEY C E. Thermodynamic changes in mechanochemically synthesized magnesium hydride nanoparticles [J]. Journal of the American Chemical Society, 2010, 132(14): 5077–5083.
28. CHO E S, RUMINSKI A M, ALONI S, et al. Graphene oxide/metal nanocrystal multilaminates as the atomic limit for safe and selective hydrogen storage [J]. Nature Communications, 2016, 7(1): 1–8.
29. CHO E S, RUMINSKI A M, LIU Y S, et al. Hierarchically controlled inside-out doping of Mg nanocomposites for moderate temperature hydrogen storage [J]. Advanced Functional Materials, 2017, 27(47): 1704316.
30. HIGUCHI K, YAMAMOTO K, KAJIOKA H, et al. Remarkable hydrogen storage properties in three-layered Pd/Mg/Pd thin films [J]. Journal of Alloys and Compounds, 2002, 330: 526–530.
31. ZHANG J, ZHOU Y, MA Z, et al. Strain effect on structural and dehydrogenation properties of MgH_2 hydride from first-principles calculations [J]. International Journal of Hydrogen Energy, 2013, 38(9): 3661–3669.
32. ZHANG J, MAO C, CHEN J, et al. Strain tuned dehydrogenation thermodynamics of magnesium based hydride: A first principle study [J]. Computational Materials Science, 2015, 105: 71–74.
33. PENG B, LI L, JI W, et al. A quantum chemical study on magnesium (Mg)/magnesium–hydrogen (Mg–H) nanowires [J]. Journal of Alloys and Compounds, 2009, 484(1–2): 308–313.
34. LI W, LI C, MA H, et al. Magnesium nanowires: enhanced kinetics for hydrogen absorption and desorption [J]. Journal of the American Chemical Society, 2007, 129(21): 6710–6711.
35. LI L, PENG B, JI W, et al. Studies on the hydrogen storage of magnesium nanowires by density functional theory [J]. The Journal of Physical Chemistry C, 2009, 113(7): 3007–3013.
36. ZHANG X, SHEN Z, JIAN N, et al. A novel complex oxide TiVO3. 5 as a highly active catalytic precursor for improving the hydrogen storage properties of MgH2 [J]. International Journal of Hydrogen Energy, 2018, 43(52): 23327–23335.
37. BARKHORDARIAN G, KLASSEN T, BORMANN R. Fast hydrogen sorption kinetics of nanocrystalline Mg using Nb_2O_5 as catalyst [J]. Scripta Materialia, 2003, 49(3): 213–217.
38. CUI J, WANG H, LIU J, et al. Remarkable enhancement in dehydrogenation of MgH_2 by a nano-coating of multi-valence Ti-based catalysts [J]. Journal of Materials Chemistry A, 2013, 1(18): 5603–5611.
39. MALKA I, PISAREK M, CZUJKO T, et al. A study of the ZrF_4, NbF_5, TaF_5, and $TiCl_3$ influences on the MgH_2 sorption properties [J]. International Journal of Hydrogen Energy, 2011, 36(20): 12909–12917.
40. FAN M-Q, LIU S-S, ZHANG Y, et al. Superior hydrogen storage properties of MgH_2–10 wt.% TiC composite [J]. Energy, 2010, 35(8): 3417–3421.

41. LILLO-RóDENAS M, GUO Z, AGUEY-ZINSOU K, et al. Effects of different carbon materials on MgH_2 decomposition [J]. Carbon, 2008, 46(1): 126–137.
42. ULLAH RATHER S, TAIMOOR A A, MUHAMMAD A, et al. Kinetics of hydrogen adsorption on MgH_2/CNT composite [J]. Materials Research Bulletin, 2016, 77: 23–28.
43. HOU X, HU R, ZHANG T, et al. Hydrogenation thermodynamics of melt-spun magnesium rich Mg-Ni nanocrystalline alloys with the addition of multiwalled carbon nanotubes and TiF_3 [J]. Journal of Power Sources, 2016, 306: 437–447.
44. LUO F P, WANG H, OUYANG L Z, et al. Enhanced reversible hydrogen storage properties of a Mg–In–Y ternary solid solution [J]. International Journal of Hydrogen Energy, 2013, 38(25): 10912–10918.
45. OUYANG L, CAO Z, WANG H, et al. Dual-tuning effect of in on the thermodynamic and kinetic properties of Mg_2Ni dehydrogenation [J]. International Journal of Hydrogen Energy, 2013, 38(21): 8881–8887.
46. XIE D H, LI P, ZENG C X, et al. Effect of substitution of Nd for Mg on the hydrogen storage properties of Mg_2Ni alloy [J]. Journal of Alloys and Compounds, 2009, 478(1): 96–102.
47. SONG W, LI J, ZHANG T, et al. Microstructure and tailoring hydrogenation performance of Y-doped Mg2Ni alloys [J]. Journal of Power Sources, 2014, 245: 808–815.
48. SONG W, LI J, ZHANG T, et al. Dehydrogenation behavior and microstructure evolution of hydrogenated magnesium–nickel–yttrium melt-spun ribbons [J]. RSC Advances, 2015, 5(67): 54258–54265.
49. LIANG G, HUOT J, BOILY S, et al. Hydrogen storage in mechanically milled Mg–LaNi5 and MgH2–LaNi5 composites [J]. Journal of Alloys and Compounds, 2000, 297(1–2): 261–265.
50. WANG P, ZHANG H, DING B, et al. Structural and hydriding properties of composite Mg–ZrFe1. 4Cr0. 6 [J]. Acta Materialia, 2001, 49(5): 921–926.
51. ARES J R, LEARDINI F, DíAZ-CHAO P, et al. Non-isothermal desorption process of hydrogenated nanocrystalline Pd-capped Mg films investigated by ion beam techniques [J]. International Journal of Hydrogen Energy, 2014, 39(6): 2587–2596.
52. SCHULZ R, HUOT J, LIANG G, et al. Recent developments in the applications of nanocrystalline materials to hydrogen technologies [J]. Materials Science and Engineering: A, 1999, 267(2): 240–245.
53. ZALUSKA A, ZALUSKI L, STRöM–OLSEN J. Nanocrystalline magnesium for hydrogen storage [J]. Journal of Alloys and Compounds, 1999, 288(1–2): 217–225.
54. FRIEDRICHS O, AGUEY-ZINSOU F, FERNANDEZ J A, et al. MgH_2 with Nb_2O_5 as additive, for hydrogen storage: chemical, structural and kinetic behavior with heating [J]. Acta Materialia, 2006, 54(1): 105–110.
55. SCHIMMEL H, JOHNSON M, KEARLEY G, et al. Structural information on ball milled magnesium hydride from vibrational spectroscopy and ab-initio calculations [J]. Journal of Alloys and Compounds, 2005, 393(1–2): 1–4.
56. JEON K-J, MOON H R, RUMINSKI A M, et al. Air-stable magnesium nanocomposites provide rapid and high-capacity hydrogen storage without using heavy-metal catalysts [J]. Nature Materials, 2011, 10(4): 286–290.
57. NORBERG N S, ARTHUR T S, FREDRICK S J, et al. Size-dependent hydrogen storage properties of Mg nanocrystals prepared from solution [J]. Journal of the American Chemical Society, 2011, 133(28): 10679–10681.
58. ZHANG S, GROSS A F, VAN ATTA S L, et al. The synthesis and hydrogen storage properties of a MgH_2 incorporated carbon aerogel scaffold [J]. Nanotechnology, 2009, 20(20): 204027.
59. QU J, WANG Y, XIE L, et al. Superior hydrogen absorption and desorption behavior of Mg thin films [J]. Journal of Power Sources, 2009, 186(2): 515–520.
60. QU J, LIU Y, XIN G, et al. A kinetics study on promising hydrogen storage properties of Mg-based thin films at room temperature [J]. Dalton Transactions, 2014, 43(15): 5908–5912.
61. SANDROCK G. A panoramic overview of hydrogen storage alloys from a gas reaction point of view [J]. Journal of Alloys and Compounds, 1999, s 293–295(2): 877–888.
62. LISHUAI X. Microstructure, hydrogen absorption/desorption behaviour and tissue stability of magnesium-based Alloys [D]: Xi'an Northwestern Polytechnical University, 2020.

63. GARNER S, CHAISE A, RANGO P D, et al. MgH_2 intermediate scale tank tests under various experimental conditions [J]. International Journal of Hydrogen Energy, 2011, 36(16): 9719–9726.
64. DELHOMME B, RANGO P D, MARTY P, et al. Large scale magnesium hydride tank coupled with an external heat source [J]. International Journal of Hydrogen Energy, 2012, 37(11): 9103–9111.
65. ANONYMOUS. Hydrogen Maple Energy:Magnesium-based solid-state hydrogen storage device phase results released [EB/OL]. 2020–07–14. https://baijiahao.baidu.com/s?id=1672132618955500354&wfr=spider&for=pc
66. NETWORK G N E. Hydrofon Energy's 100 hydrogen heavy trucks land in Jining [Z]. 2020–08–11.
67. NISHIMIYA N, SUZUKI A, ONO S. A novel batch-type hydrogen transmitting system using metal hydrides [J]. International Journal of Hydrogen Energy, 1982, 7(9): 741–750.
68. QINGHUI C, YUCHONG Z. The world's only magnesium-based solid-state hydrogen storage vehicle debuts at the 22nd Industrial Fair [EB/OL]. 2020–09–16. https://news.dayoo.com/gzrbyc/202009/16/158752_53568452.htm
69. HYDREXIA. Hydrexia provides hydrogen storage equipmen, metal hydride, Mg-based solid-state hydrogen storage trailer [EB/OL]. 2022–05–20. https://hydrexia.com/equipment/applications
70. ANONYMOUS. Shanghai Magnesium Source Power Technology Co Ltd [EB/OL]. 2022–05–20. http://www.mgpower.com.cn
71. CHEN Q, LIAO J. An ordered layered magnesium-based alloy hydrogen storage material and preparation method for fuel cells, CN108832138A [P].
72. ANONYMOUS. Hydrogen fuel cell electric two-wheeled vehicle" project won the first prize of Changzhou Innovation and Entrepreneurship Competition [J]. New Energy Technology, 2021, (8): 1.
73. QIAN L, LUO Q. Thermodynamics and Kinetics of Metal Hydrides [M]. Shanghai: Shanghai University Press, 2021.
74. ZOU J. Power station system based on magnesium-based hydrogen storage material, CN207664150U [P].
75. TAISHO O. Properties and Applications of Metal Hydrides [M]. Beijing: Chemical Industry Press, 1990.
76. WEIQIANG Z, SHAOCHUN Z. Development and analysis of high purity hydrogen from chemical plant tailgases and hydrogen storage tanks for various metal hydrides [C]. Proceedings of the 2006 Annual Conference of the Chinese Society of Environmental Science (Volume II). 2006: 227–230.
77. ZHOU C, LIU Y, LIU B. Application of a nano-magnesium-based hydrogen storage alloy powder, CN111217327A [P].
78. WANG W, ZHOU X, LONG X. Progress in the separation of hydrogen isotopes by metal hydride method [J]. Isotopes, 2011, 24(S1): 15–20.
79. BAO Z, WU Z, SERGE N N, et al. Research progress of metal hydride high-temperature thermal storage technology [J]. Advances in Chemical Engineering, 2012, 31(08): 1665–1670+1676.
80. BOGDANOVIĆ B, HOFMANN H, NEUY A, et al. Ni-doped versus undoped Mg–MgH2 materials for high temperature heat or hydrogen storage [J]. Journal of Alloys and Compounds, 1999, 292(1–2): 57–71.
81. BOGDANOVIC B, SPLIETHOFF B. Active MgH_2-Mg-systems for hydrogen storage [J]. International Journal of Hydrogen Energy, 1987, 12(12): 863–873.
82. KAWAMURA M, ONO S, HIGANO S. Experimental studies on the behaviours of hydride heat storage system [J]. Energy Conversion and Management, 1982, 22(2): 95–102.
83. BOGDANOVIE B, SPLIETHOFF B, RITTER A. The magnesium hydride system for heat storage and cooling [J]. Zeitschrift Für Physikalische Chemie, 1989, 164(Part2): 1497–1508.
84. REISER A, BOGDANOVIC B, SCHLICHTE K. The application of Mg-based metal-hydrides as heat energy storage systems [J]. International Journal of Hydrogen Energy, 2000, 25(5): 425–430.

85. FELDERHOFF M, BOGDANOVIE B. High temperature metal hydrides as heat storage materials for solar and related applications [J]. International Journal of Molecular Sciences, 2009, 10(1): 325–344.
86. BOGDANOVIĆ B, RITTER A, SPLIETHOFF B. Active MgH_2-Mg systems for reversible chemical energy storage [J]. Angewandte Chemie International Edition, 2010, 29(3): 223–234.
87. WIERSE M, WERNER R, GROLL M. Magnesium hydride for thermal energy storage in a small-scale solar-thermal power station [J]. Journal of the Less Common Metals, 1991, 172(part-P3): 1111–1121.
88. GROLL M, ISSELHORST A, WIERSE M. Metal hydride devices for environmentally clean energy technology [J]. International Journal of Hydrogen Energy, 1994, 19(6): 507–515.
89. URBANCZYK R, PEINECKE K, PEIL S, et al. Development of a heat storage demonstration unit on the basis of Mg_2FeH_6 as heat storage material and molten salt as heat transfer media [J]. International Journal of Hydrogen Energy, 2017, 42(19): 13818–13826.
90. WAN Q, JIANG L, LI Z, et al. Current research status of Mg-based thermal storage materials for solar collector power generation [J]. New Material Industry, 2016, (5): 58–62.
91. LI G, LIU H, LI L, et al. Research progress of metal hydride heat pump air conditioning [J]. Journal of Refrigeration, 2005, (02): 1–7.
92. NI J, YANG T, CHEN J, et al. Metal hydride heat pump and air conditioning [J]. Journal of Solar Energy, 2006, (03): 314–320.
93. BOGDANOVI B, RITTER A, SPLIETHOFF B, et al. A process steam generator based on the high temperature magnesium hydride/magnesium heat storage system [J]. International Journal of Hydrogen Energy, 1995, 20(10): 811–822.
94. ZHANG Z, WANG Y. Hydrogen storage materials and their applications inenergetic materials materials [J]. Progress of Fine Petrochemicals, 2006, (11): 28–31.
95. CHEN X, ZOU J, ZENG X, et al. Application of magnesium-based hydrogen storage materials inenergetic materials materials [J]. Journal of Fire Explosives, 2016, 39(03): 1–8.
96. SCHAPBACH L, ZUTTEL A. Hydrogen-storage materials for mobile applications [J]. Nature, 2001, 414(6861): 353–358.
97. REILLY JR J J, WISWALL JR R H. Reaction of hydrogen with alloys of magnesium and nickel and the formation of Mg_2NiH_4 [J]. Inorganic Chemistry, 1968, 7(11): 2254–2256.
98. LIU L, LI F, ZHI C, et al. Effects of magnesium-based hydrogen storage materials on the performance of AP/Al/HTPB composite solid propellants [J]. Energetic Materials, 2009, 17(05): 501–504.
99. LIU J, YUNJUN L. Compatibility study of hydrogen storage alloy combustibles withenergetic materials components commonly used in solid propellants [J]. Journal of Military Engineering, 2008, (09): 1133–1136.
100. LIU J, LUO Y. Study on curing porosity of butyl hydroxyl propellant containing hydrogen storage alloy [J]. Solid Rocket Technology, 2011, 34(01): 92–96.
101. ZHANG Y, XU S, ZHAO F, et al. Experimental study on the effect of MgH_2 on the ignition and combustion performance ofenergetic materials materials [J]. Journal of Fire Explosives, 2021, 44(04): 504–513.
102. LIU L, LI F, ZHI C, et al. Effect of Mg_2NiH_4 on the thermal decomposition of ammonium perchlorate [J]. Journal of Higher Education Chemistry, 2007, (08): 1420–1423.
103. FICHTNER M, FUHR O, KIRCHER O. Magnesium alanate—a material for reversible hydrogen storage [J]. Journal of Alloys and Compounds, 2003, 356(none): 418–422.
104. FIFER R A, COLE J E. Catalysts for nitramine propellants, 4379007 [P].
105. YAO M, CHEN L, PENG J. Effects of $MgH_2/Mg(BH_4)_2$ powders on the thermal decomposition behaviors of 2,4,6-trinitrotoluene (TNT) [J]. Propellants Explosives Pyrotechnics, 2015, 40(2): 197–202.
106. YAO M, CHEN L, SHANG P, et al. Effect of $Mg(BH4)_2$ and MgH_2 on the thermal decomposition of RDX [J]. Chinese Journal of Safety Science, 2013, 23(01): 115–120.
107. LIU Y, CAO Y, HUANG L, et al. Rare earth–Mg–Ni-based hydrogen storage alloys as negative electrode materials for Ni/MH batteries [J]. Journal of Alloys and Compounds, 2011, 509(3): 675–86.

108. ZHU M. Research progress of Mg-based hydrogen storage alloys and their application in Ni-MH batteries [J]. China Science Foundation, 2001, (02): 21–25.
109. LIU Y, ZHANG X. Effect of lanthanum additions on electrode properties of Mg_2Ni [J]. Journal of Alloys and Compounds, 1998, 267(1–2): 231–234.
110. CUI N, LUO J L. Electrochemical study of hydrogen diffusion behavior in Mg_2Ni-TYPE hydrogen storage alloy electrodes [J]. International Journal of Hydrogen Energy, 1999, 24(1): 37–42.
111. XU Y, MULDER F M. Non-alloy Mg anode for Ni-MH batteries: Multiple approaches towards a stable cycling performance [J]. International Journal of Hydrogen Energy, 2021, 46(37): 19542–19553.
112. OUMELLAL Y, ROUGIER A, NAZRI G, et al. Metal hydrides for lithium-ion batteries [J]. Nature Materials, 2008, 7(11): 916–921.
113. HUANG L, AYMARD L, BONNET J-P. MgH 2–TiH 2 mixture as an anode for lithium-ion batteries: synergic enhancement of the conversion electrode electrochemical performance [J]. Journal of Materials Chemistry A, 2015, 3(29): 15091–15096.
114. MECHELEN J, NOHEDA B, LOHSTROH W, et al. Mg-Ni-H films as selective coatings: tunable reflectance by layered hydrogenation [J]. Applied Physics Letters, 2004, 84(18): 3651–3653.
115. HUIBERTS J N, GRIESSEN R, RECTOR J H, et al. Yttrium and lanthanum hydride films with switchable optical properties [J]. Nature, 1996, 380(6571): 231–234.
116. LEE Y A, KALANUR S S, SHIM G, et al. Highly sensitive gasochromic H_2 sensing by nano-columnar WO_3-Pd films with surface moisture [J]. Sensors and Actuators B Chemical, 2017, 238: 111–119.
117. REZAEI S D, SHANNIGRAHI S, RAMAKRISHNA S. A review of conventional, advanced, and smart glazing technologies and materials for improving indoor environment [J]. Solar Energy Materials and Solar Cells, 2017, 159: 26–51.
118. ZHANG Y N, PENG H, QIAN X, et al. Recent advancements in optical fiber hydrogen sensors [J]. Sensors and Actuators B Chemical, 2017, 244: 393–416.
119. PENG L, LIU Y, CHEN J, et al. Progress in the study of hydrochromic magnesium-based functional films [J]. Progress of Materials in China, 2018, 37(12): 970–977+993.
120. YAMADA Y, MIURA M, TAJIMA K, et al. Optical switching durability of switchable mirrors based on magnesium–yttrium alloy thin films [J]. Solar Energy Materials and Solar Cells, 2013, 117: 396–399.
121. YAMADA Y, MIURA M, TAJIMA K, et al. Influence on optical properties and switching durability by introducing Ta intermediate layer in Mg–Y switchable mirrors [J]. Solar Energy Materials and Solar Cells, 2014, 125: 133–137.
122. RICHARDSON T J, SLACK J L, ARMITAGE R D, et al. Switchable mirrors based on nickel-magnesium films [J]. Applied Physics Letters, 2001, 78(20): 3047–3049.
123. LOHSTROH W, WESTERWAAL R J, NOHEDA B, et al. Self-organized layered hydrogenation in black Mg_2NiH_x switchable mirrors [J]. Physical Review Letters, 2004, 93(19): 197404.
124. BAO S, TAJIMA K, YAMADA Y, et al. Color-neutral switchable mirrors based on magnesium-titanium thin films [J]. Applied Physics A, 2007, 87(4): 621–624.
125. YAMADA Y, BAO S, TAJIMA K, et al. Optical properties of switchable mirrors based on magnesium-calcium alloy thin films [J]. Applied Physics Letters, 2009, 94(19): 191910.
126. LIU Y, CHEN J, PENG L, et al. Improved optical properties of switchable mirrors based on Pd/Mg-TiO_2 films fabricated by magnetron sputtering [J]. Materials and Design, 2018, 144: 256–262.
127. SLAMAN M, DAM B, SCHREUDERS H, et al. Optimization of Mg-based fiber optic hydrogen detectors by alloying the catalyst [J]. International Journal of Hydrogen Energy, 2008, 33(3): 1084–1089.
128. SLAMAN M, DAM B, PASTUREL M, et al. Fiber optic hydrogen detectors containing Mg-based metal hydrides [J]. Sensors and Actuators B Chemical, 2007, 123(1): 538–545.

129. ZHAO Q, LI Y, SONG Y, et al. Fast hydrogen-induced optical and electrical transitions of Mg and Mg-Ni films with amorphous structure [J]. Applied Physics Letters, 2013, 102(16): 161901.
130. HAN S, CHEN Y, GANG Z, et al. Optical fiber hydrogen sensor based on an annealing-stimulated Pd–Y thin film [J]. Sensors and Actuators B Chemical, 2015, 216(9): 11–16.
131. IMAI H, TAGAWA T, NAKAMURA K. Catalytic activities of hydrogen storage alloys for decomposition of alcohols [J]. Applied Catalysis, 1990, 62(1): 348–352.
132. DOLE M, WILSON F R, FIFE W P. Hyperbaric hydrogen therapy: a possible treatment for cancer [J]. Science, 1975, 190(4210): 152–154.
133. OHSAWA I, ISHIKAWA M, TAKAHASHI K, et al. Hydrogen acts as a therapeutic antioxidant by selectively reducing cytotoxic oxygen radicals [J]. Nature Medicine, 2007, 13(6): 688–694.
134. LIU H, CHEN C, HE S, et al. Characteristics of magnesium materials and progress of medical research [J]. Research in Dentistry, 2013, 29(09): 874–876.
135. SUN X. Hydrogen molecular biology [M]. Shanghai: Shanghai: Second Military Medical University Press, 2013.
136. LIU Z, ZHANG W, HAN X, et al. Analysis of hydrogen and hydrogen-enriched water in agricultural production [J]. Seed Science and Technology, 2020, 38(10): 102–103.
137. CENG J. Method and application for the preparation of slow and controlled release hydrogen fertiliser or compound hydrogen fertiliser, CN106699330A [P].
138. LI L, LIU Y, WANG S, et al. Magnesium hydride-mediated sustainable hydrogen supply prolongs the vase life of cut carnation flowers via hydrogen sulfide [J]. Frontiers in Plant Science, 2020, 11: 595376.

The manufacturer's authorised representative in the EU is Springer Nature Customer Service Centre GmbH, Europaplatz 3, 69115 Heidelberg, Germany. If you have any concerns regarding our products, please contact ProductSafety@springernature.com

Printed and bound by CPI Group (UK) Ltd, Croydon, CR0 4YY
30/09/2026
02225298-0003